Ecology and Biogeography of Marine Benthos

Ecology and Biogeography of Marine Benthos

Editor

Renato Mamede

Basel • Beijing • Wuhan • Barcelona • Belgrade • Novi Sad • Cluj • Manchester

Editor
Renato Mamede
Department of Biology
ECOMARE, CESAM
University of Aveiro
Aveiro
Portugal

Editorial Office
MDPI AG
Grosspeteranlage 5
4052 Basel, Switzerland

This is a reprint of articles from the Special Issue published online in the open access journal *Diversity* (ISSN 1424-2818) (available at: https://www.mdpi.com/journal/diversity/special_issues/V66DMSHWH5).

For citation purposes, cite each article independently as indicated on the article page online and as indicated below:

Lastname, A.A.; Lastname, B.B. Article Title. *Journal Name* **Year**, *Volume Number*, Page Range.

ISBN 978-3-7258-2020-7 (Hbk)
ISBN 978-3-7258-2019-1 (PDF)
doi.org/10.3390/books978-3-7258-2019-1

© 2024 by the authors. Articles in this book are Open Access and distributed under the Creative Commons Attribution (CC BY) license. The book as a whole is distributed by MDPI under the terms and conditions of the Creative Commons Attribution-NonCommercial-NoDerivs (CC BY-NC-ND) license.

Contents

About the Editor vii

Renato Mamede
Ecology and Biogeography of Marine Benthos
Reprinted from: *Diversity* **2024**, *16*, 502, doi:10.3390/d16080502 1

Antía Lourido, Santiago Parra and Francisco Sánchez
Soft-Bottom Infaunal Macrobenthos of the Avilés Canyon System (Cantabrian Sea)
Reprinted from: *Diversity* **2023**, *15*, 53, doi:10.3390/d15010053 6

Seyed Ehsan Vesal, Federica Nasi, Rocco Auriemma and Paola Del Negro
Effects of Organic Enrichment on Bioturbation Attributes: How Does the Macrofauna Community Respond in Two Different Sedimentary Impacted Areas?
Reprinted from: *Diversity* **2023**, *15*, 449, doi:10.3390/d15030449 20

Binbin Shan, Zhenghua Deng, Shengwei Ma, Dianrong Sun, Yan Liu, Changping Yang, et al.
A New Record of *Pinctada fucata* (Bivalvia: Pterioida: Pteriidae) in Mischief Reef: A Potential Invasive Species in the Nansha Islands, China
Reprinted from: *Diversity* **2023**, *15*, 578, doi:10.3390/d15040578 41

Diego Carreira-Flores, Regina Neto, Hugo R. S. Ferreira, Edna Cabecinha, Guillermo Díaz-Agras, Marcos Rubal and Pedro T. Gomes
Colonization in Artificial Seaweed Substrates: Two Locations, One Year
Reprinted from: *Diversity* **2023**, *15*, 733, doi:10.3390/d15060733 52

Helen Grace P. Bangi and Marie Antonette Juinio-Meñez
Feeding and Reproductive Phenotypic Traits of the Sea Urchin *Tripneustes gratilla* in Seagrass Beds Impacted by Eutrophication
Reprinted from: *Diversity* **2023**, *15*, 843, doi:10.3390/d15070843 67

Alice Guzzi, Maria Chiara Alvaro, Matteo Cecchetto and Stefano Schiaparelli
Echinoids and Crinoids from Terra Nova Bay (Ross Sea) Based on a Reverse Taxonomy Approach
Reprinted from: *Diversity* **2023**, *15*, 875, doi:10.3390/d15070875 85

Sulia Goeting, Huan Chiao Lee, László Kocsis, Claudia Baumgartner-Mora and David J. Marshall
Diversity and Distribution of the Benthic Foraminifera on the Brunei Shelf (Northwest Borneo): Effect of Seawater Depth
Reprinted from: *Diversity* **2023**, *15*, 937, doi:10.3390/d15080937 105

Moira Buršić, Andrej Jaklin, Milvana Arko Pijevac, Branka Bruvo Mađarić, Lucija Neal, Emina Pustijanac, et al.
Seasonal Variations in Invertebrates Sheltered among *Corallina officinalis* (Plantae, Rodophyta) Turfs along the Southern Istrian Coast (Croatia, Adriatic Sea)
Reprinted from: *Diversity* **2023**, *15*, 1099, doi:10.3390/d15101099 128

Rafael Bañón, Juan Fariña and Alejandro de Carlos
New Data on Exotic Muricid Species (Neogastropoda: Muricidae) from Spain Based on Integrative Taxonomy
Reprinted from: *Diversity* **2023**, *15*, 1185, doi:10.3390/d15121185 142

Jianfeng Mou, Xuebao He, Kun Liu, Yaqin Huang, Shuyi Zhang, Yongcan Zu, et al.
Benthic Biodiversity by Baited Camera Observations on the Cosmonaut Sea Shelf of East Antarctica
Reprinted from: *Diversity* **2024**, *16*, 277, doi:10.3390/d16050277 . . . . **152**

Cynthia Mariana Hernández-Casas, Ángela Catalina Mendoza-González, Deisy Yazmín García-López and Luz Elena Mateo-Cid
Spatio-Temporal Structure of Two Seaweeds Communities in Campeche, Mexico
Reprinted from: *Diversity* **2024**, *16*, 344, doi:10.3390/d16060344 . . . . **162**

About the Editor

Renato Mamede

Renato Mamede holds a Ph.D. in Marine Biology from the University of Aveiro, specializing in benthic habitat mapping and modeling, which he completed in 2018. He is currently a Ph.D. researcher at ECOMARE, Center for Environmental and Marine Studies (CESAM), Department of Biology, University of Aveiro. For the last five years, Renato has been deeply involved in marine ecology and biotechnology, focusing on the study of benthic habitats and the traceability of economically important seafood species. His research has contributed to the understanding of marine benthic ecosystems as well as to food safety, conservation of natural stocks, and the economic valorization of marine resources. In addition, Renato is an invited member of the ICES Working Group on Fisheries Benthic Impact and Trade-offs (WGFBIT), where he is actively involved in assessing the sensitivity of Portuguese seabed habitats to disturbances, such as bottom fishing.

Editorial

Ecology and Biogeography of Marine Benthos

Renato Mamede

ECOMARE, CESAM—Centre for Environmental and Marine Studies, Department of Biology, University of Aveiro, Santiago University Campus, 3810-193 Aveiro, Portugal; renatomamede@ua.pt

The ocean floor, which spans approximately 71% of the Earth's surface, stretches from sunlit shallow waters to the profound depths of ocean trenches around 11,000 m deep, encompassing regions from the polar ice to tropical seas [1]. This immense and recondite realm is home to an extraordinary variety of habitats, including soft sediments, rocky and coral reefs, hydrothermal vents, mud volcanoes, and seagrass meadows [2]. The conservation of these benthic ecosystems is not merely important; it is vital, as they provide indispensable resources and services such as pollution control, food supply, raw materials, nutrient cycling, nurseries for diverse species, water quality maintenance, and the preservation of biodiversity [3].

In the face of escalating challenges like global warming and other human-induced and natural threats—including illegal fishing or invasive species [4]—it has never been more crucial to adopt ecosystem-based strategies for conserving and managing the marine environment [5,6]. However, despite the importance of these ecosystems and the threats they face, only about a quarter of the ocean floor has been mapped with (only) bathymetric information [7]. In alignment with the goals of the United Nations Decade of Ocean Sciences for Sustainable Development (2021–2030) [8] and Sustainable Development Goal 14, "Life below water" [9], this Special Issue was launched to deepen our understanding of the ecology and biogeography of these vital marine ecosystems.

In the first article of this Special Issue, Lourido et al. [10] investigated the Avilés Canyon System (ACS) on the northern Atlantic coast of Spain to identify areas of special conservation interest. Their study involved three surveys aimed at understanding the composition and distribution of organisms and habitats within this region. These researchers observed predominantly sandy sediments, with finer, organic-rich sediments in deeper areas and coarser sediments in shallower stations. Polychaetes were found to be the most abundant and diverse faunal group, followed by crustaceans and mollusks. This study identified five major macrobenthic communities, which were primarily influenced by depth and sediment type, offering considerable insight into the infaunal assemblages of the ACS.

Vesal et al. [11] examined the effects of different sources of organic matter—terrigenous/freshwater allochthonous inputs and sewage—on the activity of soft-bottom communities. They sampled macrofauna from two areas: in front of the Po River delta and a sewage discharge zone in the Gulf of Trieste (Adriatic Sea, Italy). This study revealed higher bioturbation indices at the northern sampling stations of the prodelta and at stations located 25 m from the main sewage outfall. Species diversity was notably higher in the prodelta and increased with distance from the outfall. These authors concluded that the observed differences in bioturbation across these areas were primarily attributable to variations in sediment type and levels of organic enrichment. This study provides substantial insights into the effective management and sustainable utilization of coastal resources, particularly in regions significantly affected by human activities.

Shan et al. [12] reported the arrival of the bivalve *Pinctada fucata* at Mischief Reef (Nansha Islands, South China Sea), a species previously unrecorded in this region, likely due to human activities. Genetic analysis of seven South China Sea populations suggested that Lingshui (Hainan Island) could be a source of *P. fucata* at Mischief Reef. This finding

Citation: Mamede, R. Ecology and Biogeography of Marine Benthos. *Diversity* **2024**, *16*, 502. https://doi.org/10.3390/d16080502

Received: 13 August 2024
Accepted: 15 August 2024
Published: 19 August 2024

Copyright: © 2024 by the author. Licensee MDPI, Basel, Switzerland. This article is an open access article distributed under the terms and conditions of the Creative Commons Attribution (CC BY) license (https://creativecommons.org/licenses/by/4.0/).

highlights the impact of human activities on marine ecosystems and underscores the urgent need for management strategies to prevent the introduction of additional invasive species.

Carreira-Flores et al. [13] investigated the efficacy of artificial substrates, specifically the Artificial Seaweed Monitoring System (ASMS), for assessing natural variability in assemblages over large spatial scales. This study employed the ASMS to monitor macrofauna along a 200-km stretch of the Galician Coast over 3, 6, 9, and 12 months. The results revealed differences in macrofauna assemblages between locations, supporting the theory that benthic community succession is not linear but involves a blend of different successional stages. The ASMS demonstrated its effectiveness in capturing scale-dependent patterns and temporal variability, proving to be a valuable tool for non-destructive environmental monitoring.

Bangi and Juinio-Meñez [14] investigated the sea urchin *Tripneustes gratilla*, a major grazer, to assess its response to habitat changes near fish farms in Bolinao, Philippines. These researchers compared the feeding and reproductive traits of sea urchins in three seagrass beds located at different distances from the fish farms. Sea urchins closer to the farms exhibited smaller gut and feeding structures, lower consumption rates, and reduced gonadal indices, which were associated with reduced seagrass productivity and water movement. In contrast, the increased size of Aristotle's lantern observed at more distant stations was not attributed to food limitations. This study underscores the sensitivity of *T. gratilla* to environmental changes that may influence food quality variability.

Identifying species within an ecosystem is vital for ecological research and conservation efforts, and DNA barcoding can significantly speed up this process while complementing traditional methods. Guzzi et al. [15] employed a "Reverse Taxonomy" approach, in which DNA barcoding is followed by morphological analysis, to identify 70 echinoid and 22 crinoid specimens from two Antarctic regions: Terra Nova Bay (TNB, Ross Sea) and the Arctic Peninsula (Weddell Sea). This study identified 13 sea urchin species, including six new records for TNB and four crinoids, and corrected a previous misidentification of an echinoderm species. Notably, all identified crinoids were new records for TNB. This research sets a baseline for future biodiversity studies, especially within the TNB region, now part of the Ross Sea Marine Protected Area.

Goeting et al. [16] investigated the largely unexplored marine benthic diversity of the Palawan/North Borneo ecoregion, known for its high species richness. This study focused on benthic foraminifera from the Brunei shelf, identifying 99 species across 31 families and 56 genera from depths ranging from 10 to 200 m. Notably, 52 of these species were recorded for the first time in the region. Oxygen isotope analysis of selected species indicated cooler temperatures at greater depths, while carbon isotope data revealed species differences related to habitat preferences, food sources, and biomineralization. This study observed that species diversity increased with depth, suggesting greater disturbance in shallow waters and more pronounced specialization in deeper waters. Additionally, the results of this study indicate the presence of (at least) three biotopes on the Brunei shelf, each associated with different depth classes, thereby providing valuable insights into the regional diversity and distribution of foraminifera.

Previous research on *Corallina officinalis* (Plantae, Rhodophyta) has underscored its significant role in supporting high invertebrate biodiversity, attributed to the complex habitat structure it provides. Buršić et al. [17] examined seasonal variations in invertebrate assemblages at nine sites in the coastal region of southern Istria and the Brijuni National Park (Croatia, Adriatic Sea). These researchers recorded 29,711 invertebrates in winter, when algal growth is at its peak, and 22,292 in summer, when algal growth is minimal. Invertebrate abundance increased with algal biomass, with the highest density—586,000 individuals per square meter—observed in the Premantura area during the winter. The dominant invertebrate groups were amphipods, polychaetes, bivalves, and gastropods. This study deepened our understanding of *C. officinalis* turfs and confirmed the high invertebrate richness associated with these settlements.

In 2023, two exotic gastropod species, *Ocinebrellus inornatus* and *Rapana venosa* (Neogastropoda: Muricidae), were detected in Galician waters (Spain). Bañón et al. [18] reported new data on the occurrences of these species in this region, confirmed through morphological analysis and DNA barcoding. *O. inornatus* is now established in the Illa de Arousa (Ría de Arousa), while *R. venosa* was confirmed in this Ria, with evidence that its range extends beyond this area. These species probably arrived through imported shellfish, with this study emphasizing the need for a monitoring program to track the spread of exotic marine species.

In 2022, a baited camera was deployed for the first time in the Cosmonaut Sea, East Antarctica, capturing images at a depth of 694 m. Mou et al. [19] identified 31 species, including 23 invertebrates and 8 fish across eight phyla, from a collection of 2403 pictures and 40 videos, with the Antarctic jonasfish *Notolepis coatsi* being the most frequent fish species. This study documented ten vulnerable marine species and characterized the macrobenthic community as sessile suspension feeders, associated with fauna inhabiting muddy substrates interspersed with rocks. This research provides essential insights into macro- and megafauna communities and their habitats, establishing a baseline of Antarctic biodiversity that will support future ecosystem monitoring and conservation efforts in the Cosmonaut Sea.

In the final paper of this Special Issue, Hernández-Casas et al. [20] investigated the spatio-temporal variability of macroalgae communities in Xpicob and Villamar (Mexico) during three climatic seasons. This study identified 74 taxa from three classes: Phaeophyceae (3), Florideophyceae (36), and Ulvophyceae (35). Filamentous algae dominated the intertidal zone, while fleshy and calcareous algae were more prevalent in deeper waters. Twenty-eight species were found in both sites, with 46 taxa being site-specific: 13 in Xpicob and 33 in Villamar. Results indicated that species richness and distribution varied with environmental factors, with winter rains promoting algal growth in Xpicob and the dry season favoring Villamar. These findings underscore the considerable impact of local conditions on macroalgae communities and greatly enhance our understanding of these vital marine resources, which are valuable for marine management and conservation efforts.

In summary, this Special Issue compiles eight research articles, two communications, and one brief report, providing a comprehensive exploration of marine ecosystems ranging from shallow coastal waters to the depths of the ocean. These studies examine a diverse array of benthic habitats, including soft sediments, seagrass meadows, and rocky and artificial substrates, with research conducted across four continents—Antarctica, Asia, Europe, and North America—and within all major climate zones: polar, temperate, and tropical. The primary focus has been on macrofauna, examined at community and species levels, with some studies addressing megafauna, foraminifera, and macroalgae. These works employed a blend of traditional and remote sensing techniques for data collection, along with taxonomic, molecular, and biogeochemical methods for laboratory analysis. By utilizing a range of diversity and physiological indices, as well as advanced statistical techniques, the studies within this Special Issue have succeeded in describing and characterizing previously unexplored areas, identifying invasive species, monitoring pollution sources, and testing innovative substrates to enhance monitoring efforts. Collectively, this Special Issue makes a significant and inspiring contribution to our understanding of marine ecology and biogeography, offering valuable insights to support the conservation and management of marine benthos.

However, given the vastness of marine benthos, there is still much to explore and understand. Future benthic studies should aim to be more cost-effective and innovative in maximizing data utilization—critical steps toward achieving an ecosystem-based approach to the sustainable use of the marine environment. This will require a harmonious blend of traditional methods with cutting-edge techniques such as remote sensing, statistical modeling, and mapping. Additionally, integrating diverse types of data, including morphological, environmental, molecular, and isotopic information, will be key to unlocking a deeper understanding of these complex ecosystems [21–23].

Acknowledgments: As guest editor of the Special Issue, "Ecology and Biogeography of Marine Benthos", I would like to express my gratitude to the authors whose valuable works have been published in this issue and have contributed to its success. My sincere appreciation also goes to all academic co-editors and reviewers, whose volunteer efforts were essential to the completion of this Special Issue.

Conflicts of Interest: The author declares no conflicts of interest.

List of Contributions

1. Lourido, A.; Parra, S.; Sánchez, F. Soft-Bottom Infaunal Macrobenthos of the Avilés Canyon System (Cantabrian Sea). *Diversity* 2023, 15, 53. https://doi.org/10.3390/d15010053.
2. Vesal, S.; Nasi, F.; Auriemma, R.; Del Negro, P. Effects of Organic Enrichment on Bioturbation Attributes: How Does the Macrofauna Community Respond in Two Different Sedimentary Impacted Areas? *Diversity* 2023, 15, 449. https://doi.org/10.3390/d15030449.
3. Shan, B.; Deng, Z.; Ma, S.; Sun, D.; Liu, Y.; Yang, C.; Wu, Q.; Yu, G. A New Record of *Pinctada fucata* (Bivalvia: Pterioida: Pteriidae) in Mischief Reef: A Potential Invasive Species in the Nansha Islands, China. *Diversity* 2023, 15, 578. https://doi.org/10.3390/d15040578.
4. Carreira-Flores, D.; Neto, R.; Ferreira, H.R.S.; Cabecinha, E.; Díaz-Agras, G.; Rubal, M.; Gomes, P.T. Colonization in Artificial Seaweed Substrates: Two Locations, One Year. *Diversity* 2023, 15, 733. https://doi.org/10.3390/d15060733.
5. Bangi, H.; Juinio Meñcz, M. Feeding and Reproductive Phenotypic Traits of the Sea Urchin Tripneustes gratilla in Seagrass Beds Impacted by Eutrophication. *Diversity* 2023, 15, 843. https://doi.org/10.3390/d15070843.
6. Guzzi, A.; Alvaro, M.; Cecchetto, M.; Schiaparelli, S. Echinoids and Crinoids from Terra Nova Bay (Ross Sea) Based on a Reverse Taxonomy Approach. *Diversity* 2023, 15, 875. https://doi.org/10.3390/d15070875.
7. Goeting, S.; Lee, H.; Kocsis, L.; Baumgartner-Mora, C.; Marshall, D. Diversity and Distribution of the Benthic Foraminifera on the Brunei Shelf (Northwest Borneo): Effect of Seawater Depth. *Diversity* 2023, 15, 937. https://doi.org/10.3390/d15080937.
8. Buršić, M.; Jaklin, A.; Arko Pijevac, M.; Bruvo Mađarić, B.; Neal, L.; Pustijanac, E.; Burić, P.; Iveša, N.; Paliaga, P.; Iveša, L. Seasonal Variations in Invertebrates Sheltered among *Corallina officinalis* (Plantae, Rodophyta) Turfs along the Southern Istrian Coast (Croatia, Adriatic Sea). *Diversity* 2023, 15, 1099. https://doi.org/10.3390/d15101099.
9. Bañón, R.; Fariña, J.; de Carlos, A. New Data on Exotic Muricid Species (Neogastropoda: Muricidae) from Spain Based on Integrative Taxonomy. *Diversity* 2023, 15, 1185. https://doi.org/10.3390/d15121185.
10. Mou, J.; He, X.; Liu, K.; Huang, Y.; Zhang, S.; Zu, Y.; Liu, Y.; Cao, S.; Lan, M.; Miao, X.; et al. Benthic Biodiversity by Baited Camera Observations on the Cosmonaut Sea Shelf of East Antarctica. *Diversity* 2024, 16, 277. https://doi.org/10.3390/d16050277.
11. Hernández-Casas, C.; Mendoza-González, Á.; García-López, D.; Mateo-Cid, L. Spatio-Temporal Structure of Two Seaweeds Communities in Campeche, Mexico. *Diversity* 2024, 16, 344. https://doi.org/10.3390/d16060344.

References

1. Weatherall, P.; Marks, K.M.; Jakobsson, M.; Schmitt, T.; Tani, S.; Arndt, J.E.; Rovere, M.; Chayes, D.; Ferrini, V.; Wigley, R. A New Digital Bathymetric Model of the World's Oceans. *Earth Space Sci.* 2015, 2, 331–345. [CrossRef]
2. Montefalcone, M.; Tunesi, L.; Ouerghi, A. A Review of the Classification Systems for Marine Benthic Habitats and the New Updated Barcelona Convention Classification for the Mediterranean. *Mar. Environ. Res.* 2021, 169, 105387. [CrossRef] [PubMed]
3. Galparsoro, I.; Borja, A.; Uyarra, M.C. Mapping Ecosystem Services Provided by Benthic Habitats in the European North Atlantic Ocean. *Front. Mar. Sci.* 2014, 1, 23. [CrossRef]
4. Harris, P.T. Anthropogenic Threats to Benthic Habitats. In *Seafloor Geomorphology as Benthic Habitat: GeoHab Atlas of Seafloor Geomorphic Features and Benthic Habitats*; Elsevier: Amsterdam, The Netherlands, 2019; pp. 35–61. [CrossRef]

5. O'Higgins, T.G.; Lago, M.; DeWitt, T.H. *Ecosystem-Based Management, Ecosystem Services and Aquatic Biodiversity. Theory, Tools, and Applications*; Springer: Cham, Switzerland, 2020; pp. 1–580.
6. Boudouresque, C.F.; Astruch, P.; Bănaru, D.; Blanchot, J.; Blanfuné, A.; Carlotti, F.; Changeux, T.; Faget, D.; Goujard, A.; Harmelin-Vivien, M.; et al. The Management of Mediterranean Coastal Habitats: A Plea for a Socio-Ecosystem-Based Approach. In *Evolution of Marine Coastal Ecosystems under the Pressure of Global Changes. Proceedings of Coast Bordeaux Symposium and of the 17th French-Japanese Oceanography Symposium*; Springer International Publishing: Bordeaux, France, 2020; pp. 297–320.
7. Release of the GEBCO_2024 Grid. Available online: https://www.gebco.net/news_and_media/release_of_gebco_2024_grid.html (accessed on 8 August 2024).
8. The Ocean Decade—The Science We Need for the Ocean We Want. Available online: https://oceandecade.org/ (accessed on 8 August 2024).
9. UN. *Transforming Our World: The 2030 Agenda for Sustainable Development*; United Nations: New York, NY, USA, 2015; pp. 1–41.
10. Lourido, A.; Parra, S.; Sánchez, F. Soft-Bottom Infaunal Macrobenthos of the Avilés Canyon System (Cantabrian Sea). *Diversity* **2023**, *15*, 53. [CrossRef]
11. Vesal, S.E.; Nasi, F.; Auriemma, R.; Del Negro, P. Effects of Organic Enrichment on Bioturbation Attributes: How Does the Macrofauna Community Respond in Two Different Sedimentary Impacted Areas? *Diversity* **2023**, *15*, 449. [CrossRef]
12. Shan, B.; Deng, Z.; Ma, S.; Sun, D.; Liu, Y.; Yang, C.; Wu, Q.; Yu, G. A New Record of *Pinctada fucata* (Bivalvia: Pterioida: Pteriidae) in Mischief Reef: A Potential Invasive Species in the Nansha Islands, China. *Diversity* **2023**, *15*, 578. [CrossRef]
13. Carreira-Flores, D.; Neto, R.; Ferreira, H.R.S.; Cabecinha, E.; Díaz-Agras, G.; Rubal, M.; Gomes, P.T. Colonization in Artificial Seaweed Substrates: Two Locations, One Year. *Diversity* **2023**, *15*, 733. [CrossRef]
14. Bangi, H.G.P.; Juinio-Meñez, M.A. Feeding and Reproductive Phenotypic Traits of the Sea Urchin *Tripneustes gratilla* in Seagrass Beds Impacted by Eutrophication. *Diversity* **2023**, *15*, 843. [CrossRef]
15. Guzzi, A.; Alvaro, M.C.; Cecchetto, M.; Schiaparelli, S. Echinoids and Crinoids from Terra Nova Bay (Ross Sea) Based on a Reverse Taxonomy Approach. *Diversity* **2023**, *15*, 875. [CrossRef]
16. Goeting, S.; Lee, H.C.; Kocsis, L.; Baumgartner-Mora, C.; Marshall, D.J. Diversity and Distribution of the Benthic Foraminifera on the Brunei Shelf (Northwest Borneo): Effect of Seawater Depth. *Diversity* **2023**, *15*, 937. [CrossRef]
17. Buršić, M.; Jaklin, A.; Arko Pijevac, M.; Bruvo Mađarić, B.; Neal, L.; Pustijanac, E.; Burić, P.; Iveša, N.; Paliaga, P.; Iveša, L. Seasonal Variations in Invertebrates Sheltered among *Corallina officinalis* (Plantae, Rodophyta) Turfs along the Southern Istrian Coast (Croatia, Adriatic Sea). *Diversity* **2023**, *15*, 1099. [CrossRef]
18. Bañón, R.; Fariña, J.; de Carlos, A. New Data on Exotic Muricid Species (Neogastropoda: Muricidae) from Spain Based on Integrative Taxonomy. *Diversity* **2023**, *15*, 1185. [CrossRef]
19. Mou, J.; He, X.; Liu, K.; Huang, Y.; Zhang, S.; Zu, Y.; Liu, Y.; Cao, S.; Lan, M.; Miao, X.; et al. Benthic Biodiversity by Baited Camera Observations on the Cosmonaut Sea Shelf of East Antarctica. *Diversity* **2024**, *16*, 277. [CrossRef]
20. Hernández-Casas, C.M.; Mendoza-González, Á.C.; García-López, D.Y.; Mateo-Cid, L.E. Spatio-Temporal Structure of Two Seaweeds Communities in Campeche, Mexico. *Diversity* **2024**, *16*, 344. [CrossRef]
21. Misiuk, B.; Brown, C.J. Benthic Habitat Mapping: A Review of Three Decades of Mapping Biological Patterns on the Seafloor. *Estuar. Coast. Shelf Sci.* **2024**, *296*, 108599. [CrossRef]
22. Brandt, A.; Gutt, J.; Hildebrandt, M.; Pawlowski, J.; Schwendner, J.; Soltwedel, T.; Thomsen, L. Cutting the Umbilical: New Technological Perspectives in Benthic Deep-Sea Research. *J. Mar. Sci. Eng.* **2016**, *4*, 36. [CrossRef]
23. Schenone, S.; Azhar, M.; Delmas, P.; Thrush, S.F. Towards Time and Cost-Efficient Habitat Assessment: Challenges and Opportunities for Benthic Ecology and Management. *Aquat. Conserv.* **2023**, *33*, 1603–1614. [CrossRef]

Disclaimer/Publisher's Note: The statements, opinions and data contained in all publications are solely those of the individual author(s) and contributor(s) and not of MDPI and/or the editor(s). MDPI and/or the editor(s) disclaim responsibility for any injury to people or property resulting from any ideas, methods, instructions or products referred to in the content.

Article

Soft-Bottom Infaunal Macrobenthos of the Avilés Canyon System (Cantabrian Sea)

Antía Lourido [1,*], Santiago Parra [1] and Francisco Sánchez [2]

[1] Centro Oceanográfico de A Coruña, Instituto Español de Oceanografía, (IEO-CSIC), 15001 A Coruña, Spain
[2] Centro Oceanográfico de Santander, Instituto Español de Oceanografía, (IEO-CSIC), 39004 Santander, Spain
* Correspondence: antia.lourido@ieo.csic.es

Abstract: The Aviles Canyon System (Northern Atlantic coast of Spain) is one of the ten marine regions studied in the Spanish seas by the LIFE+ INDEMARES project, which aims to identify special areas of conservation within the Natura 2000 Network. This study aims to characterize the composition and distribution of the macrobenthic fauna in order to provide baseline data to obtain a basic knowledge of the environment. Three oceanographic surveys were carried out to investigate species and habitats of this deep-sea ecosystem. The stations were sampled using a box corer, in order to evaluate the distribution and biodiversity of the macroinfauna, and to analyse the granulometric composition and the organic matter content. Sediments were mainly sandy in nature, the finest sediments with the highest organic matter content were found in the deepest areas, while coarser sediments were located in shallow stations. Polychaetes were the best represented group in total number of species and individuals, followed by crustaceans and molluscs. Five major macrobenthic assemblages were determined through multivariate analyses. Bathymetry and sedimentary composition were the main factors structuring the benthic community separating shallow and coarser stations from deeper and finer ones.

Keywords: infaunal macrobenthos; soft-bottom; community structure; Aviles Canyon System

Citation: Lourido, A.; Parra, S.; Sánchez, F. Soft-Bottom Infaunal Macrobenthos of the Avilés Canyon System (Cantabrian Sea). *Diversity* **2023**, *15*, 53. https://doi.org/10.3390/d15010053

Academic Editor: Renato Mamede

Received: 22 November 2022
Revised: 28 December 2022
Accepted: 29 December 2022
Published: 2 January 2023

Copyright: © 2023 by the authors. Licensee MDPI, Basel, Switzerland. This article is an open access article distributed under the terms and conditions of the Creative Commons Attribution (CC BY) license (https://creativecommons.org/licenses/by/4.0/).

1. Introduction

The deep-sea, the largest ecosystem on earth which comprises more than 65% of the surface of the world, is usually defined as deeper than 200 m, where the shelf break begins [1], and where the shallow fauna of the shelf is replaced by the deep-sea fauna of the deep ocean basins. Deep-sea was believed to be faunistically very poor, based on the lack of light, cold temperature and the immense pressure that characterize this area. However, this idea was wrong: this environment, formed by hard and soft bottoms [1], is constituted by a variety of ecosystems and crossed by submarine canyons. Submarine canyons are irregular incisions ("V" or "U" shaped) that cut down the continental slope, connecting continental shelves to deep ocean basins [2]. These complex and heterogeneous systems, that are abundant and ubiquitous along continental margins, are very productive areas [3]. Canyons present different bottom types, are usually recognized as organically enriched environments, and can also be highly unstable due to high sediment loads, internal tides or episodic strong down canyon flows [4,5]. Moreover, these submarine structures are pathways for the transport of sediments and organic matter from the continental shelf to deep basins, often providing high quality food supply [6]. Benthic infaunal communities depend on this allochthonus organic matter, and consequently, canyons maintain higher density of benthic assemblages than open slopes at comparable depths [5,7].

Nowadays, the understanding of the deep-sea and the topographical features that intersect them are changing with the development of new technologies for sampling and study these areas [8]. Submarine canyons are major sources of habitat heterogeneity in continental margin settings [2,9] by change local and regional conditions, and this sediment can be periodically resuspended by natural causes such as internal waves or tidal currents. Consequently, the increased habitat heterogeneity in canyons enhances benthic biodiversity on hard substrata as well as mobile sediments, and they are also often recognized as 'hotspots' of biological activity [2].

This work is focused on the study of the Avilés Canyon System (ACS), a complex system of interrelated canyons located in the central Cantabrian Sea (North Iberian continental margin) that was declared SIC (Site of Community Interest) in 2015 after exhaustive sampling campaigns carried out in the area. The ACS is constituted by three main canyons and some other minor tributaries [10]. There is also a marginal platform and a relevant rocky outcrop in the area. The ACS is impacted by a range of natural and anthropogenic activities such as fisheries, intense sea traffic, degradation of the coast due to excessive industrial, urban, and tourism development, etc. Therefore, the managing of the fisheries that take place in the area, and the protection of these marine habitats are very important targets in the ACS. The canyon ecosystems (benthic and pelagic) and the physical processes that support them, together with geology and geophysics, were the focus of a major work program: INDEMARES (LIFE+) project "Inventory and designation of marine Natura 2000 areas in Spanish sea" whose main objective "is to contribute to the protection and sustainable use of the biodiversity in the Spanish seas through the identification of valuable areas for the Natura 2000 Network" (www.indemares.es; accessed on 10 October 2022), and include the study of ten possible new areas that accomplish the necessary requirements for being established as MPAs (Marine Protected Areas). Although the investigation was multidisciplinary, this paper specifically deals with the infaunal macrobenthos of the area, to increase our knowledge on these marine species and communities. The macrofauna are defined as small-sized organisms (0.25–0.50 mm), living buried within the sediment of the ocean floor, and consisting primarily of polychaetes, peracarid crustaceans and bivalve molluscs [1].

In spite of numerous studies conducted in submarine canyons around the world, little is known about the benthic macroinfauna inhabiting these sediments at the present time. Previous ecological works in the submarine canyons areas have been centered on the study of large organisms, such as corals and sponges, or upon fish populations of commercial value and marine top predators (e.g., cetaceans, ...), but there are relatively few studies focused upon the community structure of infaunal organisms [9]. Therefore, the objective of this study was to characterize the composition and distribution of the soft-bottom macrofaunal communities of the Aviles Canyon System in order to provide baseline data to contribute to the knowledge of this particular environment. The results of the present research (i) will provide information about endobenthic organisms, (ii) working at the species level will allow us to better understand the interrelationships between the species that structure each community, and (iii) will allow us to describe patterns of biodiversity, abundance and community structure of soft-bottom infaunal macrobenthos of the ACS as well as its currently relationship with environmental variables.

2. Materials and Methods

2.1. Study Area

The ACS is a complex canyon and valley system located in the Cantabrian Sea (southern Bay of Biscay) (Figure 1), and is constituted by three major canyons: La Gaviera Canyon, El Corbiro Canyon and Avilés Canyon, and by two conspicuous morphologic features: El Canto Nuevo marginal platform and the Agudo de Fuera rocky outcrop. The Avilés Canyon is located 7 miles from the Spanish coast, from 128 to 4700 m depth, and it is characterized by a sedimentary and V-shaped bottom that is 75 km long. The El Corbiro Canyon is 23 km long, and also has a sedimentary V-shaped floor, while La Gaviera Canyon presents a U-shaped bottom with two narrow differentiated flanks: the east side shows a sedimentary character, whereas the west side is characterized by the incision of different gullies [10]. The continental shelf shows a flat slope with rock outcrops and limited sedimentary zones, showing thin unconsolidated sedimentary cover [10]. These submarine canyons play an important role as high productivity systems transporting sediments and organic matter from the continental shelf to the Biscay Abyssal Plain [10]. Moreover, in this area exists a high diversity of habitats and biological communities resulting from a high gradient of environmental variables, and from the existence of strong hydrodynamic activity associated to the topography [10,11]. Additionally, this canyon system is known to provide Essential Fish Habitats (EFH) for important commercial species, such as hake and monkfish, as well as habitats for sharks, cetaceans and giant squid [12]. The problem in this area is that the intense fishing activity is causing several damaging interactions, as well as the uses of the coast, or the extreme industrial, urban, and tourism expansion are also impacting offshore seas [13].

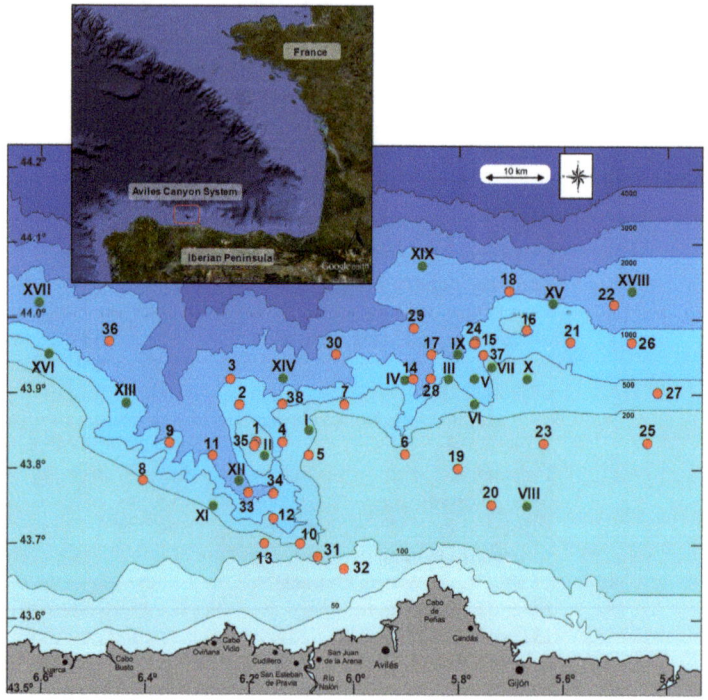

Figure 1. Location of the ACS, and position of the stations sampled on this study area. Green circles with roman numerals mark the stations where only sediment was collected. Red symbols with cardinal numerals mark the stations where sediment and infauna were collected.

2.2. Sampling Program

Within the framework of the INDEMARES (LIFE+) project, three cruises were carried out in 2010 and 2011 in the ACS to study the infaunal communities of the area, among other multidisciplinary objectives.

During the three surveys, a total of 57 stations (83–1881 m depth) were sampled using an USNEL box corer (sampling area = 0.09 m^2) (Figure 1): 57 samples were used to study the sediment characteristics (surveys 0410, 0710 and 0511), and 38 samples were used to study the endobenthic organisms (surveys 0710 and 0511). Previously, sampling stations were selected after confirmation of the presence of soft sediments with the multibeam echosounder (low backscatter values), and a single sample was taken in each station. The stations were not replicated temporally. The infaunal samples were washed carefully through a 0.5 mm mesh size sieve. The retained material was anesthetized with a MgCl$_2$ solution, and then preserved with 8% buffered formaldehyde stained with Rose Bengal solution for later sorting and identification of the fauna. The sorted fauna were classified to the lowest possible taxonomic level and counted, after which they were transferred to a 70% ethanol solution. Additionally, they were assigned to feeding categories (WoRMS, http://www.marinespecies.org, accessed on 10 October 2022), and five trophic groups were considered: carnivores, surface deposit feeders, subsurface deposit-feeders, suspension-feeders and others (the others group includes organisms with very heterogeneous food characteristics such as omnivores, herbivores and scavengers). An additional sediment sample was taken at each station within the box corer to analyze the granulometric composition and the organic matter content of the superficial substratum. These sediment samples were frozen onboard until the later processing in the laboratory. Particle size was analyzed by a combination of dry sieving and sedimentation techniques [14] in order to estimate the median grain size (Q_{50}) and the sorting coefficient (S_0). The organic matter content was estimated as the weight loss of dried samples (100 °C, 24 h) after combustion (500 °C, 24 h).

2.3. Data Analysis

Macrofaunal species richness (S), the Shannon–Wiener diversity index (H', as $\log_2$) and Pielou evenness index (J') were calculated for each sampling station using the community analysis PRIMER v6 software [15]. This software was also used for multivariate analyses. Abundances were calculated and expressed as number of individuals per m^2 of sampled area (ind.m^{-2}), and these data were organized into a sample vs. species matrix. Cluster analysis (based on the group-average sorting algorithm) and non-metric multidimensional scaling (MDS) ordination was performed using Bray–Curtis similarity measure, after fourth root transformation of the biological data to classify and order the stations into macrobenthic assemblages. These analyses were carried out to identify the groups of sampling sites that do not differ significantly in the macrofaunal composition. Values of selected abiotic features were further superimposed to visually detect any related pattern in that ordination.

The one-way ANOSIM test (PRIMER) was used to test for differences among samples from different sites with the null hypothesis that there are no significant differences. In addition, the SIMPER (Similarity Percentage) routine was then run to identify species that greatly contributed to the differentiation of station groups. The species present in each group of stations were further classified according to the constancy and fidelity indexes [16,17], which evaluates the constancy and the numerical importance of each species within a group of stations (Supplementary Material, Table S1). Five categories of constancy index were considered according to the number of times the species was found in the total of samples: constant (>76%), very common (51–75%), common (26–50%), uncommon (13–25%) and rare (<12%). According to the ratio between this index and the total constancy in the considered area (or group of stations), the fidelity index classifies the species as accidental (<10%), occasional (11–33%), accessory (34–50%), preferential (51–66%), elective (67–90%) and exclusive (>91%). The BIO-ENV routine of the PRIMER package was used to study the association of environmental factors with species abundances. This was carried out using the species abundances dataset and environmental factors dataset. All variables expressed in percentages were previously transformed by log (x + 1), and all the abiotic variables were standardised. The abiotic variables used in these analyses were water depth (m), total organic matter content (TOM; %), and sediment characteristics, including the weight percentage of coarse sands (>500 µm), fine sands (62–500 µm) and mud (<62 µm), the mean particle diameter (Q_{50}; mm), and the sorting coefficient (S_0).

3. Results

3.1. Sediments

The soft bottoms sampled in the ACS are dominated by sandy sediments, mainly by fine and very fine sands. The finest sediments with the highest organic matter content are found in the deepest areas of the continental shelf and slope, while coarser sediments are located in the shallower stations of the continental shelf (see Supplementary Material, Table S2 with the main characteristics of the sampling stations and Figure S1 with the PCA analysis representation).

Specifically, the shallowest bottoms of the continental shelf (<500 m depth) are characterized by the coarser sediments of the study area (very coarse, coarse, medium and fine sands). The organic matter content is low (1.2–5%), the sorting coefficient moderate, and the median grain size ranges between 0.12 and 1.15 mm. It is significant to highlight the presence of coarse sand and very coarse sand in the SW region of the continental shelf-break, in front of the Avilés Canyon; at medium depths (500–1000 m depth), fine and very fine sands dominate the sampled bottoms. The medium grain size range between 0.08 and 0.20 mm, the sorting coefficient between poor and moderate, and the organic matter content between 1.8% and 7.6%; finest and muddy sediments are located in the deepest stations (>1000 m depth), and they are characterized by the highest organic matter content (1.8–11.5%), by a lower sorting coefficient (from bad to moderate), and by a median grain size between 0.03 and 0.42 mm.

This general pattern of the spatial distribution of different types of sediment is more or less altered on a smaller scale by the structural complexity of the area.

3.2. Macrofaunal Composition and Structure

A total of 5053 individuals belonging to 505 taxa in 176 families were collected from the infaunal samples. Polychaetes (56.8% of total abundance; 263 species) were by far the most abundant group, followed by Arthropoda (Crustacea Subphyllum: 17.7%; 151 species). Molluscs (13.9%; 60 species), Echinodermata (6.3%; 24 species), and Others group were less common. Figure 2 shows the density (ind/m^2) of the major faunal groups ordered by depth. Macrofaunal abundance decrease with depth: the highest abundance values appeared in shallow stations (<500 m depth): sites 27, 25, 6 and 20 (2811, 2511, 2433 and 2311 indv/m^2, respectively) at a depth of 457, 157, 168 and 144 m; while the lowest values were recorded at deeper stations: sites 2, 3, and 22 (544, 567 and 733 indv/m^2, respectively) at a depth of 637, 1033 and 1184 m. There is also a decrease in the species richness with depth, ranging from 81 species at station 25 (157 m) to 29 species at station 2 (637 m). The highest Shannon H' diversity index was observed at stations 25 and 37 (H' = 5.8; medium and fine sand; 157 and 499 m depth, respectively), while the lowest diversity index was found at station 34 (H' = 4.5; very fine sand; 1017 metres depth) (Supplementary Material, Figure S2).

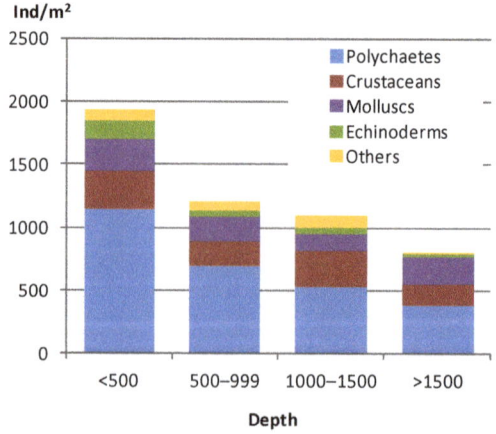

Figure 2. Abundance indices (indv/m^2) of the major macrofauna taxa on the ACS according to depth (m).

Thirteen families (or higher taxonomic level) accounted for more than 50% of all individuals: the polychaetes Spionidae, Paraonidae, Ampharetidae, Sabellidae, Capitellidae, Syllidae, Maldanidae, Onuphidae and Cirratulidae, the bivalves Bivalvia indet. and Thyasiridae, the nemerteans, and the Echinoidea indet. The most speciose families were Syllidae (31 species), Paraonidae (29), Spionidae (22), Ampharetidae (20), Onuphidae (12), Ampeliscidae (10), Lumbrineridae (10) and Terebellidae (10). Regarding the species, the dominant ones were the polychaetes *Eclysippe vanelli*, *Levinsenia flava*, *Prionospio cirrifera*, *Aricidea (Aricidea) wassi*, *Euchone incolor*, *Glycera lapidum* and *Prionospio* sp., the sipunculid *Onchnesoma steenstrupii steenstrupii*, and the bivalves *Timoclea ovata* and *Kelliella* sp., which accounted for more than 25% of all individuals (Table 1). Six species were present on more than 20 stations: *Glycera lapidum*, *Levinsenia flava*, *Prionospio* sp., *Euchone incolor*, *Onchnesoma steenstrupii steenstrupii*, and *Notomastus latericeus* (Table 1).

Table 1. Dominant species which accounted for more than 25% of abundance of all individuals, together with species present on more than 20 stations.

	Dominance (%)	Presence (N° Stations)
Eclysippe vanelli	3.0	
Levinsenia flava	2.8	25
Onchnesoma steenstrupii steenstrupii	1.8	23
Prionospio cirrifera	1.5	
Aricidea (Aricidea) wassi	1.4	
Euchone incolor	1.4	24
Glycera lapidum	1.4	26
Timoclea ovata	1.4	
Kelliella sp.	1.4	
Prionospio sp.	1.2	25
Notomastus latericeus		23

Polychaetes were the best represented group in total number of organisms at the three bathymetric levels considered (less than 500 m; 500–1000 m; more than 1000 m). At shallow, medium and deep stations, polychaetes dominated the infaunal community with values of 33.5–80.9%, 39–83.6%, and 28.5–82.4%. The same polychaete families, Spionidae, Paraonidae, and Ampharetidae, were the most abundant ones at these three bathymetric levels. This infaunal group was followed by crustaceans (5.9–39.9%; 6.0–28.6%; 11.8–53.8%), mainly dominated by Ampeliscidae, Melitidae, and Phoxocephalidae amphipods, and by molluscs (3.3–30.7%; 5.6–33.8%; 2.0–26.4%), with Thyasiridae and Veneridae as more abundant families. Echinoderms and the Others group were less abundant.

3.3. Multivariate Analyses

The ANOSIM test revealed significant differences in faunistic composition between all sites (global $R = 0.779$, $p = 0.001$). The dendrogram obtained by cluster analysis showed the presence of four major biological groups of stations at a similarity level of 25% (Figures 3 and 4): group A (140 ± 40 m depth), B (1676 ± 291 m depth), C (161 ± 38 m depth), and D (D1: 489 ± 84 m depth; D2: 970 ± 225 m depth). Stations 3, 10, 11, 18, 21, 22 and 33 did not belong to group A, B, C or D (see Supplementary Material, Table S3 with the results of ANOSIM test for these groups). nMDS ordination (Figure 5) showed similar results to those of the dendrogram (stress: 0.21). The summary of characteristics for each association is shown in Table 2.

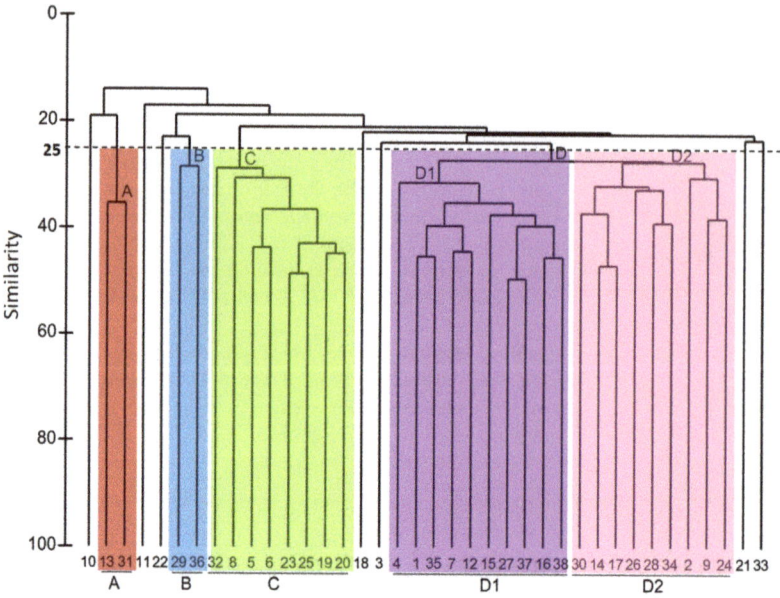

Figure 3. Hierarchical cluster analysis for benthic macrofauna at the ACS.

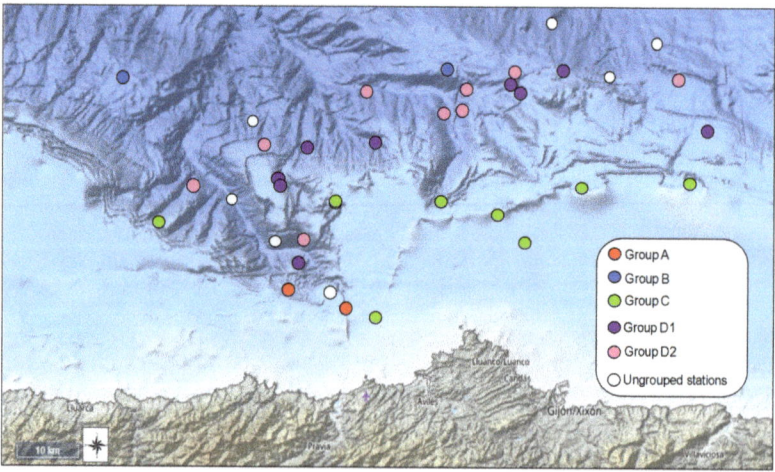

Figure 4. Location of sampling sites in the ACS showing the groups determined by cluster analysis.

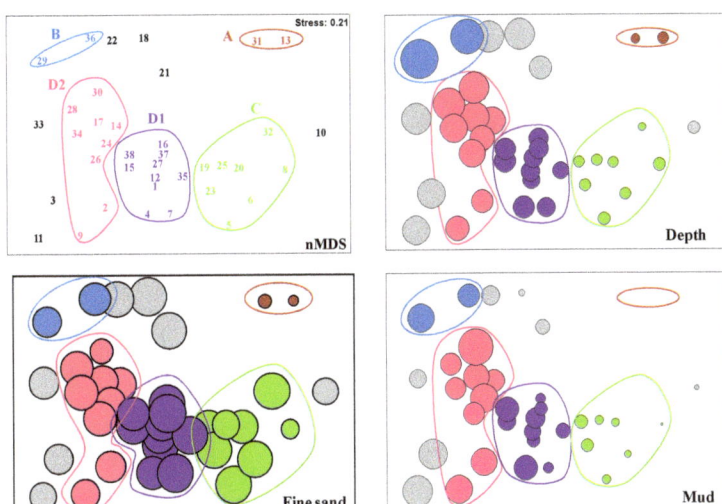

Figure 5. Non-metric multidimensional scaling (nMDS) ordination of sampling sites showing groups determined by cluster analysis, and ordination of sites with values of some environmental variables superimposed (depth; fine sand; mud).

Table 2. Abiotic and biotic characteristics of the main infaunal macrobenthic assemblages discriminated by the multivariate analysis of abundance data.

		Depth (m)	Q$_{50}$ (mm)	S$_0$	TOM(%)	CS	FS	Mud	Sediment Type	S	N
Group A	Mean	140 ± 40	1.14 ± 0.01	Mod	2.49 ± 0.45	90.6 ± 2.8	9.4 ± 2.8	0	MS (1 site),	56 ± 18	2067 ± 47
	Range	112–168	1.13–1.15		2.17–2.81	88.7–92.6	7.4–11.3	0	VCS (1)	43–68	2033–2100
Group B	Media	1676 ± 291	0.14 ± 0.05	Poor	5.01 ± 0.47	4.8 ± 1.9	64.0 ± 6.9	31.3 ± 8.8	VFS (1),	44 ± 5	978 ± 251
	Range	1470–1881	0.11–0.17		4.67–5.34	3.4–6.1	59.1–68.9	25.0–37.4	FS (1)	40–47	800–1156
Group C	Media	161 ± 38	0.37 ± 0.22	Mod	2.75 ± 0.78	26.3 ± 23.6	70.0 ± 22.6	3.7 ± 3.1	FS (3), MS (3),	63 ± 14	1903 ± 536
	Range	83–208	0.23–0.88		1.34–3.75	8.1–76.4	23.3–88.0	0–8.6	CS (1), VCS (1)	39–81	1122–2511
Group D1	Media	489 ± 84	0.15 ± 0.04	Mod	3.48 ± 1.48	3.2 ± 3.4	82.2 ± 9.2	14.6 ± 8.6	VFS (3),	61 ± 12	1673 ± 581
	Range	356–612	0.11–0.23	Poor	1.22–6.60	0.7–12.1	68.1–91.7	5.3–28.7	FS (7)	39–75	744–2811
Group D2	Media	970 ± 225	0.10 ± 0.03	Poor	5.47 ± 2.22	1.4 ± 1.2	64.9 ± 12.8	33.6 ± 13.0	Mud (1),	51 ± 13	1217 ± 418
	Range	637–1318	0.04–0.12	Mod	1.84–8.25	0–3.1	37.2–76.9	22.5–62.8	VFS (8)	29–66	544–1800

Notes: Q$_{50}$: mean grain size; S$_0$: sorting coefficient; TOM: total organic matter; CS: coarse sand; FS: fine sand; MS: medium sand; VCS: very coarse sand; VFS: very fine sand; S: number of species; N: number of individuals per m^2; Mod: moderate sorted.

Group A was made up of two shallow stations (medium and very coarse sand), and was characterized by the lowest organic matter content, and the highest number of individuals. The most abundant species of this group (accounting for 25% of all the fauna; Supplementary Material, Table S4) were the bivalves *Limatula* sp. (constant/elective), the polychaetes *Pisione* sp. (constant/exclusive), and the amphipods family Aoridae, while the most characteristics species were the constant and exclusive polychaetes *Caulleriella bioculata*, *Goniadella* sp., *Pisione* sp., *Sphaerosyllis bulbosa* and the bivalve *Spisula* sp.

Group B comprises the deepest stations of the sampling program. These stations were composed of fine and very fine sediments, with relatively high contents of organic matter. The caprellids (Caprellidae indet.; common/preferential), the polychaetes *Aurospio dibranchiata* (constant/preferential), and the scaphopods *Antalis agilis* were the most abundant species of this group (accounting for 25% of all the fauna). The polychaetes *Nothria* sp. (constant/exclusive), *Levinsenia* sp. 1 (constant/elective), and *Notoproctus* sp. (constant/elective), and the isopod Anthuridae sp. (constant/elective) characterized this group.

Group C is defined by the shallowest stations of the study area. This group is composed by very coarse, coarse, medium and fine sands, with low organic matter content. The most abundant species (accounting for 25% of all the fauna) were the nemerteans (Nemertea indet.), the bivalves Bivalvia indet. and *Timoclea ovata* (constant/preferential), the echinoids Echinoidea indet., and the polychaetes *Aricidea (Aricidea) wassi* (constant/preferential), *Aonides paucibranchiata*, *Euchone incolor*, and *Galathowenia oculata*. The most characteristic species of this group were the polychaetes *Aponuphis bilineata* (very common/exclusive), *Pista* sp. (very common/elective), and *Spiochaetopterus* sp. 1 (very common/elective), and the decapods (Decapoda indet.; very common/elective).

Group D was further subdivided into group D1 and group D2. Group D1 was made up of fine sand stations located at medium depths. The most abundant taxa of this group (accounting for 25% of all the fauna) were the polychaetes *Eclysippe vanelli* (constant/preferential), *Levinsenia flava*, *Magelona filiformis*, and *Prionospio cirrifera*, and the bivalves Bivalvia indet., and *Kelliella* sp., while the most characteristics species were the polychaetes *Ophelina abranchiata* (constant/elective), the maldanids Maldanidae gen. sp. 2 (very common/elective), and *Spiophanes wigleyi* (very common/elective). Group D2 comprised sediments deeper than D1, with the highest organic matter content of the study area, and with the finest granulometric fractions (mainly, very fine sands). The polychaetes *Levinsenia flava* and *Aurospio dibranchiata*, the caudofoveata (Caudofoveata indet.), the sipunculid *Onchnesoma steenstrupii steenstrupii*, the bivalves *Genaxinus eumyarius* (very common/exclusive) and *Axinulus croulinensis* (very common/preferential), and the nemerteans (Nemertea indet.) were the most abundant taxa of this group (accounted for 25% of all the fauna). The amphipods *Metaphoxus simplex* (constant/elective) and the polychaetes *Paradiopatra* sp. 2 (very common/elective) also characterized this group.

SIMPER analysis showed that the bivalve *Limatula* sp., and the polychaetes *Pisione* sp., *Protodorvillea kefersteini* and *Prionospio cirrifera* explained most of the dissimilarity between groups A and C (average dissimilarity = 80.1%), while the bivalves *Limatula* sp. and *Tellina* sp. together with the polychaete *Pisione* sp., contributed greatly to the differentiation of A from D1 (average dissimilarity = 86.9%). The bivalve *Timoclea ovata*, and the polychaetes *Eclysippe vanelli*, *Ophelina abranchiata*, and *Levinsenia gracilis* differentiated group C from D1 (average dissimilarity = 73.4%). Group D1 differed from D2 (average dissimilarity = 72.8%) due to the polychaetes *Eclysippe vanelli*, *Prionospio cirrifera*, and *Aricidea (Aricidea) wassi*, whereas the polychaetes *Nothria* sp., *Levinsenia* sp. 1, and the cirratulids (Cirratulidae indet.), the sipunculid *Onchnesoma steenstrupii steenstrupii*, and the cumaceans (Cumacea indet.) greatly contributed to separating D2 from B (average dissimilarity = 77.0%).

3.4. Relationship between Biotic and Environmental Variables

The BIO-ENV procedure analysis suggests that the variables depth, fine sand and mud content are the major structuring factors of the benthic community (ρ = 0.671). Depth and mud content show the highest correlation values if they are considered separately (ρ_w = 0.500 and 0.473, respectively).

The nMDS ordination of sites with superimposed values of these three abiotic variables showed that stations appeared distributed from left to right following decreasing values of depth, mud and fine sand content (Figure 5).

3.5. Trophic Structure

The macrofaunal community of the ACS was dominated by deposit feeders, specifically by surface deposit feeders which represent more than 40% of total abundance on average. At most stations (33 stations), surface deposit feeders predominated and represented between 25.6% and 54.7% of the populations. The second most strongly represented trophic group were the subsurface deposit feeders which accounted for more than 25% of total abundance on average (from 4.2 to 53.2%); this is the major feeding mode at four stations. Suspensivores (0–31.6%) dominated only in one station, while Carnivores and the Others group did not dominate any station. Moreover, all trophic groups showed the same pattern with depth, decreasing their abundances at greater bathymetries. However, the pattern of the proportion of each trophic group is different (Figure 6). In this case, carnivores and surface deposit feeders decrease their proportion with depth, whereas subsurface deposit feeders and the Others group increase this proportion. Suspensivores show a light decrease with depth.

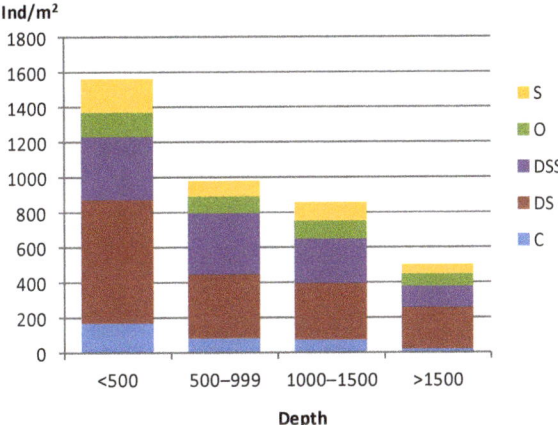

Figure 6. Relative abundance (%) of macrofaunal feeding types at each strata. O: Others group; C: carnivores; S: suspensivores; DSS: subsurface deposit feeders; DS: surface deposit feeders.

4. Discussion

Previously to this study, [11] characterized the habitat of the deep-water coral reefs in La Gaviera Canyon (ACS), and [10] described the geomorphology of the ACS. However, the only preliminary descriptions focused on the species composition of the benthic assemblages of the ACS were made by [18] using an unpublished historical dataset obtained in 1987–1988, and by [19], with the data of this study but only considering the family taxonomic level. [18] described the macrobenthic community assemblages in the ACS (N Iberian Shelf) on the basis of 42 anchor dredge and/or epibenthic sledge collected in 1987–1988, sieved on a 2 mm mesh, collected at water depths of 31–1400 m. They identified 810 macrofaunal taxa belonging to eleven Phyla, and, in the present work, were identified 505 taxa and six Phyla (38 stations from 83 to 1881 m). Other [18] work focused on larger animals (megafaunal communities) than this study does, such as epifauna settling on big stones, corals or pennatulaceans, and comprised phyla not considered in the present study, such as Cnidarians, Brachiopods or Bryozoans. Therefore, these factors, together with the different sampling methods used in both studies (gears, sieves, etc.), make comparisons difficult, and reveal that this work adds new information with actualized data about the ACS, more focused in the infaunal community (gear: box corer; sieve: 0.5 mm; maximum depth: 1881 m). In spite of that, there are general patterns that can be observed in both works, such as the decrease in sediment grain size with depth. As was stated in [19], median grain-size decreased and sediment organic matter content increased with increasing water depth in the ACS. The coarser sediments (very coarse, coarse and medium sands) are located at the shallowest stations (continental shelf), while fine and very fine sands prevail below 300 m. This gradient in grain size from the shallowest sites down to the deeper areas has also been recorded in other deep-sea areas [1,20,21]. Moreover, [18] suggested depth as the major structuring agent of the benthic community, and that sediment characteristics were also an important factor influencing this community structure. The same results are shown in this work, where the statistical analyses select depth, fine sand and mud as the major structuring agents of the community, and where the cluster analyses also discriminated between shallow from deep stations.

The distribution and diversity of deep-sea infauna have mainly been related to depth and sediment [6], and canyons systems are extremely variable in the terms of sediment and organic matter [5]. The presence of three canyons and rocky outcrops on the continental shelf has a strong influence on the sedimentary processes, generating the different deposits found [10]. Different sediment types can allow the settlement of different benthic fauna, and moreover, sediment composition changes with depth, and this fact limits the distribution of some species [1]. The depth distribution of a species may also be controlled by food availability [22]: the abundances of the macrofaunal organisms decrease to the abyss because of the reduction in nutrient input [23]. Furthermore, there are other factors besides sediment type and nutrient input that vary with depth and that can also influence infaunal organisms' distribution, such as temperature, hydrostatic pressure, light intensity or current dynamics besides biological interactions (competition, predation) [23]. Therefore, macrofaunal abundance decreases with depth in many deep-sea environments [1], and the results of this work agree with this pattern that has already been recorded in other deep-sea areas ([24] in the Gay Head-Bermuda transect; [25] in the tropical Northeast Atlantic; [20] in the Goban Spur in the NE Atlantic Ocean; [21] in the Northwest Atlantic; [1] in the Gulf of Mexico; [5] in three Portuguese submarine canyons in the NE Atlantic).

According to the studies that appear in [23], species richness pattern follows a parabolic curve along depth gradient: diversity increases to intermediate depths but then decreases toward the abyss. In the ACS this pattern does not take place because the existence of the submarine canyons could affect this pattern, and in this work the species richness decreases regularly with depth, giving rise to a similar decrease in the diversity index according to depth. One possible explanation for this pattern is that species diversity may also be related to the sediment particle diversity that provides great habitat complexity, so if the particle diversity decreases with depth, the same is going to happen with the species diversity [23].

Polychaetes are the numerically dominant macrofaunal taxon in the deep-sea [26,27]. In the ACS, these annelids accounted for 56.8% of the total macrofaunal abundance, attaining similar relative abundances to other deep-sea areas [5,18,21,26,28]. The most numerous families of this faunistic group were mostly the same as have already been recorded in these other studies, where Spionidae, Paraonidae and Cirratulidae stand out among the rest of the families of these deep-sea areas. Our results also show that the next most abundant faunistic group of deep-sea macrofauna are the peracarid crustaceans, followed by the molluscs, as also found by [26].

Regarding the trophic groups, deposit feeders species are largely dominant. Deposit feeders ingest sediment so they rely on sediments for nutrition [23]. They comprise the majority of species, and specifically in this work, surface deposit feeders, that feed from the sediment surface, are more numerous than subsurface deposit feeders, that feed as they burrow through the sediment, at most of the stations in the ACS. Deposit feeding is the dominant feeding mode in the deep-sea [5,6], with surface deposit feeders as the most abundant macrofaunal feeding mode in deep-sea sediments [26,29]. The dominance of deposit feeding may arise because the habitat is unsuitable for suspension feeders, and also due to a possible limitation of resources and prey for macrophagous feeders. Moreover, the proportion of surface deposit feeders and carnivores, the latter typically more abundant at shallow stations [5,30,31], tended to decrease with water depth, whereas the proportion of subsurface deposit feeders and the Others group tended to increase. Suspension feeders, which feed on material they collect from the water column, decrease in overall importance with increasing water depth [27] given the low suspended-particle in the deep sea. Therefore, in this work suspension feeding is rare [23,26].

Matches with Habitat Characterization

During the cruises carried out in the INDEMARES project, the sea floor of the margin was mapped to locate and map possible Vulnerable Marine Ecosystems, including continental shelf, continental slope and a narrow band of abyssal plain attached to the base of the continental slope [10]. The investigation was multidisciplinary, involving geology–geophysics, biology (benthic and pelagic), ecology and physical oceanography. The INDEMARES surveys used multidisciplinary sampling gear in an attempt to reach these objectives, such as otter trawl and beam trawl for communities that inhabit sedimentary areas; ROV, benthic photolander platform and photogrammetric sled for complex and vulnerable areas; box-corer to study sedimentological characteristics, organic content data and infaunal communities (this work); multibeam echosounder, high-resolution seismic profiles (TOPAS system), etc. Thus, the study of the geomorphology and shallow structure of the area will provide important information about the configuration of the sea floor, which is essential in understanding the distribution of benthic communities and bottom circulation patterns.

The results obtained in those studies shown that the particular habitats found in the area do not seem to match with the criteria for the classification of the habitats according to EUNIS (discrepancies in their design, the same habitat can be classified into different levels, etc.) [32]. However, there are 24 habitats identified already in the ACS after the study carried out by the INDEMARES (LIFE+) project [33], which collected different kinds of data: bathymetry and geomorphology, substratum, hydrography and dynamics, biological samples, and fishery impacts. Some of these benthic habitats identified match with the infaunal groups determined in this work through the multivariate analyses. Thus, group A is located on the "Faunal communities in Atlantic offshore circalittoral coarse sediment" (MD321), group C on the "Faunal communities in Atlantic offshore circalittoral sand" (MD521), group D1 mainly on the "Sparse communities on Atlantic upper bathyal sand" (ME521) and group D2 on the "Atlantic upper bathyal sand" (ME52). Some of these habitats, particularly the habitats located in sedimentary grounds, and their biological communities can be altered and disturbed by fishing activities [13].

The present study provides an important insight into infaunal assemblages of the ACS. Benthic infauna are mostly sedentary, so they are frequently examined to track the impact of disturbance on marine environments. Exploration and investigation on canyons macrobenthic assemblages provide important information for understanding the ecosystem processes in those environments. This work based on the ACS and its ecosystems tries to address the actual condition of the canyons system as well as to gain important environmental baseline information, because there is a lack of studies in deep infaunal communities as well as a taxonomic bias toward larger animals [34]. Over the last years, significant progress on deep sea macroecology has been made, but much still remains to be done. Therefore, future studies should investigate the distribution of individual species in the deep sea, as well as to test specific hypotheses and patterns, and to increase our information of specific taxa, basins and oceans [35].

Supplementary Materials: The following supporting information can be downloaded at: https://www.mdpi.com/article/10.3390/d15010053/s1, Figure S1: Principal component analysis (PCA) showing sampling sites and the environmental variables for the ACS (Q50: median grain size; S0: sorting coefficient; TOM: total organic matter content); Figure S2: Density (ind/m^2), richness (numer of taxa) and Shannon index (H', $\log_2$) of macroinfauna vs. depth; Table S1: Constancy and fidelity indices; Table S2: Depth, location, sedimentary type and sorting coefficient (S_0) of sampling stations in the ACS; Table S3: Results of the ANOSIM test; Table S4: Constancy and Fidelity indices of the more abundant species of each group of stations.

Author Contributions: Conceptualization, A.L., S.P. and F.S.; methodology, A.L., S.P. and F.S.; software, A.L.; validation, A.L., S.P. and F.S.; formal analysis, A.L.; investigation, A.L., S.P. and F.S.; resources, A.L., S.P. and F.S.; data curation, A.L., S.P. and F.S.; writing—original draft preparation, A.L., S.P. and F.S.; writing—review and editing, A.L., S.P. and F.S.; visualization, A.L., S.P. and F.S.;

supervision, A.L., S.P. and F.S.; project administration, F.S.; funding acquisition, F.S. All authors have read and agreed to the published version of the manuscript.

Funding: This study was partially funded by the European Commission LIFE+ 'Nature and Biodiversity' (INDEMARES Project, 07/NAT/E/000732) and by the Spanish Science and Technology Ministry (ECOMARG3 Project).

Institutional Review Board Statement: Not applicable.

Informed Consent Statement: Not applicable.

Data Availability Statement: Not applicable.

Acknowledgments: The authors would like to thank all the participants of the oceanographic surveys, and the crew of the R/V Vizconde de Eza and Thalassa.

Conflicts of Interest: The authors declare no conflict of interest.

References

1. Thistle, D. The deep-sea floor: An overview. In *Ecosystems of the Deep Oceans. Ecosystems of the World*; Tyler, P.A., Ed.; Elsevier: Amsterdam, The Netherlands, 2003; Volume 28, pp. 5–37.
2. De Leo, F.C.; Smith, C.R.; Rowden, A.A.; Bowden, D.A.; Clark, M.R. Submarine canyons: Hotspots of benthic biomass and productivity in the deep sea. *Proc. R. Soc. Lond.* **2010**, *277*, 2783–2792. [CrossRef] [PubMed]
3. Coelho, H.; Villarreal, M.R.; Leitão, P.C.; del Río, G.D. The low frequency circulation over Aviles canyon: Observations and modeling. *Geophys. Res. Abstr.* **2004**, *6*, 07206.
4. Schlacher, T.A.; Schlacher-Hoenlinger, M.A.; Williams, A.; Althaus, F.; Hooper, J.N.A.; Kloser, R. Richness and distribution of sponge megabenthos in continental margin canyons off southeastern Australia. *Mar. Ecol. Prog. Ser.* **2007**, *340*, 73–88. [CrossRef]
5. Cunha, M.R.; Paterson, G.L.J.; Amaro, T.; Blackbird, S.; Stigter, H.C.; Ferreira, C.; Glover, A.; Hilario, A.; Kiriakoulakis, K.; Neal, L.; et al. Biodiversity of macrofaunal assemblages from three Portuguese submarine canyons (NE Atlantic). *Deep Sea Res. II* **2011**, *58*, 2433–2447. [CrossRef]
6. Mamouridis, V.; Cartes, J.E.; Parra, S.; Fanelli, E.; Salinas, J.I.S. A temporal analysis on the dynamics of deep-sea macrofauna: Influence of environmental variability off Catalonia coasts (western Mediterranean). *Deep Sea Res. Part I* **2011**, *58*, 323–337. [CrossRef]
7. Paterson, G.L.J.; Glover, A.G.; Cunha, M.R.; Neal, L.; de Stitger, H.; Kiriakoulakis, K.; Billett, D.S.M.; Wolff, G.; Tiago, A.; Ravara, A.; et al. Disturbance, productivity and diversity in deep-sea canyons: A worm's eye view. *Deep Sea Res. II* **2011**, *58*, 2448–2460. [CrossRef]
8. Tyler, P.A. *Ecosystems of the Deep Oceans. Ecosystems of the World*; Elsevier: Amsterdam, The Netherlands, 2003.
9. De Leo, F.C.; Vetter, E.W.; Smith, C.R.; Rowden, A.A.; McGranaghan, M. Spatial scale-dependent hábitat heterogeneity influences submarine canyon macrofaunal abundance and diversity off the Main and Northwest Hawaiian Islands. *Deep Sea Res. II* **2014**, *104*, 267–290. [CrossRef]
10. Gómez-Ballesteros, M.; Druet, M.; Muñoz, A.; Arrese, B.; Rivera, J.; Sánchez, F.; Cristobo, J.; Parra, S.; García-Alegre, A.; González-Pola, C.; et al. Geomorphology of the Avilés Canyon System, Cantabrian Sea (Bay of Biscay). *Deep Sea Res. II* **2014**, *106*, 99–117. [CrossRef]
11. Sánchez, F.; González-Pola, C.; Druet, M.; García-Alegre, A.; Acosta, J.; Cristobo, J.; Parra, S.; Ríos, P.; Altuna, A.; Gómez-Ballesteros, M.; et al. Habitat characterization of deep-water coral reefs in La Gaviera Canyon (Avilés Canyon System, Cantabrian Sea). *Deep Sea Res. Part II* **2014**, *106*, 118–140. [CrossRef]
12. Sánchez, F.; Olaso, I. Effects of fisheries on the Cantabrian Sea shelf ecosystem. *Ecol. Model.* **2004**, *172*, 151–174. [CrossRef]
13. Punzón, A.; Arronte, J.C.; Sánchez, F.; García-Alegre, A. Spatial characterization of the fisheries in the Avilés canyon system (Cantabrian Sea, Spain). *Cienc. Mar.* **2016**, *42*, 237–260. [CrossRef]
14. Buchanan, J.B. Sediment analysis. In *Methods for the Study of Marine Benthos*; Holme, N.A., McIntyre, A.D., Eds.; Blackwell Scientific Publications: Oxford, UK, 1984; pp. 41–65.
15. Clarke, K.; Gorley, R. *PRIMER v6: User Manual/Tutorial*; Primer-E Ltd.: Plymouth, UK, 2006.
16. Glémarec, M. Bionomie benthique de la partie orientale du Golfe de Morbihan. *Cah. Biol. Mar.* **1964**, *5*, 33–96.
17. Cabioch, L. Contribution a la connaissance des peuplements benthiques de la Manche occidentale. *Cah. Biol. Mar.* **1968**, *9*, 493–720.
18. Louzao, M.; Anadon, N.; Arrontes, J.; Alvarez-Claudio, C.; Fuente, D.M.; Ocharan, F.; Anadon, A.; Acuna, J.L. Historical macrobenthic community assemblages in the Avilés Canyon, N Iberian Shelf: Baseline biodiversity information for a marine protected area. *J. Mar. Syst.* **2010**, *80*, 47–56. [CrossRef]
19. Lourido, A.; Parra, S.; Sánchez, F. A comparative study of the macrobenthic infauna of two bathyal Cantabrian Sea areas: The Le Danois Bank and the Avilés Canyon System (S Bay of Biscay). *Deep Sea Res. II* **2014**, *106*, 141–150. [CrossRef]
20. Flach, E.; Muthumbi, A.; Heip, C. Meiofauna and macrofauna community structure in relation to sediment composition at the Iberian margin compared to the Goban Spur (NE Atlantic). *Prog. Oceanogr.* **2002**, *52*, 433–457. [CrossRef]
21. Levin, L.A.; Gooday, A.J. The deep Atlantic Ocean. In *Ecosystems of the Deep Oceans. Ecosystems of the World*; Tyler, P.A., Ed.; Elsevier: Amsterdam, The Netherlands, 2003; Volume 28, pp. 111–178.
22. Gage, J.D. Food inputs, utilization, carbon flow and energetics. In *Ecosystems of the Deep Oceans. Ecosystems of the World*; Tyler, P.A., Ed.; Elsevier: Amsterdam, The Netherlands, 2003; Volume 28, pp. 315–380.

23. Stuart, C.T.; Rex, M.A.; Etter, R.J. Large-scale spatial and temporal patterns of deep-sea benthic species diversity. In *Ecosystems of the Deep Oceans. Ecosystems of the World*; Tyler, P.A., Ed.; Elsevier: Amsterdam, The Netherlands, 2003; Volume 28, pp. 295–311.
24. Hessler, R.R.; Sanders, H.L. Faunal diversity in the deep sea. *Deep Sea Res.* **1967**, *14*, 65–78. [CrossRef]
25. Galeron, J.; Sibuet, M.; Mahaut, M.; Dinet, A. Variation in structure and biomass of the benthic communities at three contrasting sites in the tropical Northeast Atlantic. *Mar. Ecol. Prog. Ser.* **2000**, *197*, 121–137. [CrossRef]
26. Gage, J.D.; Tyler, P.A. *Deep-Sea Biology. A Natural History of Organisms at the Deep-Sea Floor*; Cambridge University Press: Cambridge, UK, 1991.
27. Ramirez-Llodra, E.; Brandt, A.; Danovaro, R.; De Mol, B.; Escobar, E.; German, C.R.; Levin, L.A.; Martinez Arbizu, P.; Menot, L.; Buhl-Mortensen, P.; et al. Deep, diverse and definitely different: Unique attributes of the world's largest ecosystem. *Biogeosciences* **2010**, *7*, 2851–2899. [CrossRef]
28. Escobar-Briones, E.; Estrada-Santillán, E.L.; Legendre, P. Macrofaunal density and biomass in the Campeche canyon, southwestern Gulf of Mexico. Tropical studies in oceanography. The deep Gulf of Mexico benthos Program. *Deep-Sea Res. II* **2008**, *55*, 2679–2685. [CrossRef]
29. Shields, M.A.; Blanco-Perez, R. Polychaete abundance, biomass and diversity patterns at the Mid-Atlantic Ridge, North Atlantic Ocean. *Deep Sea Res. II* **2013**, *98B*, 315–325. [CrossRef]
30. Stora, G.; Bourcier, M.; Arnoux, A.; Gerino, M.; Le Campion, J.; Gilbert, F.; Durbec, J.P. The deep-sea macrobenthos on the continental slope of the northwestern Mediterranean Sea: A quantitative approach. *Deep Sea Res. I* **1999**, *46*, 1339–1368. [CrossRef]
31. Probert, P.K.; Glasby, C.J.; Grove, S.L.; Paavo, B.L. Bathyal polychaete assemblages in the region of the Subtropical Front, Chatham Rise, New Zealand. *N. Z. J. Mar.* **2009**, *43*, 1121–1135. [CrossRef]
32. Davies, C.E.; Moss, D.; Hill, M.O. *EUNIS Habitat Classification Revised 2004*; Report of European Environment Agency (EEA); European Nature Information System (EUNIS): Copenhagen, Denmark, 2004.
33. Sánchez, F.; Gómez-Ballesteros, M.; González-Pola, C.; Punzón, A. *Sistema de Cañones Submarinos de Avilés. Proyecto LIFE +INDEMARES*; Fundación Biodiversidad del Ministerio de Agricultura, Alimentación y Medio Ambiente: Madrid, Spain, 2014.
34. Clark, M.R.; Rowden, A.A.; Schlacher, T.; Williams, A.; Consalvey, M.; Stocks, K.I.; Rogers, A.D.; O'Hara, T.D.; White, M.; Shank, T.M.; et al. The ecology of seamounts: Structure, function, and human impacts. *Ann. Rev. Mar. Sci.* **2010**, *2*, 253–278. [CrossRef] [PubMed]
35. McClain, C.R.; Rex, M.A.; Etter, R.J. Patterns in deep-sea macroecology. In *Marine Macroecology*; Witman, J.D., Roy, K., Eds.; The University of Chicago Press: Chicago, IL, USA, 2009; Chapter 3.

Disclaimer/Publisher's Note: The statements, opinions and data contained in all publications are solely those of the individual author(s) and contributor(s) and not of MDPI and/or the editor(s). MDPI and/or the editor(s) disclaim responsibility for any injury to people or property resulting from any ideas, methods, instructions or products referred to in the content.

Article

Effects of Organic Enrichment on Bioturbation Attributes: How Does the Macrofauna Community Respond in Two Different Sedimentary Impacted Areas?

Seyed Ehsan Vesal [1,2,*], Federica Nasi [1], Rocco Auriemma [1] and Paola Del Negro [1]

[1] National Institute of Oceanography and Applied Geophysics-OGS, Via A. Piccard, 54, 34151 Trieste, Italy
[2] Department of Life Sciences, University of Trieste, Via A. Valerio, 4/1, 34127 Trieste, Italy
* Correspondence: ehsan.vesal.64@gmail.com

Abstract: We assessed the influence of different organic matter (OM) inputs associated with terrigenous/freshwater allochthonous and sewage derive on bioturbation and irrigation potential community indices (BP_c and IP_c) of the soft-bottom macrofauna community. The macrofauna was sampled from two different sedimentary impacted areas, in front of the Po River Delta (northern Adriatic Sea) and sewage discharge diffusion zone (Gulf of Trieste). The highest values of BP_c and IP_c were observed at the northward sampling stations of the prodelta and the stations 25 m distance in front of the main sewage outfall. Species richness showed high values in the prodelta likely due to the OM positive effect from the delta, and it increased with increasing distance from the pipeline due to the effect of OM from the sewage discharge. The bioturbation indices differed due to the presence of surface deposit feeders and the injection depth (from 2 to 5 cm) with limited movement at the station located northwards in the prodelta and 25 m distance in the diffusion zone. We infer that the difference in bioturbation indices was likely due to the effects of grain-size composition and the degree of organic enrichment in both study areas.

Keywords: macrofauna community; organic enrichment; bioturbation potential; bio-irrigation potential; coastal areas; northern Adriatic sea

Citation: Vesal, S.E.; Nasi, F.; Auriemma, R.; Del Negro, P. Effects of Organic Enrichment on Bioturbation Attributes: How Does the Macrofauna Community Respond in Two Different Sedimentary Impacted Areas? *Diversity* **2023**, *15*, 449. https://doi.org/10.3390/d15030449

Academic Editors: Renato Mamede and Marcos Rubal

Received: 22 January 2023
Revised: 27 February 2023
Accepted: 14 March 2023
Published: 17 March 2023

Copyright: © 2023 by the authors. Licensee MDPI, Basel, Switzerland. This article is an open access article distributed under the terms and conditions of the Creative Commons Attribution (CC BY) license (https://creativecommons.org/licenses/by/4.0/).

1. Introduction

Macrofauna organisms are considered a key biological component due to high biomass and biodiversity [1], further, most of them are sessile and play a critical role in cycling nutrients, oxygenation of deeper sediment layers, and sediment reworking. They show marked responses to environmental changes depending on their species-specific sensitivity/tolerance levels [2–4]. Biogenic activities such as bioturbation and bio-irrigation by benthic organisms are fundamental due to not only to mixing the substrate as sediment particle preservation and sediment reworking (i.e., burrow and mound construction, particle ingestion, food caching, prey excavation, etc., as bioturbation activities), but also burrow-dwelling can ventilate the sediments, creating a rapid exchange of water between the overlying water and subsurface sediment (i.e., bio-irrigation activities), which relates directly to species activities; food availability and geochemical composition within the substrate are all affected [5–7]. These fundamental processes also affect sediment turnover, diffusive and advective processes that transport elements in dissolved and particulate among sediments [8,9], and consequently, have implications for ecosystem-related functions [10–12].

The coastal marine ecosystems are subjected to several impacts from natural to anthropogenic origins, which depose huge amounts of organic matter (OM). The major natural point of OM sources are rivers which largely contribute to accumulating allochthonous OM in the area interested by their plumes [13,14]. Besides, the human impacts and growing urban development of coastal areas entail increased anthropogenic pressures such as domestic and municipal wastewater disposal into marine environments. Sewage-derived materials present a widespread environmental problem in coastal waters which often release a high amount of OM into shallow subtidal habitats [15,16] and contain organic contaminants, fecal sterols, heavy metals, bacteria, nutrients, and large amounts of suspended and particulate organic matter [17].

The coastal sediments act as a sink for the accumulation of allochthonous-OM [18]. Despite this, some studies have documented the importance of allochthonous, terrestrial/riverine resource supply for marine communities, e.g., [19,20]. However, a high amount of allochthonous OM could cause the most pervasive threat to the diversity, structure, and functioning of marine coastal ecosystems [21–23].

Macrofauna communities adapt to environmental disturbances, and the anthropogenic impact factors have to be measured against the background of natural forces; an anthropogenic factor can be detected if its impact exceeds the intensity and frequency of natural physical disturbance [24]. Detrimental effects of sewage discharge are evidenced [25,26], but it is challenging to disentangle and quantify the relative importance of anthropogenic and natural organic matter in environments with competing activities, permanent alterations, and persistent usage [27,28]. Community responses to anthropogenic disturbances are rarely compared to natural disturbance patterns. Such comparisons increase our ability to predict the responses of organisms to future disturbances and help place human activity in a more realistic perspective of natural history [29]. Hence, biogenic activities such as bioturbation and bioirrigation can diminish the possible negative effects of organic contaminants in the sediment, if contamination does not reach high levels causing partial or total defaunation [8,30]. However, in some cases, these can be influenced by the different amounts of OM and its allochthonous origin [31].

Generally, the degree of bioturbation increases with decreasing subaerial exposure time of a deposit. The strength of this association can vary greatly at a local scale and is determined by the organism, the consistency of the substrate, and grain size [32]. Many of the environmental stresses affecting organisms within shallow subtidal to supratidal environments vary as a function of grain size [33]. In fact, the lability of OM affects the exchange rate of dissolved compounds between the sediment and water column [34]. Muddy sediments generally have more reduced conditions than sandy sediments, which affect the mineralization rates of OM in the sediments, and therefore, their metabolic capacity [35].

Bioturbation has been quantified by a series of modeled simulations and calculated with metrics from benthic quantitative data, such as bioturbation potential community-BPc [36] and irrigation potential community-IPc [37] calculations. Bioturbation potential calculations are linked to the adoption of a trait-based approach and can be quantitatively estimated from benthic quantitative data using the metric of bioturbation indices (BPc and IPc) and it is useful when trying to categorize and understand ecosystem functions conducted by benthic communities [12,36,37].

The consequences of environmentally driven changes in biodiversity to BP_c, and its relation to ecosystem functioning, have been explored in terrestrial [38], marine habitats [36,39], local scales [40,41], regional scales [12,42,43], for different contexts [42,44,45]; for a variety of ecosystem functions including productivity [43], nutrient cycling [36], carbon storage [40,43], and decomposition of plant pigments in surface sediments [46].

Besides, another important feature is the sediment irrigation derived from animals that affect the different biogeochemical processes on the seafloor. Bio-irrigation is mainly caused by burrow-dwelling organisms that can ventilate the sediments, creating a fast water interchange between the overlying water and subsurface sediments [31,47]. The latter process is mostly induced by suspension deposit-feeding activities and ventilation rates of benthic organisms [48]. Accordingly, bio-irrigation is predominantly related to body mass and feeding type [49]. Ref. [37] modified the BP_c index suggested by [36] into community irrigation potential (IP_c), as a new index, whereas in the bioturbation potential calculation-BP_c, the mobility trait presumably underrates the contribution of sessile organisms with low mobility rate but high bio-irrigation efficiency. In this context, [37] tried to replace the reworking and mobility traits with the feeding types, burrow, and depth pocket injection of burrows (as bio-irrigation functional characteristics).

So far, the BP_c index has been usefully applied in many marine studies and by calculating BP_c over time, or for different locations or scenarios, changes in the efficiency of the organism-sediment couple can be monitored for compliance in support of management and policy objectives [50,51]; the new IP_c index has been less adopted in ecological surveys [31,52].

The environmental effects of OM enrichment depend on origin-specific conditions including the prevailing physicochemical and biological features of the receiving environment [53,54]. The results presented in this study will help to understand the relationship between macrofauna invertebrates and their bioturbation processes and the spatial distribution of OM with two different origins (natural and anthropogenic), which has been poorly investigated, especially in coastal marine environments. In addition, they will provide how macrofauna bioturbation attributes can play a key role in protecting and managing marine coastal areas. Therefore, we provide the different biological responses of macrofauna community to perturbation impact by evaluating changes in the metric of bioturbation indices (BP_c and IP_c) caused by natural and anthropogenic organic enrichment in two different areas. We focused on the Po River Delta (northern Adriatic Sea) and Servola pipelines (Gulf of Trieste, northern Adriatic Sea).

Specifically, this study aims to investigate the effects of organic enrichment by natural and anthropogenic impacts on the macrofauna community by applying bioturbation and bio-irrigation indices (BP_c and IP_c) in two different impacted areas. We hypothesized that the macrofauna community, inhabiting the coastal area in front of the Po River Delta and nearby sewage outfalls, respond differently in terms of bioturbation attributed to uneven amounts of OM. We aimed to answer the following specific questions: (1) Does the structure of the benthic macrofauna community affect bioturbation processes in different sedimentary environments? (2) Do macrofauna bioturbation attributes show spatial variability associated with different OM inputs of terrigenous/freshwater allochthonous and sewage-derived materials? (3) Are the bioturbation attribute patterns driven by specific sediment physicochemical parameters?

2. Material and Methods

2.1. Study Area

The study was performed in coastal areas located in the northern Adriatic Sea subjected to a high amount of organic enrichment from natural (Po River Delta) and anthropogenic origins (sewage discharges in the Gulf of Trieste) (Figure 1 and Table S1).

Among European transitional systems, the Po River Delta is considered the major one, which is characterized by multiple physical-chemical and biological processes favoring natural organic enrichment and sedimentation [55], and wide seasonal, daily variability in chemical-physical parameters and fluvial inputs [56]. The Po River, with a drainage basin of 71,000 km^2 and a length of 673 km, is the most important river in Italy and one of the largest in Europe. It extends over 685 km^2 and most of the drainage basin runs through a wide low-gradient alluvial plain, with seven river branches, several lagoons, and wetlands [57], it is characterized by two annual floods (>5000 m^3 s^{-1}) associated with rainfall in autumn and snowmelt in spring [58]. Its total discharge is not equally distributed along the coast of the delta, where only 20% flows into the northern coast, 30% to the Pila tip (then driven southward by coastal currents), and the remaining 50% into the southern coast. During normal flow conditions, transported fine-grained sediments undergo a relatively rapid deposition nearby the mouths (~6 cm year^{-1} near the Pila distributary; [57]). Conversely, during flood events, these particles may cover a wide distance before reaching the sea bottom. The plume is principally transported southward along the shelf due to the predominant cyclonic Western Adriatic Coastal Current–WAC (driven chiefly by the pressure gradient established between interior dense water and coastal freshwater set up by the Italian rivers), and it is subjected to wind-induced resuspension events promoted principally by the north-easterly Bora wind [59]. Furthermore, the latter tends to confine the plume along the Italian coastline [60], especially during winter when this katabatic wind is stronger; the south-easterly Scirocco drives riverine water northward [61].

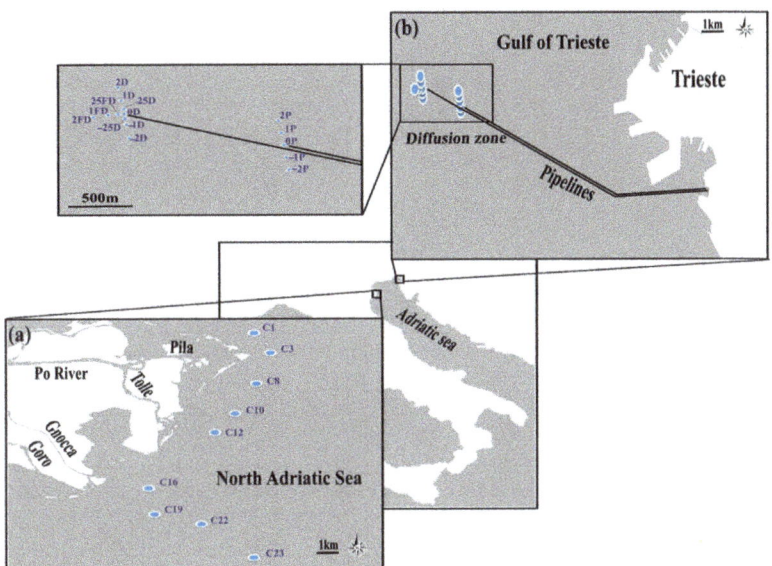

Figure 1. Map showing the location of both study areas: (**a**) the Po River prodelta and (**b**) the area nearby the two outfalls of Servola sewage plant (Gulf of Trieste).

The Gulf of Trieste is a shallow embayment of about 100 km coastline and vastness of about 600 km^2, located in the northern Adriatic Sea (Italy) with an average depth of 17 m. Geographical, hydrological, and sedimentological features and the physical and chemical features of the Gulf were exhaustively described by [62–64]. The Gulf hosts the main sewage treatment plant of Trieste city which is carrying out organic enrichment from anthropogenic origin. The plant is located at the foot of the Servola hill, serving up to 200,000 inhabitants of Trieste with a maximum flow of 6000 L sec^{-1} [64,65]. In this plant, the wastewaters are subjected to a treatment of the physical-chemical type since 1992 (while for the sludge anaerobic digestion heated with the recovery of the biogas produced is used). The Servola sewage disposal plant is composed of two adjacent pipelines (6.5 and 7.5 km) leading to the sea at a depth of 20 to 23 m with 600 sewage diffusion towers and a length of dispersion zone of about 1.5 km (1 km longest and 0.5 km shortest pipe, respectively) [63,66] by type of mixture collected and treating both wastewaters and meteoric within 35 million m^3 per year [16]. There is a greater flow of wastewater through pipe ends than through diffusion towers because fluid parts of wastewater only flow upward through diffusion towers due to the difference in density, which is distributed by currents throughout the area [64].

2.2. Sampling Design and Samples Processing

Sediments in front of the Po River delta were collected in December 2014 (after flood events; [14]); whereas the sampling nearby sewage outfalls (in the Gulf of Trieste) was performed in April 2018 before the improvement of sewage treatments; [16] in the spatial scales, in order to consider the best time to conduct sampling in both study areas which most probably contained the highest possible amounts of organic enrichment discharged (natural and anthropogenic) and could change the community structures to the highest possible degrees. Therefore, according to the seawater currents [14], nine sampling sites in the Po River prodelta area were located at increasing depths (between 9 and 21 m) and distance from the main distributary mouth (Po di Pila) along the southward river plume (Figure 1a, Table S1). In the Gulf of Trieste, to expose the best coverage of the whole diffusion area, as well as consider the seawater currents in the gulf and the largest amount of organic enrichment discharged from the main outfall [63], 15 stations were sampled following an increasing distance from the pipelines (<5, 100, and >200 m) for each outfall, the shortest pipe ('proximal' transect) and the longest and main one ('distal' transect). Additionally, in the 'distal' transect were sampling stations at 25 m from the duct (Figure 1b, Table S1).

In both areas, sediments were collected by a van Veen grab (0.1 m^2). The macrofauna was collected in three replicates and sieved with a 1 mm mesh and the organisms were instantly fixed with ethanol (80%). In the laboratory, the organisms were separated from the sediment and identified into the lowest possible taxonomical level employing a stereomicroscope (Model; Zeiss Discovery V.12, 8–110× final magnification) and counted for each station separately, and species names were updated wherever needed.

Weight estimate (Wet Weight-WW) was measured for each taxon [64]. Subsequently, to obtain the Dry Weight (DW), samples were placed in an oven at 100 °C for 24 h, cooled in a lab desiccator to the normal temperature of the room, and then weighed. The organisms were heated to 500°C for 24 h in an oven and cooled in a lab desiccator to the room temperature to obtain their ash quantity, and then weighed. Ash weight was subtracted from DW to obtain Ash Free Dry Weight (AFDW) [67]. The environmental variables considered in this study (i.e., shells, sand, silt, clay, Total Organic Carbon-TOC, Total Nitrogen-TN, and carbon and nitrogen molar ratio-C:N) were determined in the same samples, which were thoroughly described by [31] for the Po River coastal area, whereas by [63] for the Gulf of Trieste.

2.3. Estimation of the Bioturbation Potential (BP) and Irrigation Potential (IP)

Community bioturbation potential (BP$_c$) is a metric first described by [36], which combines abundance and biomass data with information about the life traits of individual species or taxonomic groups. This information describes modes of sediment reworking and mobility of taxa in a dataset, two traits known to regulate biological sediment mixing, a key component of bioturbation [39,68].

The bioturbation potential-BP [36] was computed according to the following equation:

$$BP_c = \sum_{i=1}^{n} BP_i, \text{ whereas } BP_i = (B_i/A_i)^{0.5} * A_i * M_i * R_i$$

where B$_i$ and A$_i$ metrics are biomass AFDW in (gr m^{-2}), and the number of individuals (m^2), respectively. M$_i$ and R$_i$ are categorical scores of species that represent increasing mobility (M$_i$) and increasing sediment reworking (R$_i$). Community-level bioturbation potential (BP$_c$) and individual taxa (BP$_i$) were calculated across the whole sampling species. This study used the list of mobility (M) and reworking (R) scores from literature, i.e., [69,70] and expert knowledge [31,63,64,71] (Supplementary Table S2 provides the category of species scores for M$_i$ and R$_i$).

The irrigation potential (IP) of [37] is defined by traits including burrow type (BT$_i$), feeding type (FT$_i$) and injection pocket depth (ID$_i$) which are the irrigation behaviors of benthic macrofauna species and their effects on ecosystem functioning. The irrigation potential equation of each species taking into account the different sampled stations is given by:

$$IP_c = \sum_{i=1}^{n} IP_i, \text{ where } IP_i = (B_i/A_i)^{0.75} * A_i * BT_i * FT_i * ID_i$$

The mean individual biomass of each species (i) is expressed by the ratio B$_i$/A$_i$, where B$_i$ is the biomass of species (AFDW) grams per m^2, while A$_i$ is abundance per m^2. As described by [37], the categorical trait demonstrates the species-specific occurrence of the relevant trait and is assigned by numerical scores. We considered a bit of modification for the categorical trait scores due to the lack of the sub-suspension/funnel habit in the sampling area and the prevailing occurrence of the deposit-feeding type. The categorical traits were adjusted by different scores considering deposit feeder type in surface and subsurface deposit feeder moods (Table S2 provides the categorical taxa scores for BT_i, FT_i and ID_i). The data collection on BT_i was obtained from previous studies [47,52,72]. We acquired the scores on FT_i mood based on the literature, e.g., [73] and the databases (www.polytraits.lifewatchgreece.eu; www.marlin.ac.uk/biotic, accessed on 1 April 2020). Moreover, we implemented the information on ID_i moods from the literature, e.g. [74–76]. If no information could be available for the categorical trait scores at the species level, a score was used from the next highest taxonomic rank or indicated as not available information (n.a. in Table S2) and deleted from the index calculation.

2.4. Statistical Analyses

Before all the analyses, data were explored and checked for normality and collinearity following Shapiro–Wilk's and Spearman's rank correlation coefficient, respectively [77]. The differences in species number (richness), BP_c, IP_c, and environmental factors among groups of stations in both study areas (for factors see the 'transects' namely in Table S1) were computed by Mann–Whitney U tests. A one-way PERMANOVA test was used to check for significant differences in BP_i and IP_i values for every single species among groups of stations in both study areas, where factors (reported in Table S1) were selected as fixed factors. When significant differences were noticed, PERMANOVA pairwise tests were performed. Unrestricted permutations of raw data and 9999 permutations were applied.

To observe any spatial patterns in bioturbation attribute values, a non-metric multidimensional scaling analysis (nMDS) was applied for two matrices (i.e., BP_i and IP_i, for both study areas, separately). The environmental variables (i.e., sand, silt, clay, TOC, and C:N) were overlaid as supplementary variables (vectors) onto ordination spaces to investigate their relations in this distribution.

In addition, we measured the relative frequencies of scores for the factors 'north and south' in the prodelta and 'distance gradient' in the diffusion zones in order to assess the variation in species scores in sampling areas.

Further, to indicate the significance covaried coherently on the BP_i and IP_i values, i in both study areas and Similarity Profiles (SIMPROF) analysis was applied. To detect which taxa were mainly responsible for bioturbation and irrigation activities (BP_i and IP_i data, respectively) at stations gathered into different transects in both study areas, SIMPER analysis was used and different factors (see Table S1) were determined. A cut-off at <70 % was applied.

Additionally, distance-based redundancy analysis (dbRDA) was used to detect the relationships between bioturbation indices (BP_i and IP_i values) and selected species by the SIMPER test and environmental variables. Before analysis, environmental data were normalized.

To highlight the spatial relationship between predictor variables (the considered environmental parameters) and response variables (BP_c and IP_c values and scores frequencies), linear regression and Spearman's correlations were computed for each area separately. By doing so, the predictive power of environmental parameters for each bioturbation attribute was discriminated via the coefficient of determination r_s (Spearman's correlation) and R^2 (linear regression).

For the multivariate analyses, the matrices BP_i and IP_i for sampling areas were square root, and the Bray–Curtis similarity was applied. The Mann–Whitney test was computed using STATISTICA 7 software and the multivariate analyses were performed using PRIMER 7 (PRIMER-E Ltd., Plymouth, UK) software.

3. Results

3.1. Taxonomic Composition of the Macrobenthic Community

In both study areas, the macrofauna abundance varied from 7.3 ± 1.4 ind. m^{-2} at C23 to 532.3 ± 62.0 ind. m^{-2} at C8 and ranged from 36.0 ± 0.6 ind. m^{-2} at 0P to 343.6 ± 9.0 ind. m^{-2} at -1D in the Po River delta and the Gulf of Trieste, respectively. Regarding biomass, the lowest value was observed at C12 (0.03 ± 0.0 g m^{-2}), whereas at -25D the highest biomass was measured (2.6 ± 0.1 g m^{-2}). Polychaetes were the dominant taxa (41.75%), mollusks (32.92%), crustaceans (10.18%), echinodermata (12.65%), and other groups (anthozoa and sipuncula together 2.50%) were found in the Po River prodelta, whereas polychaetes (74.30% of the total abundance) followed by mollusks (16.29%), crustaceans (7.09%), echinodermata (2.31%), and other groups (sipuncula = 0.01%) in the diffusion zone (Gulf of Trieste) (Figure 2). In both sampling areas, a total of 253 taxa were found. In the Po River delta, C23 had the lowest species number (six species) and C1 was the highest one (39 species), whereas, the species number showed the minimum value at 0D (35 species) and maximum value at the 2P (75 species), in the Gulf of Trieste (Figure 3).

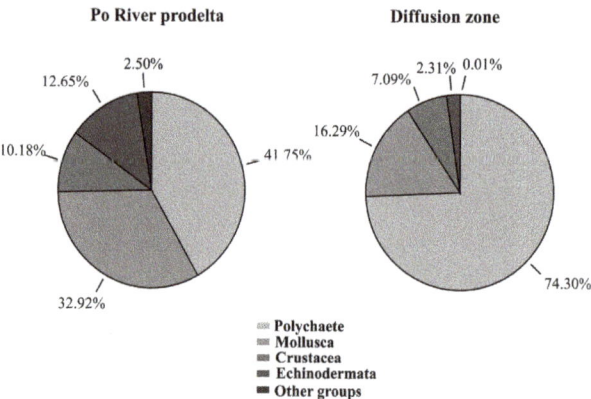

Figure 2. Pie charts representing the proportion of polychaetes, mollusks, crustaceans, echinodermata, and other groups for both study areas. The values close to the pie charts indicate total abundance.

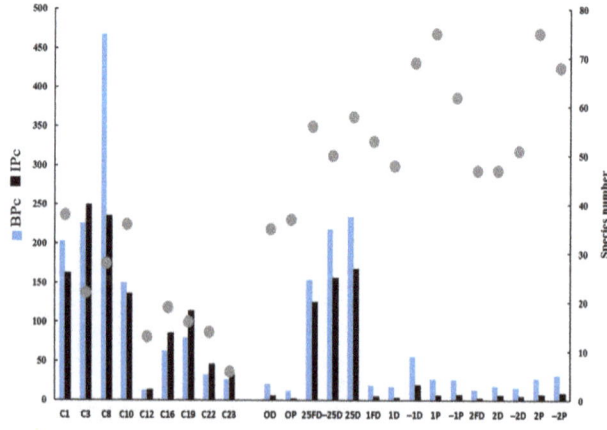

Figure 3. Total bioturbation potential (BP_c) and irrigation potential (IP_c) of community and species number in different sampled areas.

Overall, in the diffusion zone area, significantly higher species numbers were recorded if compared to the Po River prodelta (U test, z = 3.8; $p < 0.01$) (Figure 3).

3.2. Macrofauna Bioturbation Attributes

The community bioturbation potential (BP_c) values were lower at C12 (12.8) and 0P (11.8), while the highest ones were estimated at C8 (468.01) and 25D (234.8) in the Po River delta and the Gulf of Trieste, respectively. Similarly, the irrigation potential (IP_c) showed lower values at C12 (13.2) and 0P (1.82), whereas this index was higher at C3 (250.0) and 25D (167.9) in the coastal area nearby the Po River mouth and pipelines, respectively. In addition, BP_c and IP_c did not follow variation patterns with species richness in both sampling areas (Figure 3). Overall, higher values of irrigation potential were noticed at the Po River prodelta if compared to the area nearby the pipelines, as corroborated by the U test ($z = -2.9$; $p < 0.01$). Further, higher values of the indices were noticed at the stations placed in the north part of the Po River (U test 'north' vs 'south': $z = 2.3$; $p < 0.05$) for BP_c and IP_c (Figure 4). In addition, significantly major values of BP_c and IP_c were measured at stations located 25 m from the sewage pipeline (H = 7.8; $p < 0.05$ for both indices). No differences were measured between stations gathered in 'distal' and 'proximal' transects for diffusion area (Figure 4).

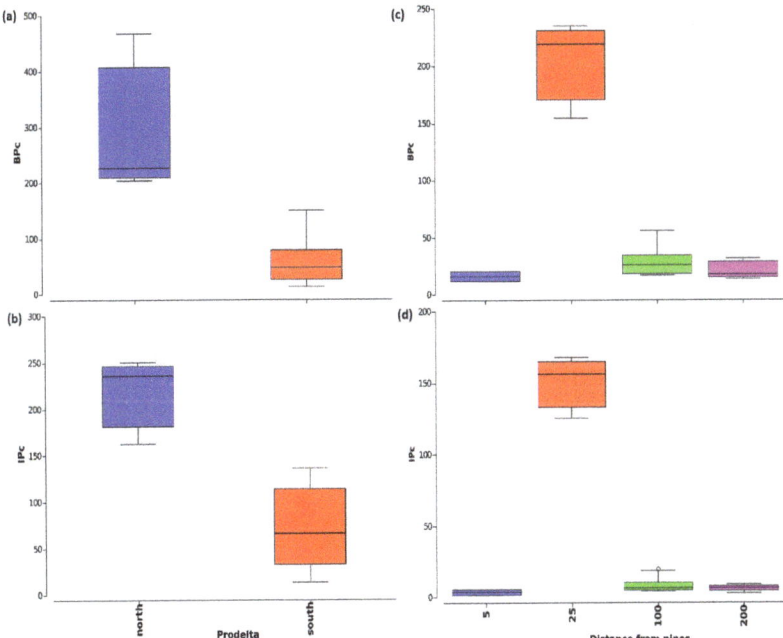

Figure 4. Boxplots showing the variability of bioturbation potential (BP_c) and irrigation potential (IP_c) of community at north and south for Po River coastal area (**a**,**b**), and at increasing distance from pipelines (<5, 25, 100, and >200 m) in the Gulf of Trieste (**c**,**d**).

The PERMANOVA tests on bioturbation attribute values (BP_i and IP_i) for every single species showed differences between both study areas. According to the PERMANOVA main test, a highly significant difference was observed at increasing distances from the main distributary mouth (Po di Pila) along the southward river plume (Pseudo-F = 2.5 and 2.6; $p < 0.05$ for BP_i and IP_i, respectively). Similarly, significant differences were noticed in bioturbation indices among stations placed at increasing distances from the pipes (Pseudo-F

= 3.02 and 3.03; $p < 0.01$, respectively). In addition, the bioturbation indices of species significantly differed between stations gathered in 'distal' and 'proximal' transects (Pseudo-F = 2.0; $p < 0.05$ for BP_i and IP_i).

Additionally, SIMPER analysis performed on BP_i and IP_i values showed differences among the sampling stations in the Po River prodelta area and the Gulf of Trieste. For the first sampling area, the higher BP_i values of the bivalve *Striarca lactea* (Contrib% = 7.4) and the polychaete *Owenia fusiformis* (Contrib% = 6.9) were responsible for the main difference from the stations placed on the northward part and southward of the Po River delta (north vs south: 77.8 %, average dissimilarity). Similarly, *O. fusiformis* mostly contributed to the dissimilarity between 'north vs south' (Contrib% = 11.8) due to high values of IP_i at northern sites (average dissimilarity = 75.6). By SIMPER analysis on BP_i and IP_i values, the dissimilarity in the Gulf of Trieste between stations placed in <5 m and 25 m distance groups (<5 vs 25: 77.8 and 87.8 % average dissimilarity for BP_i and IP_i values, respectively) were mainly due to the high BP_i and IP_i values of the mollusk *Polititapes aureus* sampled at 25 m from the main pipe (Contribution% = 5.7 and 8.6, respectively). The dissimilarity between stations located in 25 and 100 m groups (25 vs 100: 77.1% and 84.9 average dissimilarity for BP_i and IP_i values, respectively) was mostly due to the high value of polychaetes *Capitella capitata* (25 m group), with contribution% = 5.65 and 8.35 for BP_i and IP_i values, respectively. SIMPER analysis showed the dissimilarity between stations placed at 25 vs >200 m (25 vs >200: 85.9 and 90.5%, average dissimilarity for BP_i and IP_i values, respectively) were characterized by polychaetes *C. capitata* that were highly present at 25 m station group with contribution 6.1 and 8.7% for BP_i and IP_i values, respectively.

Considering the relative frequencies (%) of scores belonging to the BP_c values, it showed higher occurrences for reworking scores representing the superficial modifiers (64.0%) for the stations located in the northern and southern part of the prodelta (64.0 and 49.0%, respectively). In addition, towards the southern stations, a higher% of biodiffusers was calculated (29.0%). In the diffusion area, we measured slightly increasing percentages of 'superficial modifiers' score coupled with decreasing values of 'biodiffusors' from the stations located nearby the main outfall towards the farther ones (Figure 5a). Regarding the mobility scores (Figure 5b), in the coastal area of Po River, major values of relative frequencies of organisms with 'limited movements' were measured in northern stations compared to southern ones. In addition, in the Gulf of Trieste, we observed great changes in score % belonging to 'movements through the sediment matrix' and 'free movements via 'burrow types'. The latter was noticed with a higher value nearby the outfalls (34.5%) and remarkably decreased towards the stations far from the diffusion zone. On the contrary, we measured increasing values of 'movements through 'the sediment matrix' and 'organism that lives in fixed tubes' scores at stations placed at 100 and >200 m from the main pipe.

According to the IP_c values, relative frequencies (%) indicated the scores belonging to burrow type varied in the coastal area of prodelta. A higher % of 'infauna with internal irrigation' score in northern stations (34.0%), whereas the increasing value of the 'blind-ended irrigation' score (41.8%) was noticed at southern ones. A decreasing value of the 'open irrigation' score was observed moving away from the stations' nearby diffusion zone towards the farther ones in the Gulf of Trieste (Figure 6a). 'Surface deposit feeder' and 'injection depth from 2 to 5 cm' were highly expressed northward and 25 m distance from the main outfall in the prodelta and Gulf of Trieste, respectively (Figure 6b,c). Lastly, the stations gathered in 25 m were characterized by a higher percentage of 'depth over 10 cm' score when compared with the later stations in both study areas.

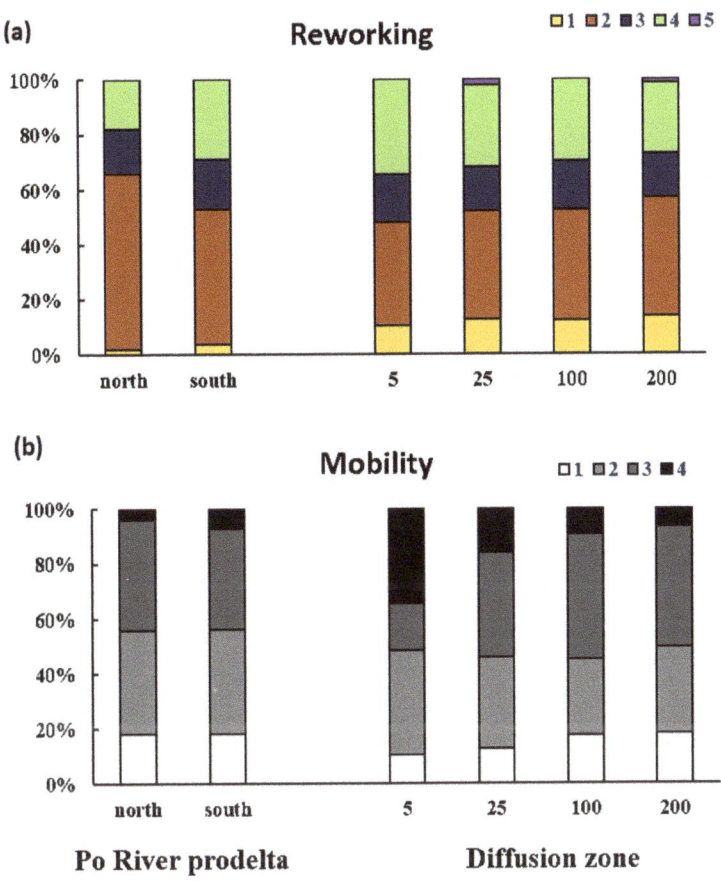

Figure 5. Relative frequencies (%) of scores belonging to reworking (a) and mobility (b) of bioturbation activities of community at north and south for Po River coastal area and at increasing distance from pipelines (<5, 25, 100, and >200 m) in the Gulf of Trieste. See Table S2 for scores.

3.3. Relation between Bioturbation Indices and Environmental Factors

Considering both sampling areas, TN and TOC were highly covaried (r_s = 0.906), therefore, we deleted TN from the subsequent analyses. Table 1 summarizes the linear regression analysis for environmental parameters as predictors and bioturbation indices (BP_c and IP_c) as response variables.

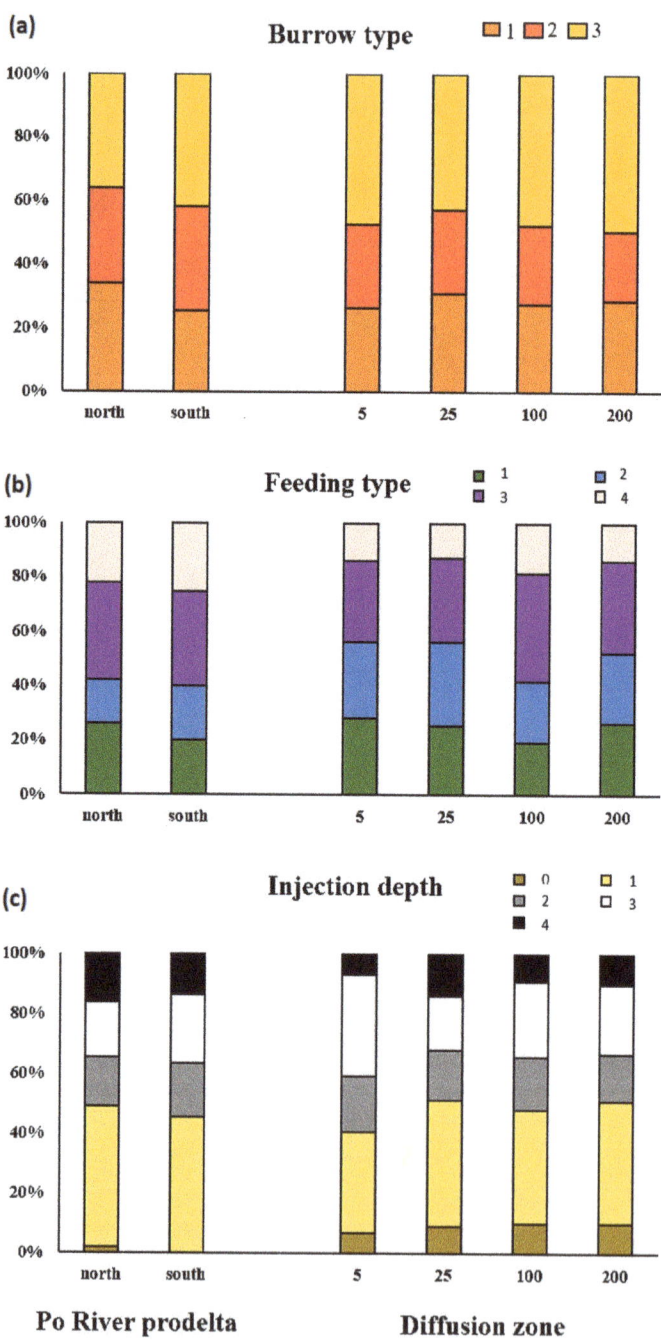

Figure 6. Relative frequencies (%) of scores belonging to burrow type (**a**), feeding type (**b**), and injection depth (**c**) of bio-irrigation activities of community at north and south for Po River coastal area and at increasing distance from pipelines (<5, 25, 100, and >200 m) in the Gulf of Trieste. See Table S2 for scores.

Table 1. Linear regression of bioturbation indices (BP$_c$ and IP$_c$) as response variables and environmental parameters as predictor. R^2: coefficient of determination; df: degrees of freedom; F and P give the calculated F value (F-test) and the corresponding probability associated with the null hypothesis. The negative linear regressions are in bold. TOC-Total Organic Carbon; C:N-carbon and nitrogen ratio.

Area	Response Variables	Predictor Variables	R^2	df	F	P
Po River prodelta	BP$_c$	Clay	0.74	1.7	20.71	<0.01
	IP$_c$	Sand	0.75	1.7	21.12	<0.01
		Clay	0.79	1.7	26.65	<0.01
		C:N	0.49	1.7	6.77	0.03
Diffusion zone	BP$_c$	Sand	0.56	1.13	17.07	<0.01
		Silt	0.57	1.13	17.28	<0.01
		Clay	0.53	1.13	14.94	<0.01
		TOC	0.29	1.13	5.44	0.03
	IP$_c$	**Clay**	**0.55**	**1.13**	**16.3**	**<0.01**
		TOC	0.31	1.13	5.96	0.02
		Silt	**0.61**	**1.13**	**20.78**	**<0.01**
		Sand	0.6	1.13	19.62	<0.01

The nMDS was performed on BP$_i$ and IP$_i$ values of the Po River coastal area and confirmed the PERMANOVA results (Figure S1a,b). Further, the differences among groups of stations (i.e., 'north and south') were enhanced by SIMPROF analyses, particularly for IP$_i$ values. The stations located in the northern part of prodelta (C1, C3, and C8) were plotted on the right side of the nMDS plot-based BP$_i$ (Figure S1a) and on the left side of the nMDS plot-based IP$_i$ values (Figure S1b). Figure S1a,b explains that clay is only responsible for the difference among the stations mainly those increasing away from the main distributary mouth (Po di Pila) along the southward river plume (i.e., C12, C19, C22, and C23). In addition, the nMDS performed on BPi and IPi composition in the Gulf of Trieste did not follow (by SIMPROF test) the same results obtained by the PERMANOVA test. Furthermore, the nMDS showed the stations located at a 25 m distance from the sewage duct (i.e., stations -25D, 25FD, and 25D) were placed on the left side of the nMDS plot-based BP$_i$ (Figure S2a) and the right side of the nMDS plot-based IP$_i$ (Figure S2b) values at the maximum distance (Bray–Curtis maximum dissimilarity) from stations located at 100 and >200 m. To compare with those stationed at 5 m away from the duct, these results indicated that species at 25 m away from the duct had different compositions in reworking and bioirrigation attributes. The latter differences could be due to higher values of sand and TOC at stations gathered at 25 m from the duct, whereas a major % of silt and clay was noticed at the 100 and >200 m group of stations.

The distance-based redundancy analysis (dbRDA) performed using BP$_i$ and IP$_i$ values of species selected by SIMPER and environmental variables for both sampling areas, explained the 54.9 and 52.2% of the total variation, respectively (Figure 7a,b). Overall, both analyses plotted the stations separately according to sampling areas and factors. Regarding the negative part of dbRDA1 (left side of Figure 7a), higher percentages of silt corresponded to high occurrences of the polychaete *Sternaspis scutata*, *O. fusiformis*, and *Heteromastus filiformis* in the prodelta area. Clay was the predominant element of the positive axis of dbRDA2, related principally with the polychaetes *Lumbrinereis lusitanica* and *Hilbigneris gracilis* that occurred in the farther stations from the diffusion zone. In addition, C:N, TOC, and sand were the predominant elements of the negative part of dbRDA2, strictly related to some bivalve species in the prodelta area (e.g., *Varicorbula gibba*, *S. lactea*, and *Peronidia albicans* and polychaetes such as *Glycera trydactila* and *Eunice vittata*). Regarding the dbRDA analyses performed on IP$_i$ values (Figure 7b), some different species were related to the second axis (e.g., the bivalve *Moerella distorta*, and the polychaetes *C. capitata* and *Maldane sarsi*).

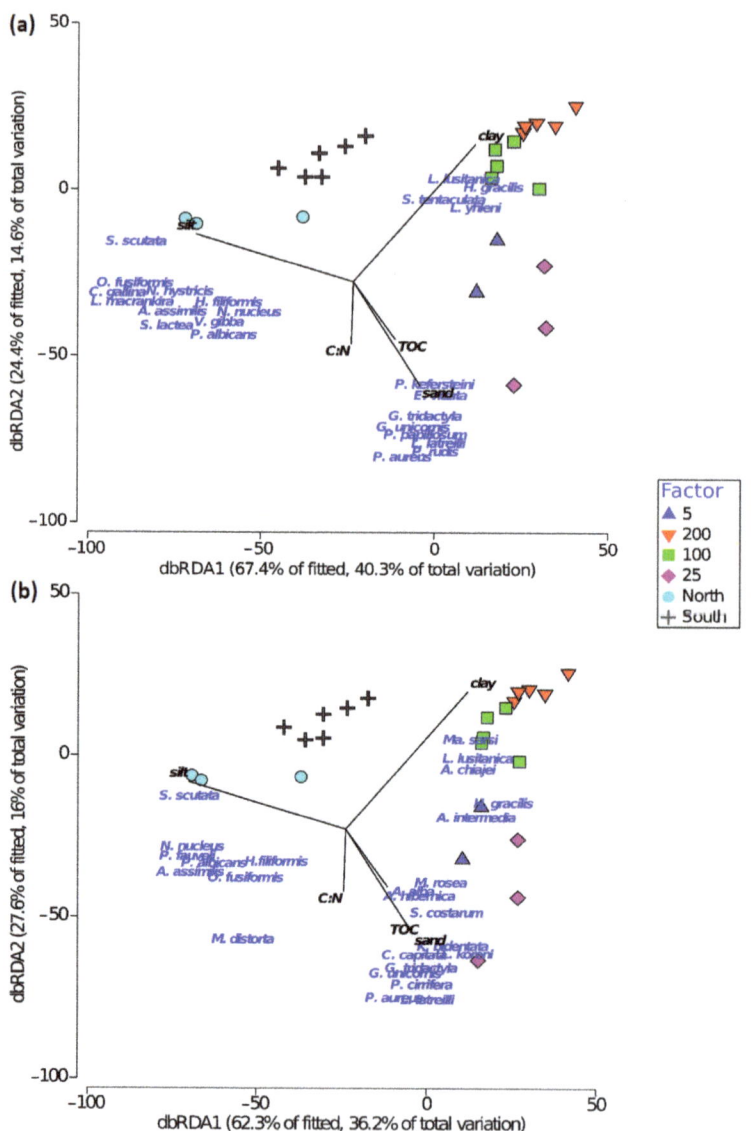

Figure 7. Distance-based redundancy analysis (dbRDA) performed using BPi (a) and IPi (b) values of species selected by SIMPER and environmental variables for both sampling areas. Arrows indicate environmental variables: sand, silt, Total Organic Carbon (TOC), Total Nitrogen (TN), and the ratio between carbon and nitrogen (C:N). See Table S2 for species names.

4. Discussion

This study shows that the variations in macrofauna bioturbation attributes among stations considering the spatial scales in both study areas were likely due to the effects of grain-size composition and the degree of organic enrichment more than the differences in the type of organic matter inputs (terrigenous/freshwater allochthonous and sewage derive). The temporal scale was not of fundamental importance here because the two study areas were far apart. In this study, two different snapshots of environmental conditions were compared in order to determine how they might influence macrofauna bioturbation characteristics in different ways. However, it was unclear how much OM affected macrofauna communities in the two sampling areas, as well as the relationship between the increased OM in the sediments and changes in benthic fauna structure and metabolism is quite complicated [78]. As a result of this study, two different macrofauna communities responded similarly to the amount of organic enrichment, which followed the model of Pearson and Rosenberg [30]. The macrofauna sampled at the stations close to the main outflow (Gulf of Trieste) and in front of the main river month (Po delta River) were characterized by a higher density and a lower diversity with the presence of opportunistic species, as compared to the macrofauna studied at the more distant stations, where the situation was the opposite.

In addition, our results show that macrofauna invertebrates with high burrowing depths and blind-ended burrows, abundant at sandier stations, enhanced bio-irrigation. Regardless of the type of organic matter sources, the high density of superficial modifiers and biodiffusors (belonging to 'reworking' traits categories) enhance the bioturbation activities, increasing the entire ecosystem functioning in both study areas, where a higher % of biodiffusers was found towards the southern stations at the coastal area of Po River, while a slightly increasing percentage of 'superficial modifiers' score coupled with decreasing values of 'biodiffusors' were observed from the stations located nearby the main outfall towards the farther ones in the diffusion area.

In both sampling areas, there was no evident variation pattern of species number and macrofaunal bioturbation attributes. In the Gulf of Trieste, the highest values of species number were observed toward the increasing distance from the pipe, whereas the maximum of BP_c and IP_c was noticed at stations located 25 m from the main outfall. Similarly, in the prodelta area, where despite the higher numbers of species at northern stations if compared to southern ones, the major values of BP_c and IP_c at northern sites did not follow richness values. Accordingly, our results evidenced that the structure of the benthic macrofauna community did not affect bioturbation processes in different sedimentary environments. Our finding is in contrast to those observed by [31]. They reported that the structure of the macrobenthic community deeply influenced the bioturbation attributes. However, they studied macrofauna invertebrates from brackish environments. The latter community is per se less structured if compared to one from coastal areas, thus few species were able to sustain the bioturbation processes [31].

We inferred that physical environmental features, mainly different grain-size fractions, determine the broad pattern of benthic organism distribution, instead of the origin of organic matters (i.e., terrigenous or sewage-derived materials). Our findings are in accord with previous studies that have reported grain size as the main driver of the spatial distribution pattern of the bioturbation attributes [70,76].

In addition, despite the lower values of species richness, the bioturbation attributes (i.e., BP_c and IP_c) in the prodelta area, were higher if compared to the diffusion zone (Gulf of Trieste), also at the stations farther from the main point of contamination (the principal distributary mouth). This is principally linked to the presence of the highest dominance of surface deposit feeders at northern sites whereas sub-surface deposit feeders and biodiffusors in the southern ones. The feeding strategy of macrofauna invertebrates is fundamental to determining the type of burrows and the modes of locomotion and defecation [47,73,79].

Our results, as observed by other authors [70,80] indicated that surface and subsurface deposit feeders mostly contributed to BP_c and IP_c of coastal marine environments. The huge amount of riverine organic matter deeply influenced the bioturbation features in prodelta areas. The continuous and high load of terrigenous material from a river is known to affect suspension-feeding animals by clogging feeding structures, interfering with particle selection, and requiring the use of energy to clear away unvented particles [81]. In fact, in the central part of the Po delta system (i.e., C8), where high current velocities and low turbidity were reported [82,83], we observed a high dominance of suspension feeder on the account of *O. fusiformis* and the bivalve *Varicorbula gibba*. The majority of these suspension feeders are superficial modifiers. Those can rework sediments but not as much as biodiffusers. The majority of these suspension feeders are superficial modifiers. Superficial modifiers are considered weak bioturbators since these animals can rework only the most superficial sediments and thus have a low impact on bioturbation processes compared to the other reworking modalities (i.e., conveyors, biodiffusors, and regenerators) [69]. The higher values of BP_c in this station (i.e., C8) were principally due to the huge amount of *O. fusiformis*.

Southern stations were characterized by high expression of biodiffuser and deposit feeders, on the account of *S. scutata* and the echinoderm *Amphiura chiajei*. Biodiffusers are dominant in muddy sediments since they can constantly and randomly biomix (both horizontally and vertically) the local sediments over a short distance, which results in particle transport [84]. Among them, gallery biodiffusers often occur in finer sediments in which they are promoters of diffusive local biomixing primarily due to burrowing activities within the upper 10–30 cm of sediments [47]. These animals, in particular polychaetes, are considered non-selective deposit feeders [85] able to feed on both fresh and aged organic matter, promoting nutrient cycling within sediment layers [31,86]. Commonly, *S. scutata* is a deep burrowing subsurface deposit feeder living below 4 cm of depth and can use the food resource provided by river floods later on and over a longer period [87].

Regarding IP_c, higher values of bio-irrigation were noticed at both groups of the site in prodelta areas, if compared to the diffusion zone, however, major bioturbation activities were reported in the southern station if compared to sediment reworking ones. These results confirmed the importance of grain-size distribution for bio-irrigation features. In finer sediments, the permeability rate is low, and therefore, open irrigation (U- or Y-shaped burrows with two or more openings to the sediment surface and radial diffusion mode) is prevalent and enhance the water interchange. These differences reflect the fact that bio-irrigation rates are species-specific and depend on the physical mechanism used [88–90]. This is supported by a shift from anoxia to air-saturated conditions at the burrow opening and increases microbial respiration faster than possible than by the molecular diffusion system [91].

The bioturbation processes in the Gulf of Trieste were lower if compared to the Po River prodelta area. We inferred that this difference could be strictly linked to the biomass but above all to the abundance of species found. In the diffusion zone, we have higher values of species but also not so many individuals for species (higher evenness values, data not reported), except for *Capitella capitata* which was observed with higher abundance but low biomass at only 0D station. On the contrary, in the prodelta coastal area, we noticed in many stations' major specimen densities [31]; the values of BP_c and IP_c were higher. The highest values of BP_c and IP_c latter, confirmed by the U test, were observed at stations placed 25 m distance away from the main sewage outfall. Regarding bioturbation processes, the latter results were due to the dominance of conveyors.

In both sampling areas, the stations close to the main outflow (Gulf of Trieste) and in front of the main river month (Po delta River) were characterized by high values of TOC and TN compared to the more distant stations [31,63]. A similar pattern was noticed for C:N ratio. Since the highest dominance of surface deposit feeders at northern sites (Po river prodelta) and the subsurface deposit feeder (e.g., *C. capitata*) at 25 m from the pipe (Gulf of Trieste), could have influenced the distribution of organic matter within sediments (i.e., TOC and TN), promoting the unexpected decreasing of organic matter at surface layers. Further, it is known that the high presence of *C. capitata* may promote the mineralization of organic matter deposited in the sea bottom within a relatively short period [92]. This result agrees well with the finding that subsurface-deposit feeders, protruding deep into the sediment and causing most of the diffusive mixing, dominate sediments containing intermediate to low-quality organic matter [93].

Reworking and ingestion of sediment particles may have contributed to modifying sediment properties and so promoted microbial population resulting in accelerated degradation of organic matter [94]. Additionally, the latter was confirmed by [95], suggesting that BP_c values could be a good predictor of oxygen consumption, denitrification, alkalinity and ammonium fluxes in fine sandy sediments. What was observed by the authors might explain the hypoxic conditions in the area nearby the outfall despite high sediment reworking. Furthermore, the higher values of IP_c at stations gathered at <5 and 25 m from the pipes were principally due to the presence of open irrigation system at sandier sites. In fact, the main role in determining how the exchange of pore water within sediments is linked to the morphology of burrows [31,79], and further, burrow irrigation is characterized by the advection of pore water in more penetrable sandy sediments, causing the building of blind-ending burrows [47]. Moreover, we observed more injection depth at the stations nearby the diffusion zone. The burrowing depth is important for the pore water exchange and local input of oxygen into anoxic sediments. In particular, the effects of deep burrowing organisms are enhanced in oxic or suboxic sediments compared to those of shallow burrowers [96].

5. Conclusions

The present study contributes to the growing body of bioturbation research, especially in the relatively less investigated coastal marine ecosystems. We provided insight into the different macrofauna responses influenced by natural and anthropogenic organic enrichment by using bioturbation indices (BP_c and IP_c) in two different sedimentary study areas. In both sampling areas, there was no evident variation pattern in species number and macrofaunal bioturbation attributes, where the highest values of species number were noticed toward the increasing distance from the pipe, whereas the maximum of BP_c and IP_c was noticed at stations located 25 m from the main outfall. Similarly, in the prodelta area, the major values of BP_c and IP_c at northern sites did not follow richness values. However, at the stations close to the main outflow (Gulf of Trieste) and in front of the main river month (Po delta River), which were characterized by high values of TOC and TN as well as C:N, two different macrofauna communities responded similarly to the amount of organic enrichment, which is characterized by a higher density and a lower diversity with the presence of opportunistic species. The difference in bioturbation indices was due to the presence of surface deposit feeders and injection depth (from 2 to 5 cm) with limited movement at the station located northwards and 25 m distance in the prodelta and diffusion zone, respectively. Such a strong dominance of these taxa in the implementation of bioturbation indices can cause instability in the ecosystem functions if these organisms disappear as a result of ecological disasters or environmental degradation. Grain-size fractions were the main drivers of the significant differences in BP_c and IP_c, and the major number of taxa in coarse sediments contributed to the highest values of bioturbation indices. The use of both BP_c and IP_c could be matched with experimental data to corroborate our findings and help introduce the application of functional traits in the assessment of the benthic ecosystem functioning, and it could be useful for monitoring programs in differing sedimentary environments, which could allow for cross-system comparisons over habitats and geographic scales. Accordingly, the deepening of the knowledge on macrofauna bioturbation attributes concerning their sediment reworking and ventilation abilities after anoxic and dystrophic events is of paramount importance in the framework of efficient management and sustainable use of coastal resources, especially areas deeply influenced by human impacts.

Supplementary Materials: The following supporting information can be downloaded at: https://www.mdpi.com/article/10.3390/d15030449/s1, Figure S1: nMDS ordination plot based on BPi (a) and IPi (b) values at sampled stations in the Po River prodelta area. The significant covaried groups of stations (by SIMPROF test) are indicated. The environmental variables vectors are overlaid; Figure S2: nMDS ordination plot based on BPi (a) and IPi (b) values at sampled stations nearby the diffusion zone in the Gulf of Trieste. The significant covaried groups of stations (by SIMPROF test) are indicated. The environmental variables vectors are overlaid; Table S1: Coordinates and depth for both sampling stations (a) along the coastal Po River Pro delta and (b) in the Gulf of

Trieste along with the sewage discharge. Indication of the transect to which the station belongs as well as the distance from the sewage pipelines are reported; Table S2: Bioturbation Potential (BPi) [69] and Irrigation Potential (IPi) [37] categorical scores of taxa observed in both sampling areas. Ri (Reworking) scores: 1-epifauna, 2-superficial modifiers, 3-upward and downward conveyors, 4-biodiffusors, 5-regenerators; Mi (Mobility) scores: 1-organism that lives in fixed tubes, 2-organism with limited movement, 3-movements through the sediment matrix, 4-free movements via burrow types. BTi (Burrow type) scores: 1-infauna with internal irrigation (e.g. siphons), 2-open irrigation (U- or Y- shaped burrows), 3-blind ended irrigation (blind-ended burrows, no burrows systems); FTi (Feeding type) scores: 1-surface filter feeder, 2-predator, 3-surface deposit feeder, 4-subsurface deposit feeder; IDi (Injection depth) scores:0-epibiont; 1-depth from 0 to 2 cm, 2-depth from 2 to 5 cm, 3-depth from 5 to 10 cm and 4-depth over 10 cm. From [37], the calculation of FTi was modified by dividing deposit feeder habit in surface and subsurface deposit feeder modalities, with different categorical scores.

Author Contributions: Conceptualization, F.N. and R.A.; Formal analysis, S.E.V. and F.N.; Funding acquisition, P.D.N.; Investigation, S.E.V., F.N. and R.A.; Methodology, S.E.V., F.N. and R.A.; Supervision, P.D.N.; Writing—original draft, S.E.V. and F.N. All authors have read and agreed to the published version of the manuscript.

Funding: This research received no external funding.

Institutional Review Board Statement: Not applicable.

Data Availability Statement: Not applicable.

Acknowledgments: In the Po River coastal area, the activities were supported by the Flagship Project RITMARE–La Ricerca ITaliana per il MARE-The Italian Research for the Sea, coordinated by the National Research Council and supported by the Ministry of Education, University and Research. In the Gulf of Trieste, the study was supported by AcegasApsAmga Hera. The authors wish to thanks Marco Segarich and Carlo Franzosini for logistical support during sampling activities, Larissa Ferrante for macrofaunal sorting activities and taxonomic identification, and Matteo Bazzaro and Federica Relitti for the grain-size and sediment organic carbon analyses.

Conflicts of Interest: The authors declare no conflict of interest.

References

1. Snelgrove, P.V. The biodiversity of macrofaunal organisms in marine sediments. *Biodivers. Conserv.* **1998**, *7*, 1123–1132. [CrossRef]
2. 2Ferraro, S.P.; Cole, F.A. Taxonomic level sufficient for assessing pollution impacts on the Southern California bight macrobenthos—Revisited. *Environ. Toxicol.* **1995**, *14*, 1031–1040. [CrossRef]
3. Lancellotti, D.A.; Stotz, W.B. Effects of shoreline discharge of iron mine tailings on a marine softbottom community in northern Chile. *Mar. Pollut. Bull.* **2004**, *48*, 303–312. [CrossRef] [PubMed]
4. Bremner, J.; Rogers, S.I.; Frid, C.L.J. Methods for describing ecological functioning of marine benthic assemblages using biological traits analysis (BTA). *Ecol. Indic.* **2006**, *6*, 609–622. [CrossRef]
5. Rhoads, D.C.; Boyer, L.F. The effects of marine benthos on physical properties of sediments. In *Animal-Sediment Relations*; Springer: Boston, MA, USA, 1982; pp. 3–52. [CrossRef]
6. Schaffner, L. Small-scale organism distributions and patterns of species diversity: Evidence for positive interactions in an estuarine benthic community. *Mar. Ecol. Prog. Ser.* **1990**, *61*, 107–117. [CrossRef]
7. Meysman, F.J.; Middelburg, J.J.; Heip, C.H. Bioturbation: A fresh look at Darwin's last idea. *Trends Ecol. Evol.* **2006**, *21*, 688–695. [CrossRef]
8. Remaili, T.A.; Simpson, S.L.; Amato, E.D.; Spadaro, D.A.; Jarolimek, C.V.; Jolley, D.F. The impact of sediment bioturbation by secondary organisms on metal bioavailability, bioaccumulation and toxicity to target organisms in benthic bioassays: Implications for sediment quality assessment. *Environ. Pollut.* **2016**, *208*, 590–599. [CrossRef]
9. Remaili, T.M.; Simpson, S.L.; Jolley, D.E.F. Effects of enhanced bioturbation intensities on the toxicity assessment of legacy-contaminated sediments. *Environ. Pollut.* **2017**, *226*, 335–345. [CrossRef]
10. Mermillod-Blondin, F. The functional significance of bioturbation and biodeposition on biogeochemical processes at the water-sediment interface in freshwater and marine ecosystems. *J. N. Am. Benthol. Soc.* **2011**, *30*, 770–778. [CrossRef]
11. Rosenberg, R.; Grémare, A.; Ducheme, J.C.; Davey, E.; Frank, M. 3D visualization and quantification of marine benthic biogenic structures and particle transport utilizing computer-aided tomography. *Mar. Ecol. Prog. Ser.* **2008**, *363*, 171–182. [CrossRef]
12. Birchenough, S.; Parker, R.; McManus, E.; Barry, J. Combining bioturbation and redox metrics: Potential tools for assessing seabed function. *Ecol. Indic.* **2012**, *12*, 8–16. [CrossRef]

13. Giani, M.; Berto, D.; Rampazzo, F.; Savelli, F.; Alvisi, F.; Giordano, P.; Ravaioli, M.; Frascari, F. Origin of sedimentary organic matter in the north-western Adriatic Sea. *Estuar. Coast. Shelf Sci.* **2009**, *84*, 573–583. [CrossRef]
14. Bongiorni, L.; Nasi, F.; Fiorentino, F.; Auriemma, R.; Rampazzo, F.; Nordström, M.C.; Berto, D. Contribution of deltaic wetland food sources to coastal macrobenthic consumers (Po River Delta, north Adriatic Sea). *Sci. Total Environ.* **2018**, *643*, 1373–1386. [CrossRef] [PubMed]
15. Koop, K.; Hutchins, P. Disposal of sewage to the ocean—A sustainable solution? *Mar. Pollut. Bull.* **1996**, *33*, 121–123. [CrossRef]
16. Nasi, F.; Vesal, S.E.; Relitti, F.; Bazzaro, M.; Teixidó, N.; Auriemma, R.; Cibic, T. Taxonomic and functional macrofaunal diversity along a gradient of sewage contamination: A three-year study. *Environ. Pollut.* **2023**, *323*, 121022. [CrossRef]
17. Moon, H.-B.; Yoon, S.-P.; Jung, R.-H.; Choi, M. Wastewater treatment plants (WWTPs) as a source of sediment contamination by toxic organic pollutants and fecal sterols in a semi-enclosed bay in Korea. *Chemosphere* **2008**, *73*, 880–889. [CrossRef]
18. Wilkinson, G.M.; Besterman, A.; Buelo, C.; Gephart, J.; Pace, M.L. A synthesis of modern organic carbon accumulation rates in coastal and aquatic inland ecosystems. *Sci. Rep.* **2018**, *8*, 15736. [CrossRef]
19. Darnaude, A.M.; Salen-Picard, C.; Polunin, N.V.C.; HarmelinVivien, M.L. Trophodynamic linkage between river runoff and coastal fishery yield elucidated by stable isotope data in the Gulf of Lions (NW Mediterranean). *Oecologia* **2004**, *138*, 325–332. [CrossRef]
20. Savage, C.; Thrush, S.F.; Lohrer, A.M.; Hewitt, J.E. Ecosystem services transcend boundaries: Estuaries provide resource subsidies and influence functional diversity in coastal benthic communities. *PLoS ONE* **2012**, *7*, e42708. [CrossRef]
21. Lotze, H.K.; Lenihan, H.S.; Bourque, B.J.; Bradbury, R.H.; Cooke, R.G.; Kay, M.C.; Kidwell, S.M.; Kirby, M.X.; Peterson, C.H.; Jackson, J.B.C. Depletion, Degradation, and Recovery Potential of Estuaries and Coastal Seas. *Science* **2006**, *312*, 1806–1809. [CrossRef]
22. Airoldi, L.; Balata, D.; Beck, M.W. The Gray Zone: Relationships between habitat loss and marine diversity and their applications in conservation. *J. Exp. Mar. Biol.* **2008**, *366*, 8–15. [CrossRef]
23. Crain, C.M.; Halpern, B.S.; Beck, M.W.; Kappel, C.V. Understanding and managing human threats to the coastal marine environment. *Ann. N. Y. Acad. Sci.* **2009**, *1162*, 39–62. [CrossRef] [PubMed]
24. Kaiser, M.J.; Clarke, K.R.; Hinz, H.; Austen, M.C.V.; Somerfield, P.J.; Karakassis, I. Global analysis of response and recovery of benthic biota to fishing. *Mar. Ecol. Prog. Ser.* **2006**, *311*, 1–14. [CrossRef]
25. Borja, A.; Galparsoro, I.; Solaun, O.; Muxika, I.; Tello, E.M.; Uriarte, A.; Valencia, V. The European Water Framework Directive and the DPSIR, a methodological approach to assess the risk of failing to achieve good ecological status. *Estuar. Coast. Shelf Sci.* **2006**, *66*, 84–96. [CrossRef]
26. Borja, A.; Dauer, D.M.; Elliott, M.; Simenstad, C. Medium and long-term recovery of estuarine and coastal ecosystems: Patterns, rates and restoration effectiveness. *Estuaries Coast.* **2010**, *33*, 1249–1260. [CrossRef]
27. Kenny, A.J.; Skjoldal, H.R.; Engelhard, G.H.; Kershaw, P.J.; Reid, J.B. An integrated approach for assessing the relative significance of human pressures and environmental forcing on the status of Large Marine Ecosystems. *Prog. Oceanogr.* **2009**, *81*, 132–148. [CrossRef]
28. Kenny, A.J.; Jenkins, C.; Wood, D.; Bolam, S.G.; Mitchell, P.; Scougal, C.; Judd, A. Assessing cumulative human activities, pressures, and impacts on North Sea benthic habitats using a biological traits approach. *ICES J. Mar. Sci.* **2018**, *75*, 1080–1092. [CrossRef]
29. Lissner, A.L.; Taghorn, G.L.; Diener, D.R.; Schroeter, S.C.; Dixon, D. Recolonization of deep-water hard substrate communities: Potential impacts from oil and gas development. *Ecol. Appl.* **1991**, *1*, 258–267. [CrossRef]
30. Pearson, T.H.; Rosenberg, R. Macrobenthic succession in relation to organic enrichment and pollution of the marine environment. *Oceanogr. Mar. Biol. Ann. Rev.* **1978**, *16*, 229–311.
31. Nasi, F.; Ferrante, L.; Alvisi, F.; Bonsdorff, E.; Auriemma, R.; Cibic, T. Macrofaunal bioturbation attributes in relation to riverine influence: What can we learn from the Po River lagoonal system (Adriatic Sea)? *Estuar. Coast. Shelf Sci.* **2020**, *232*, 106405. [CrossRef]
32. Dashtgard, S.E.; Gingras, M.K.; Pemberton, S.G. Grain-size controls on the occurrence of bioturbation. *Palaeogeogr. Palaeoclimatol. Palaeoecol.* **2008**, *257*, 224–243. [CrossRef]
33. Gingras, M.K.; Pemberton, S.G.; Saunders, T.; Clifton, H.E. The ichnology of modern and Pleistocene brackish-water deposits at Willapa Bay, Washington: Variability in estuarine settings. *Palaios* **1999**, *14*, 352–374. [CrossRef]
34. Martinez-Garcia, E.; Sanchez-Jerez, P.; Aguado-Giménez, F.; Ávila, P.; Guerrero, A.; Sánchez-Lizaso, J.L.; Fernandez-Gonzalez, V.; González, N.; Gairin, J.I.; Carballeira, C.; et al. A meta-analysis approach to the effects of fish farming on soft bottom polychaeta assemblages in temperate regions. *Mar. Pollut. Bull.* **2013**, *69*, 165–171. [CrossRef]
35. Papageorgiou, N.; Ioanna Kalantzi, I.; Karakassis, I. Effects of fish farming on the biological and geochemical properties of muddy and sandy sediments in the mediterranean sea. *Mar. Environ. Res.* **2010**, *69*, 326–336. [CrossRef]
36. Solan, M.; Cardinale, B.J.; Downing, A.L.; Engelhardt, K.A.; Ruesink, J.L.; Srivastava, D.S. Extinction and ecosystem function in the marine benthos. *Science* **2004**, *306*, 1177–1180. [CrossRef]
37. Wrede, A.; Beermann, J.; Dannheim, J.; Gutow, L.; Brey, T. Organism functional traits and ecosystem supporting services. A novel approach to predict bioirrigation. *Ecol. Indicat.* **2018**, *91*, 737–743. [CrossRef]
38. Bunker, D.E.; DeClerck, F.; Bradford, J.C.; Colwell, R.K.; Perfecto, I.; Phillips, O.L.; Naeem, S. Species loss and aboveground carbon storage in a tropical forest. *Science* **2005**, *310*, 1029–1031. [CrossRef] [PubMed]

39. Solan, M.; Wigham, B.D.; Hudson, I.R.; Kennedy, R.; Coulon, C.H.; Norling, K.; Rosenberg, R. In situ quantification of bioturbation using time-lapse fluorescent sediment profile imaging (f-SPI), luminophore tracers and model simulation. *Mar. Ecol. Prog. Ser.* **2004**, *271*, 1–12. [CrossRef]
40. Lohrer, A.; Halliday, N.; Thrush, S.; Hewitt, J.; Rodil, I. Ecosystem functioning in a disturbance-recovery context: Contribution of macrofauna to primary production and nutrient release on intertidal sandflats. *J. Exp. Mar. Biol. Ecol.* **2010**, *390*, 6–13. [CrossRef]
41. Teal, L.; Parker, E.; Solan, M. Coupling bioturbation activity to metal (Fe and Mn) profiles in situ. *Biogeosciences* **2013**, *10*, 2365–2378. [CrossRef]
42. Queirós, A.M.; Hiddink, J.G.; Johnson, G.; Cabral, H.N.; Kaiser, M.J. Context dependence of marine ecosystem engineer invasion impacts on benthic ecosystem functioning. *Biol. Invasions* **2011**, *13*, 1059–1075. [CrossRef]
43. Solan, M.; Scott, F.; Dulvy, N.K.; Godbold, J.A.; Parker, R. Incorporating extinction risk and realistic biodiversity futures: Implementation of trait-based extinction scenarios. In *Marine Biodiversity and Ecosystem Functioning: Frameworks, Methodologies, and Integration*; Oxford University Press: Oxford, UK, 2012; pp. 127–148.
44. Van Colen, C.; Rossi, F.; Montserrat, F.; Andersson, M.G.I.; Gribsholt, B.; Herman, P.M.J.; Degraer, S.; Vincx, M.; Ysebaert, T.; Middelburg, J.J. Organism-sediment interactions govern post-hypoxia recovery of ecosystem functioning. *PLoS ONE* **2012**, *7*, e49795. [CrossRef]
45. Villnäs, A.; Norkko, J.; Lukkari, K.; Hewitt, J.; Norkko, A. Consequences of increasing hypoxic disturbance on benthic communities and ecosystem functioning. *PLoS ONE* **2012**, *7*, e44920. [CrossRef] [PubMed]
46. Josefson, A.B.; Norkko, J.; Norkko, A. Burial and decomposition of plant pigments in surface sediments of the Baltic Sea: Role of oxygen and benthic fauna. *Mar. Ecol. Prog. Ser.* **2012**, *455*, 33–49. [CrossRef]
47. Kristensen, E.; Penha-Lopes, G.; Delefosse, M.; Valdemarsen, T.; Quintana, C.O.; Banta, G.T. What is bioturbation? The need for a precise definition for fauna in aquatic sciences. *Mar. Ecol. Prog. Ser.* **2012**, *446*, 285–302. [CrossRef]
48. Aller, R.C. The effects of macrobenthos on chemical properties of marine sediment and overlying water. In *Animal Sediment Relations*; McCall, P.L., Tevesz, M.J.S., Eds.; Plenum Press: New York, NY, USA, 1982; pp. 53–102. [CrossRef]
49. Christensen, B.; Vedel, A.; Kristensen, E. Carbon and nitrogen fluxes in sediment inhabited by suspension-feeding (*Nereis diversicolor*) and non-suspension feeding (*N. virens*) polychaetes. *Mar. Ecol. Prog. Ser.* **2000**, *192*, 203–217. [CrossRef]
50. Painting, S.J.; Van der Molen, J.; Parker, E.; Coughlan, C.; Birchenough, S.; Bolam, S.; Aldridge, J.N.; Forster, R.M.; Greenwood, N. Development of indicators of ecosystem functioning in a temperate shelf sea: A combined fieldwork and modelling approach. *Biogeochemistry* **2012**, *113*, 237–257. [CrossRef]
51. Van Hoey, G.; Permuy, D.C.; Vandendriessche, S.; Vincx, M.; Hostens, K. An ecological quality status assessment procedure for soft-sediment benthic habitats: Weighing alternative approaches. *Ecol. Ind.* **2013**, *25*, 266–278. [CrossRef]
52. Renz, J.R.; Powilleit, M.; Gogina, M.; Zettler, M.L.; Morys, C.; Forster, S. Community bioirrigation potential (BIPc), an index to quantify the potential for solute exchange at the sediment-water interface. *Mar. Environ. Res.* **2018**, *181*, 214–224. [CrossRef]
53. Islam, S. Nitrogen and phosphorus budget in coastal and marine cage aquaculture and impacts of effluent loading on ecosystem: Review and analysis towards model development. *Mar. Pollut. Bull.* **2005**, *50*, 48–61. [CrossRef]
54. Sweetman, A.K.; Norling, K.; Gunderstad, C.; Haugland, B.T.; Dale, T. Benthic ecosystem functioning beneath fish farms in different hydrodynamic environments. *Limnol. Oceanogr.* **2014**, *59*, 1139–1151. [CrossRef]
55. Hedges, J.I.; Keil, R.G. Sedimentary organic matter preservation: An assessment and speculative synthesis. *Mar. Chem.* **1995**, *49*, 81–115. [CrossRef]
56. Marchini, A.; Munari, C.; Mistri, M. Functions and ecological status of eight Italian lagoons examined using biological traits analysis (BTA). *Mar. Pollut. Bull.* **2008**, *56*, 1076–1085. Available online: https://www.marlin.ac.uk/biotic (accessed on 1 April 2021). [CrossRef] [PubMed]
57. Tesi, T.; Miserocchi, S.; Goñi, M.A.; Turchetto, M.; Langone, L.; De Lazzari, A.; Albertazzi, S.; Correggiari, A. Influence of distributary channels on sediment and organic matter supply in event-dominated coastal margins: The Po prodelta as a study case. *Biogeosciences* **2011**, *8*, 365. [CrossRef]
58. Boldrin, A.; Langone, L.; Miserocchi, S.; Turchetto, M.; Acri, F. Po River plume on the Adriatic continental shelf: Dispersion and sedimentation of dissolved and suspended matter during different river discharge rates. *Mar. Geol.* **2005**, *222*, 135–158. [CrossRef]
59. Frignani, M.; Langone, L.; Ravaioli, M.; Sorgente, D.; Alvisi, F.; Albertazzi, S. Fine-sediment mass balance in the western Adriatic continental shelf over a century time scale. *Mar. Geol.* **2005**, *222*, 113–133. [CrossRef]
60. Kourafalou, V.H. Process studies on the Po River plume, North Adriatic Sea. *J. Geophys. Res.* **1999**, *104*, 29963–29985. [CrossRef]
61. Pirazzoli, P.A.; Tomasin, A. Recent evolution of surge-related events in the northern Adriatic area. *J. Coast. Res.* **2002**, *18*, 537–554.
62. Lipizer, M.; De Vittor, C.; Falconi, C.; Comici, C.; Tamberlich, F.; Giani, M. Effects of intense physical and biological forcing factors on CNP pools in coastal waters (Gulf of Trieste, Northern Adriatic Sea). *Estuar. Coast. Shelf Sci.* **2012**, *115*, 40–50. [CrossRef]
63. Vesal, S.E.; Nasi, F.; Pazzaglia, J.; Ferrante, R.; Auriemma, R.; Relitti, F.; Bazzaro, M.; Del Negro, P. Assessing the sewage discharge effects on soft-bottom macrofauna through traits-based approach. *Mar. Pollut. Bull.* **2021**, *173*, 113003. [CrossRef]
64. Vesal, S.E.; Auriemma, R.; Libralato, S.; Nasi, F.; Del Negro, P. Impacts of organic enrichment on macrobenthic production, productivity, and transfer efficiency: What can we learn from a gradient of sewage effluents? *Mar. Pollut. Bull.* **2022**, *182*, 113972. [CrossRef]
65. Solis-Weis, V.; Aleffi, I.F.; Bettoso, N.; Rossini, P.; Orel, G. The benthic macrofauna at the outfalls of the underwater sewage discharges in the Gulf of Trieste (Northern Adriatic Sea, Italy). *Ann. Ser. Hist. Nat.* **2007**, *17*, 1–16.

66. Novelli, G. *Gli Scarichi a Mare Nell'alto Adriatico*; Rassegna Tecnica del Friuli Venezia Giulia: Udine, Italy, 1996; Volume 3, pp. 11–19.
67. Wetzel, M.A.; Leuchs, H.; Koop, J.H.E. PRESERVATION effects on wet weight, dry weight, and ash-free dry weight biomass estimates of four common estuarine macro-invertebrates: No difference between ethanol and formalin. *Helgol. Mar. Res.* **2005**, *59*, 206–213. [CrossRef]
68. Solan, M. The Concerted Use of 'Traditional' and Sediment Profile Imagery (SPI) Methodologies in Marine Benthic Characterisation and Monitoring. Ph.D. Thesis, Department of Zoology, National University of Ireland, Galway, Ireland, 2000. (Unpublished work).
69. Queirós, A.M.; Birchenough, S.N.; Bremner, J.; Godbold, J.A.; Parker, R.E.; Romero-Ramirez, A.; Reiss, H.; Solan, M.; Somerfield, P.J.; Van Colen, C.; et al. A bioturbation classification of European marine infaunal invertebrates. *Ecol. Evol.* **2013**, *3*, 3958–3985. [CrossRef] [PubMed]
70. Gogina, M.; Morys, C.; Forster, S.; Gräwe, U.; Friedland, R.; Zettler, M.L. Towards benthic ecosystem functioning maps: Quantifying bioturbation potential in the German part of the Baltic Sea. *Ecol. Indicat.* **2017**, *73*, 574–588. [CrossRef]
71. Nasi, F.; Nordström, M.C.; Bonsdorff, E.; Auriemma, R.; Cibic, T.; Del Negro, P. Functional biodiversity of marine soft-sediment polychaetes from two Mediterranean coastal areas in relation to environmental stress. *Mar. Environ. Res.* **2018**, *137*, 121–132. [CrossRef]
72. De-la-Ossa-Carretero, J.A.; Del-Pilar-Ruso, Y.; Giménez-Casalduero, F.; Sánchez-Lizaso, J.L. Assessing reliable indicators to sewage pollution in coastal soft-bottom communities. *Environ. Monit. Assess.* **2012**, *184*, 2133–2149. [CrossRef]
73. Jumars, P.A.; Dorgan, K.M.; Lindsay, S.M. Diet of worms emended: An update of polychaete feeding guilds. *Annu. Rev. Mar. Sci.* **2015**, *7*, 497–520. [CrossRef]
74. Levin, L.; Blair, N.; DeMaster, D.; Plaia, G.; Fornes, W.; Martin, C.; Thomas, C. Rapid subduction of organic matter by maldanid polychaetes on the North Carolina slope. *J. Mar. Res.* **1997**, *55*, 595–611. [CrossRef]
75. Atkinson, R.J.A.; Froglia, C.; Arneri, E.; Antolini, B. Observations on the burrows and burrowing behaviour of Brachynotus gemmellari and on the burrows of several other species occurring on Squilla grounds off Ancona, Central Adriatic. *Sci. Mar.* **1998**, *62*, 91–100. [CrossRef]
76. Morys, C.; Powilleit, M.; Forster, S. Bioturbation in relation to the depth distribution of macrozoobenthos in the southwestern Baltic Sea. *Mar. Ecol. Prog. Ser.* **2017**, *579*, 19–36. [CrossRef]
77. Zuur, A.F.; Ieno, E.N.; Elphick, C.S. A protocol for data exploration to avoid common statistical problems. *Methods Ecol. Evol.* **2010**, *1*, 3–14. [CrossRef]
78. Magni, P.; Vesal, S.E.; Giampaoletti, J.; Como, S.; Gravina, M.F. Joint use of biological traits, diversity and biotic indices to assess the ecological quality status of a Mediterranean transitional system. *Ecol. Ind.* **2023**, *147*, 109939. [CrossRef]
79. Kristensen, E.; Kostka, J.E. Macrofaunal burrows and irrigation in marine sediment: Microbiological and biogeochemical interactions. In *Interactions between Macro- and Microorganisms in Marine Sediments*; Kristensen, E., Haese, R.R., Kostka, J.E., Eds.; American Geophysical Union: Washington, DC, USA, 2005; pp. 125–157.
80. Breine, N.T.; De Backer, A.; Van Colen, C.; Moens, T.; Hostens, K.; Van Hoey, G. Structural and functional diversity of soft–bottom macrobenthic communities in the Southern North Sea. *Estuar. Coast. Shelf Sci.* **2018**, *214*, 173–184. [CrossRef]
81. Thrush, S.F.; Hewitt, J.E.; Cummings, V.J.; Ellis, J.I.; Hatton, C.; Lohrer, A.; Norkko, A. Muddy waters: Elevating sediment input to coastal and estuarine habitats. *Front. Ecol. Environ.* **2004**, *2*, 299–306. [CrossRef]
82. Braga, F.; Zaggia, L.; Bellafiore, D.; Bresciani, M.; Giardino, C.; Lorenzetti, G.; Maicu, F.; Manzo, C.; Riminucci, F.; Ravaioli, M.; et al. Mapping turbidity patterns in the Po river prodelta using multi-temporal Landsat 8 imagery. *Estuar. Coast. Shelf Sci.* **2017**, *198*, 555–567. [CrossRef]
83. Maicu, F.; De Pascalis, F.; Ferrarin, C.; Umgiesser, G. Hydrodynamics of the Po River-Delta-Sea system. *J. Geophys. Res. Ocean.* **2018**, *123*, 6349–6372. [CrossRef]
84. Queirós, A.M.; Fernandes, J.A.; Faulwetter, S.; Nunes, J.; Rastrick, S.P.; Mieszkowska, N.; Artioli, Y.; Yool, A.; Calosi, P.; Arvanitidis, C.; et al. Scaling up experimental ocean acidification and warming research: From individuals to the ecosystem. *Glob. Chang. Biol.* **2015**, *21*, 130–143. [CrossRef]
85. Lopez, G.R.; Levinton, J.S. Ecology of deposit feeding animals in marine sediments. *Q. Rev. Biol.* **1987**, *62*, 235–260. [CrossRef]
86. Töornroos, A.; Bonsdorff, E. Developing the multitrait concept for functional diversity: Lessons from a system rich in functions but poor in species. *Ecol. Appl.* **2012**, *22*, 2221–2236. [CrossRef]
87. Salen-Picard, C.; Arlhac, D.; Alliot, E. Responses of a Mediterranean soft bottom community to short-term (1993–1996) hydrological changes in the Rhone river. *Mar. Environ. Res.* **2003**, *55*, 409–427. [CrossRef]
88. Kristensen, E. Impact of polychaetes (*Nereis* spp. and *Arenicola marina*) on carbon biogeochemistry in coastal marine sediments. *Geochem. Trans.* **2001**, *2*, 92–103. [CrossRef] [PubMed]
89. Shull, D.H.; Benoit, J.M.; Wojcik, C.; Senning, J.R. Infaunal burrow ventilation and pore-water transport in muddy sediments. *Estuar. Coast. Shelf Sci* **2009**, *83*, 277–286. [CrossRef]
90. Kristensen, E.; Delefosse, M.; Quintana, C.O.; Flindt, M.R.; Valdemarsen, T. Influence of benthic macrofauna community shifts on ecosystem functioning in shallow estuaries. *Front. Mar. Sci.* **2014**, *1*, 41. [CrossRef]
91. Mermillod-Blondin, F.; Rosenberg, R. Ecosystem engineering: The impact of bioturbation on biogeochemical processes in marine and freshwater benthic habitats. *Aquat. Sci.* **2006**, *68*, 434–442. [CrossRef]

92. Chareopanich, C.; Montani, S.; Tsutsumi, H.; Matsuoka, S. Modification of chemical characteristics of organically enriched sediment by *Capitella* sp. I. *Mar. Poll. Bull.* **1993**, *26*, 375–379. [CrossRef]
93. Dauwe, B.P.H.J.; Herman, P.M.J.; Heip, C.H.R. Community structure and bioturbation potential of macrofauna at four North Sea stations with contrasting food supply. *Mar. Ecol. Prog. Ser.* **1998**, *173*, 67–83. [CrossRef]
94. Kinoshita, K.; Wada, M.; Kogure, K.; Furota, T. Microbial activity and accumulation of organic matter in the burrow of the mud shrimp, *Upogebia major* (Crustacea: Thalassinidea). *Mar. Biol.* **2008**, *153*, 277–283. [CrossRef]
95. Braeckman, U.; Foshtomi, M.Y.; Van Gansbeke, D.; Meysman, F.; Soetaert, K.; Vincx, M.; Vanaverbeke, J. Variable importance of macrofaunal functional biodiversity for biogeochemical cycling in temperate coastal sediments. *Ecosystems* **2014**, *17*, 720–737. [CrossRef]
96. Aller, R.C. Interactions between bioturbation and Mn cycling in marine sediments. *EOS Transact. Am. Geophys. Union* **1988**, *69*, 1106.

Disclaimer/Publisher's Note: The statements, opinions and data contained in all publications are solely those of the individual author(s) and contributor(s) and not of MDPI and/or the editor(s). MDPI and/or the editor(s) disclaim responsibility for any injury to people or property resulting from any ideas, methods, instructions or products referred to in the content.

Communication

A New Record of *Pinctada fucata* (Bivalvia: Pterioida: Pteriidae) in Mischief Reef: A Potential Invasive Species in the Nansha Islands, China

Binbin Shan [1,2,3,†], Zhenghua Deng [1,3,†], Shengwei Ma [3,4], Dianrong Sun [2,3], Yan Liu [1,2,3], Changping Yang [2,3], Qiaer Wu [3,4,*] and Gang Yu [1,3,*]

1. Tropical Aquaculture Research and Development Center, South China Sea Fisheries Research Institute, Chinese Academy of Fishery Sciences, Sanya 572000, China
2. Key Laboratory of Marine Ranching, Ministry of Agriculture Rural Affairs, Guangzhou 510300, China
3. South China Sea Fishers Research Institute, Chinese Academy of Fisheries Sciences, Guangzhou 510300, China
4. The Agriculture Department Sea Fishery Development Key Laboratory, South China Sea Fisheries Research Institute, Chinese Academy of Fishery Sciences, Guangzhou 510300, China
* Correspondence: wqe66@163.com (Q.W.); gyu0928@163.com (G.Y.)
† These authors contributed equally to this work.

Abstract: Mischief Reef is located in the eastern Nansha Islands of the South China Sea. With increasingly intense anthropogenic disturbance, *Pinctada fucata*, a previously unrecorded species in the reef, has occurred in the region. In this study, we identified and described the occurrence of *P. fucata* in Mischief Reef based on morphology and molecular markers. Furthermore, we performed a population genetics analysis of seven *P. fucata* populations of the South China Sea. All *P. fucata* populations showed significant high-level genetic diversity, but the differentiation among *P. fucata* populations was small. There was an F_{ST} value close to zero (−0.0083) between the Lingshui and Mischief Reef populations. Our results hint that Lingshui may be one of the potential sources of *P. fucata* to Mischief. In addition, we discussed the possible cause of the mass occurrence of *P. fucata*. The present study serves as a warning that anthropogenic disturbances have disrupted the local ecosystem in Mischief Reef.

Keywords: pearl oyster; South China Sea; cytochrome c oxidase subunit I; internal transcribed spacer 2; population genetics

1. Introduction

The redistribution of species is a significant outcome of anthropogenic disturbance, whether intentional or unintentional, and may occur at any scale. This process can result in the disruption of local ecosystems and even pose a threat to economies [1]. Over the last five centuries, the number of species that have been introduced to new habitats through human activities has grown at an exponential rate, with a particularly notable surge in the past two centuries [2]. The introduction of invasive aquatic species into new environments has emerged as a pressing concern for the world's oceans, representing one of the most substantial risks and ranking among the top four threats [3]. Invasive alien marine species pose a threat to both marine biodiversity and industries, such as fishing and tourism. Unlike oil spills, this situation is likely to exacerbate over time, making it an increasingly pressing issue [4]. However, due to the accidental nature of many introductions, invasion events may be linked to significant data gaps and, in some cases, can remain undetected for extended periods, ranging from years to decades or even centuries [5,6].

The genus *Pinctada* Röding, 1798, a group of pearl oysters in the class Bivalvia and family Pteriidae, is found in a broad range of environments spanning from the Indo-Pacific to Western Atlantic tropical and subtropical regions. They are predominantly associated with shallow-water habitats, particularly reef environments [7]. *Pinctada fucata* (A. Gould, 1850), also known as *Pinctada fucata martensii* (Dunker, 1880) or *Pinctada martensii* (Dunker, 1880) (https://www.marinespecies.org/aphia.php?p=taxdetails&id=397170, accessed on 8 April 2023), is an economically valuable bivalve species that is endemic to the coastal waters of the Pacific Ocean between the Tropic of Cancer and the Tropic of Capricorn [8–10]. This species is mainly cultivated for pearl production in Asia, especially in Korea, China and Japan [9,11,12]. Mischief Reef is located in the eastern Nansha Islands of China (Figure 1). In recent decades, with increased development in Mischief Reef (reclamation, aquaculture and fishery), the species of *Pinctada* have changed in this area. According to previous research by Wang and Chen [13], *P. maculata* was the sole species of the genus *Pinctada* found in the sea area surrounding the Nansha Islands. However, a recent sampling conducted at Mischief Reef revealed that *P. fucata* was also present (Figure S1). We found that *P. fucata* tended to aggregate in groups at coral reefs and attach to the nets of aquaculture cages. According to the observations of local fishermen, the presence of *P. fucata* in Mischief was initially recorded in 2016. While no substantial ecological or economic issues have been reported due to the presence of *P. fucata* in the Mischief Reef area to date, it remains unclear whether the introduction and potential proliferation of this species may cause any detrimental effects to the reef's ecosystem in the future. Because it may be considered a potential invasive species to the Nansha Islands, it is necessary to describe the new record of *P. fucata* in Mischief Reef and estimate the population genetic diversity and structure of the *P. fucata* populations.

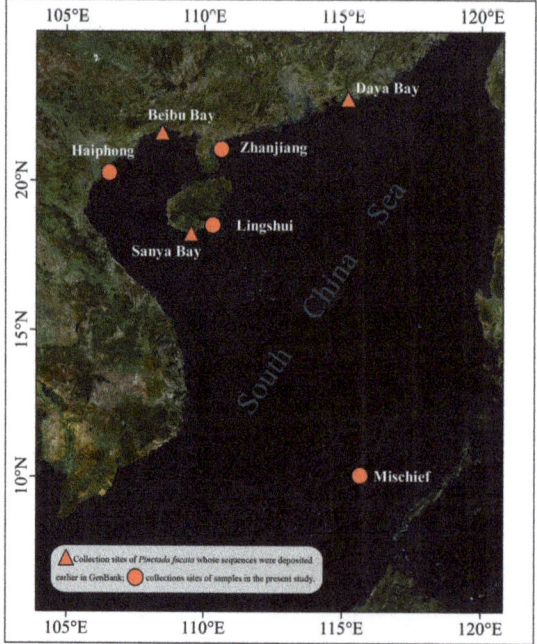

Figure 1. Sites sampled in the present study, and additional sites from which *Pinctada fucata* sequences were obtained from GenBank based on previous study [8]. For more detailed information, please refer to Supplementary Table S1. The base map is Bing Virtual Earth.

The taxonomy of pearl oyster species is primarily established based on their soft tissue and shell characteristics, including shape and colour, as outlined in previous studies [14,15]. However, the taxonomy of pearl oyster species is complex because their shells are quite similar [16] and there are not many morphological diagnosable characters available for species determination [17]. In recent years, the application of molecular sequence data has provided valuable insights into the taxonomic classification of numerous bivalve species [9,18,19]. The incursion of *P. imbricata radiata* (Leach, 1814) into the coastal waters of the eastern Adriatic Sea was reported by Gavrilović et al. [20], who relied on the analysis of the cytochrome c oxidase subunit I (COI) gene sequences. Somrup et al. [21] distinguished a new species in the genus *Pinctada* collected from Phuket, Thailand based on both morphological and COI gene sequence data. Furthermore, molecular markers also provide valuable information on nonindigenous species, facilitating the estimation of relationship between introduced populations and other geographically distinct populations [22,23]. In the present study, morphology and molecular markers were used to confirm the identity of *P. fucata* samples collected on Mischief Island. In addition, a population genetics analysis was conducted to compare the Mischief Reef population with other geographically distinct populations in the South China Sea, to estimate the potential geographic origin of the *P. fucata* introduced to Mischief Reef.

2. Materials and Methods

2.1. Sampling and DNA Extraction

Twenty *Pinctada fucata* specimens were collected in the coral reefs from Mischief Island (9°57′ N, 115°44′ E) in December 2018. Furthermore, we also sampled *P. fucata* specimens from Haiphong, Lingshui and Zhanjiang (ten individuals per population). The specimens were kept in a freezer after collection (Figure 1). Next, genomic DNA was isolated from a 50-mg sample of the adductor muscle using a phenol-chloroform extraction method following standard protocols.

2.2. Amplification and Sequencing

The specific primers PMCOI-F: 5′-TTT CTT ATC CGA ATG GAGCT-3′ and PMCOI-R: 5′-TGT ATT AAA ATG CCG ATC CG′ [24] were used to amplify a fragment of approximately 500 bp of the COI gene sequence. Internal transcribed spacer 2 (ITS2) was also amplified using the primer pair 5.8S-F and 28S-R from the study of Yu and Chu [25]. Polymerase chain reaction (PCR) amplifications were carried out in 25 reactions using the following mix: DNA template (1 µL), forwards primer (1 µL, 10 µM/L), reverse primer (1 µL, 10 µM/L), dNTPs (2 µL, 2.5 mM/L), EasyTaq DNA Polymerase (0.15 µL, 5 U/µL), and 10 × PCR buffer (2.5 µL, 25 µM/L). The PCR amplification protocol consisted of an initial denaturation at 95 °C for 5 min, followed by 35 cycles of denaturation at 94 °C for 45 s, annealing at 50 °C for 30 s, and extension at 72 °C for 1 min 45 s. A final extension step was performed at 72 °C for 10 min. The resulting PCR products of the COI gene were used as the template DNA for cycle sequencing reactions using the Big Dye Terminator Cycle Sequencing Kit. Sequencing was conducted bidirectionally on an ABI Prism 3730 automatic sequencer (Applied Biosystems, Foster City, CA, USA). The amplified products of ITS2 were purified and cloned into *Escherichia coli* using the pGEM-T Easy vector (Promega, Madison, WI, USA). The plasmids were sequenced using universal primers M13-47 and RV-M. The sequencing products were alcohol-precipitated and subsequently sequenced on an ABI 3730 automatic sequencer (Applied Biosystems, CA, USA).

2.3. Statistical Anazlysis

The obtained sequences were subjected to revision using DNASTAR software (DNASTAR, Madison, WI, USA). To confirm the classification status of our samples, eight COI sequences of *Pinctada* species were downloaded from GenBank and included in the phylogenetic tree study (Table 1). To root the tree, *Pteria penguin* (Röding, 1798) was selected as the outgroup. The neighbour-joining (NJ) tree was constructed using MEGA 5.0 [26] under the Kimura 2-parameter (K2P) model [26,27].

For population genetics analysis, definition of haplotypes was carried out using DnaSP v. 5.00, and gaps were not considered during the analysis of the sequence data [28]. To quantify the genetic diversity in each population, various measures were employed, including the number of polymorphic sites, number of haplotypes, haplotype diversity, nucleotide diversity, and mean number of pairwise differences. These parameters were calculated using Arlequin v. 3.0 [29]. By employing these measures, a comprehensive analysis of genetic variation within populations was conducted, providing valuable insights into the evolutionary processes shaping genetic diversity.

To assess the genetic differentiation between different populations, the F-statistic (Fst) was calculated using on Arlequin v. 3.0 software [30]. The Fst value was computed using the K2P method, considering different substitution rates between transitions and transversions [27]. The probability pertaining to the Fst values was assessed using random permutation techniques, wherein a minimum of 10,000 permutations were performed. Significance tests to determine differentiation between samples were carried out using exact P tests with a Markov chain procedure executed in Arlequin. Moreover, for a more comprehensive and straightforward illustration of population differentiation, we used IQ-TREE version 1.6 (http://www.iqtree.org/#download, accessed on 13 January 2022) to select the best-fit model of nucleotide substitution based on the Akaike information criterion (AIC), and we built the maximum likelihood (ML) tree based on the best-fit model [31]. Furthermore, we drew a heatmap of Fst values and exact P tests between different populations.

Table 1. Species and the GenBank accession numbers of the COI sequences used in this study.

Species	GenBank Accession No.	Sample Site
Pinctada fucata martensii *	KX669229	Sanya, China [32]
Pinctada fucata	DQ299941	China
Pinctada albina	AB261165	Kagoshima, Japan [33]
Pinctada martensii *	AB076915	Okinawa, Japan [18]
Pinctada imbricata	KX713492	Florida Keys, USA [34]
Pinctada maculata	AB076928	Okinawa, Japan [18]
Pinctada margaritifera	HM467838	China
Pinctada persica	AB777263	Hendurabi, Iran [35]
Pinctada radiata	KF284062	Ras Al Khaimah, UAE [36]
Pinctada maxima	NC018752	Not mentioned [37]
Pteria penguin	KU552127	Sanya, China [32]

* Pinctada martensii, P. fucata martensii and P. fucata were conspecific, the P. martensii and P. fucata martensii are not accepted.

3. Results

3.1. Morphological Characteristics and DNA Barcoding Identification

The body and shell are asymmetrical, the left valve is deeper than the right valve, and there is a byssal notch on the anterior side. The average length/height ratio was 0.86. Anterior and posterior auricles are located at the ends of the hinge line, and the former is larger. The hinge line is longer than the shell height. The ligament connects the two valves at the centre of the hinge line. The hinge teeth are well developed, and the posterior tooth of the left valve is above that of the right. The external colour is brown, green, red, yellow, or white, and the shell's internal surface is nacreous; the nacre is of a hard-white metallic lustre and yellow, silver, gold, or pink (Figure 2). Muscular scars are visible. According to the key characteristics for species identification based on the shell morphology of the genus *Pinctada*, the distribution pattern of processes (scales) on the external shell is different between *P. maculata* and *P. fucata* [38,39]. The processes (scales) of *P. fucata* are densely distributed, but those of *P. maculata* are sparsely distributed at regular intervals [39].

The COI gene (500 bp) was sequenced from six individuals. Two haplotypes were found. The accession numbers for the two haplotype sequences submitted to GenBank are MK748604 and MK748605. Sixteen COI sequences of *Pinctada* species, including our haplotypes, were downloaded and analysed (Table 1). The mean genetic distance between all species was calculated as 22.63%, while the genetic distance between our six specimens and *P. martensii* was only 0.26%. Furthermore, the genetic distances between our six specimens and the other *Pinctada* species (exclude the upper clade) ranged from 9.81% to 44.25%. An NJ phylogenetic tree was constructed using MEGA 5.0, with *P. penguin* chosen as the outgroup. The tree showed that our specimens, *P. fucata*, *P. fucata martensii* and *P. martensii* clustered in the same group (Figure 3).

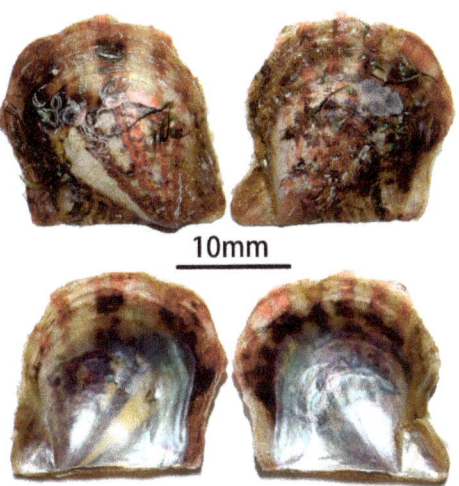

Figure 2. *Pinctada fucata* collected from in the coral reefs from the Mischief Reef.

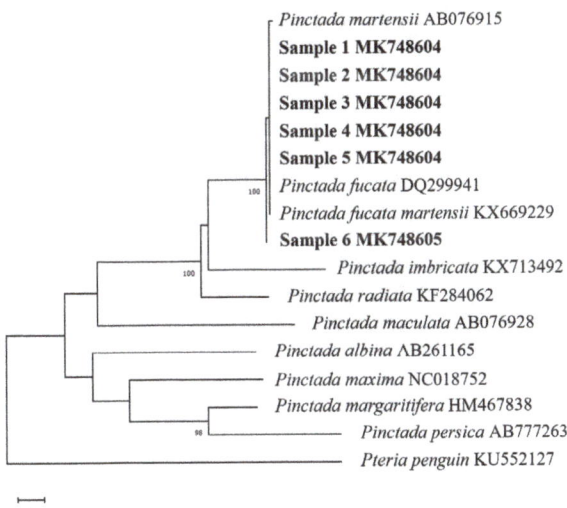

Figure 3. Neighbour-joining phylogenetic tree based on COI sequence data. The regular texts refer to sequences downloaded from GenBank, while the bold texts represent sequences that were sequenced in the present study. *Pteria penguin* was designated as the outgroup and used to root the tree.

3.2. Population Genetic Diversity

In this study, we obtained 40 ITS2 sequences with a length of 548 bp (OQ629249-OQ629288). Among these sequences, we identified a total of 31 haplotypes. Among the 31 haplotypes identified, 26 were found singly in the four locations sampled, representing 83% of the total (Supplementary Table S2). The Lingshui and Mischief populations exhibited the highest number of haplotypes with 10 each, while the Zhangjiang population had the lowest eight haplotypes (see Table 2). Notably, out of the 10 haplotypes detected in the

Mischief Reef population, only two were shared with other populations (Supplementary Table S2). Additional genetic diversity parameters are presented in Table 2, where the gene diversity ranged from 0.93 to 1.00, indicating that all *P. fucata* populations demonstrated a high level of genetic diversity.

Table 2. Population genetic diversity parameters.

	Zhanjiang	Lingshui	Mischief	Haiphong
Numbers of haplotypes	8	10	10	9
Haplotype diversity	0.9333	1.0000	1.0000	0.9778
Nucleotide diversity	0.0129	0.0148	0.0151	0.0142
Mean number of pairwise difference	7.1556	8.1556	8.3111	7.8222

3.3. Population Genetic Structure and Differentiation

Additionally, we downloaded 25 ITS2 sequences from previous study [8], and their corresponding sampling sites are shown in Figure 1. Supplementary Table S1 provides further information on these samples. The subsequent analysis was performed on a total of 65 sequences. Pairwise *Fst* were calculated based on the ITS2 sequences, and the values ranged from −0.0700 to 0.1101 (Figure 4). The Beihai and Mischief Reef populations exhibited the highest F_{ST} values, while the Shenzhen and Lingshui populations had the lowest *Fst* values. Notably, the *Fst* values between the Mischief Reef and Lingshui populations were negative (−0.0093), indicating a closer relationship between these populations than within populations. Among the populations, the Beihai population displayed greater genetic differentiation (0.0400–0.1101) than the other populations. The optimal model of nucleotide substitution was found to be HKY + F + R2. The ML tree was constructed based on 1000 replicates of ultrafast bootstrap approximation. The ML tree had a shallow topology, and there were no significant genealogical branches or sample clusters corresponding to the sampling sites (Figure 5). However, the topological configuration of the phylogenetic tree showed that the majority of individuals from Mischief Reef displayed a greater similarity to the Lingshui population in comparison to other populations, thus indicating a strong connection between the Lingshui population and the *P. fucata* population of Mischief Reef.

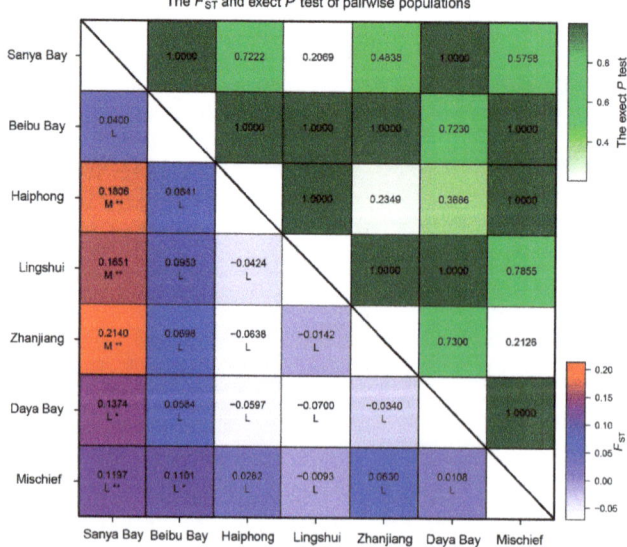

Figure 4. Heatmap of *Fst* (**right**, below diagonal) and exact *P* test (**left**, above diagonal) values of *P. fucata*. *: $p < 0.05$, **: $p < 0.01$.

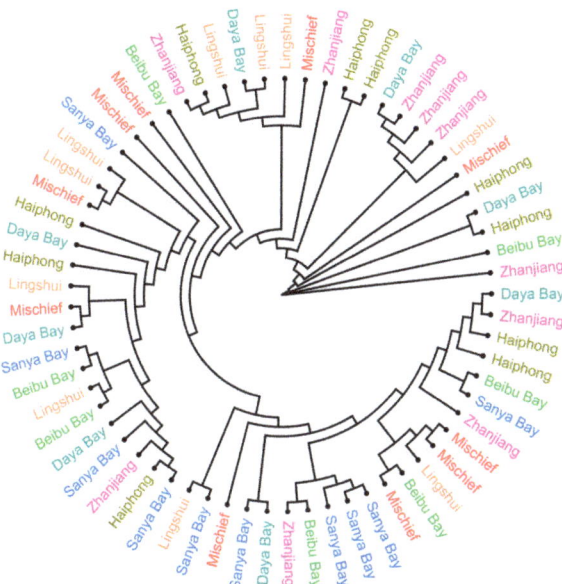

Figure 5. Maximum likelihood tree of all ITS2 sequences, with each sampling site indicated by a distinct colour. The lengths of branches represent the relative genetic difference.

4. Discussion

4.1. Identification of the Samples

Pinctada fucata and *P. imbricata* are accepted in online marine taxonomic databases such as Molluscabase, GBIF and WoRMS. Cunha et al. [10] recognized the phylogenies of *Pinctada* species, and their analyses indicated that *P. fucata* and *P. martensii* were conspecific, with the species recovered as monophyletic compared to *P. imbricata*. Furthermore, Yu and Chu [9] observed low genetic distances among *P. fucata*, *P. fucata martensii* and *P. imbricata*, and indicated they may be conspecific. According to these earlier findings, *P. fucata* is the available specific name. In the present study, we identified the samples based on molecular markers and morphology, our results confirm that the species we collected from Mischief Reef is indeed *P. fucata*.

A series of detailed investigations (investigation time: July–August 1988, May 1989, May–June 1993 and March–April 1994) of the Nansha Islands by the China Nansha expedition showed the absence of *P. fucata* [13,40]. Therefore, our study represents the first record of *P. fucata* on the Nansha Islands. While *P. fucata* is of great economic value in pearl production, the emergence of non-native species may present a potential hazard to the indigenous ecosystem. The activities of introduced bivalve species, including shell construction, bioturbation, and filter feeding, could modify the processes and functions of ecosystems, thereby exerting detrimental effects on biodiversity and the environment [3,41]. Therefore, it is important to pay close attention to the new record of *P. fucata* in Mischief Reef.

4.2. Population Genetics Analysis

In the present study, we found that all *P. fucata* populations exhibited a significant level of genetic diversity. Additionally, the high haplotype diversity (0.93–1.00) with low nucleotide diversity (0.013–0.015) observed in our study is consistent with that reported in many other bivalve species [42,43].

In our study, we investigated the population genetic differentiation of six native *P. fucata* populations and one Mischief Reef population. The Fst values and ML tree revealed that the differentiation among *P. fucata* populations was small. Previous studies have suggested that the *P. fucata* populations in China could be considered a group [44]. We also found that many pairwise Fst values were negative, indicating that differentiation between populations was very small [45]. For instance, the population sampled from Shenzhen showed negative Fst values with four populations. This phenomenon may be explained by mixed germplasm resources resulting from mass hybridization and transport during the development of *P. fucata* aquaculture [25,46]. More importantly, it is noteworthy that among the six native populations, only the Lingshui population had a negative Fst value (−0.0093) with the Mischief Reef population. Our results hinted that Lingshui had a very close relationship with *P. fucata* in Mischief Reef.

4.3. Possible Cause of Occurrence

Since we have not yet investigated the source of the *P. fucata* population in Mischief, we cannot directly address the invasion pathway and the exact origins. Thus, here we discuss the possible cause of the occurrence of *P. fucata*. Most marine invasive species are disseminated via ballast water or hull fouling, which are associated with maritime activities [4]. In North America, commercial shipping has been identified as the most significant introduction vector, responsible for 52–82% of nonindigenous species introductions over the past 30 years [47]. The pearl oyster exhibits a prolonged planktonic larval stage, which can last up to 17 days [48]. This characteristic may facilitate the entrainment of larvae in ballast tanks throughout the duration of a voyage. Other vectors, specifically hull fouling, floating ropes and aquaculture net cages, could also be introduction vectors. Previous studies have reported that aquaculture and other human activities have increased in frequency in the past decade, particularly in Lingshui and other cities in Hainan [49,50]. This raises the possibility that these activities may have facilitated the introduction of *P. fucata* from Hainan to Mischief Island, and could explain why *P. fucata* populations in Lingshui and Mischief Island exhibit a closer relationship. Marques and Breve [51] documented the presence of juvenile *P. imbricata* adhering to a floating rope found along the Uruguayan coast. Although there were no data indicating that the species was successfully and effectively settled, it could be a potentially invasive species in the Uruguayan coast. Furthermore, the expansion of native species into adjacent areas may also account for the presence of newly recorded species, as this process can produce effects similar to those generated by alien species [52]. As a result of human activities (reclamation, aquaculture, fishery), the habitat (including temperature, salinity, nutrients) in Mischief Reef may have been changed, which could benefit the migration of *P. fucata* from adjacent areas through ocean currents.

5. Conclusions

In the present study, we identified and described the occurrence of *P. fucata* in Mischief Reef. The results of the population genetics analysis hinted that the Lingshui population had a very close relationship with *P. fucata* in Mischief Reef. The presence of *P. fucata* should serve as a warning that anthropogenic disturbances have disrupted the local ecosystem in Mischief Reef. Investigating the invasion pathway of *P. fucata* is of great importance, as it highlights the necessity for management strategies to prevent the introduction of other invasive species.

Supplementary Materials: The following supporting information can be downloaded at: https://www.mdpi.com/article/10.3390/d15040578/s1, Figure S1: The occurrence of *Pinctada fucata* in Mischief Reef; Table S1: Information of the downloaded ITS2 sequences from NCBI; Table S2: Haplotypes of the four *P. fucata* populations

Author Contributions: Conceptualization, B.S., Z.D., S.M., Q.W. and G.Y.; methodology, B.S. and S.M.; software, B.S. and Z.D.; validation, B.S., S.M. and Z.D.; formal analysis, B.S.; investigation, Z.D.; resources, B.S. and S.M.; data curation, B.S., Z.D., S.M. and D.S.; writing—original draft preparation, B.S., Z.D., S.M.; writing—review and editing, B.S., Z.D., Q.W. and G.Y.; visualization, Q.W. and G.Y.; supervision, Y.L., C.Y., Q.W. and G.Y.; project administration, Q.W. and G.Y.; funding acquisition, Q.W., G.Y. and D.S. All authors have read and agreed to the published version of the manuscript.

Funding: This study was supported by the Hainan Provincial Natural Science Foundation of China under contract No. 320QN361; National Natural Science Foundation of China (42206109); Asia Cooperation Fund Project—Modern fishery cooperation between China and neighboring countries around the South China Sea; Biodiversity, germplasm resources bank and information database construction of the South China Sea Project.

Institutional Review Board Statement: Not applicable.

Data Availability Statement: All sequences obtained in present study were deposited in NCBI. The accession numbers are OQ629249-OQ629288, MK748604 and MK748605.

Conflicts of Interest: The authors declare no conflict of interest.

References

1. Tong, X.; Wang, R.; Chen, X.Y. Expansion or Invasion? A Response to Nackley et al. *Trends. Ecol. Evol.* **2018**, *33*, 234–235. [CrossRef] [PubMed]
2. Di Castri, F.; Drake, J.A.; Mooney, H.A.; Groves, R.H.; Kruger, F.J.; Rejmanek, M.; Williamson, M. History of biological invasions with emphasis on the Old World. In *Biological Invasions: A Global Perspective*; Wiley: Chichester, UK, 1989; pp. 1–30.
3. Werschkun, B.; Banerji, S.; Basurko, O.C.; David, M.; Fuhr, F.; Gollasch, S.; Grummt, T.; Haarich, M.; Jha, A.N.; Kacan, S.; et al. Emerging risks from ballast water treatment: The run-up to the International Ballast Water Management Convention. *Chemosphere* **2014**, *112*, 256–266. [CrossRef] [PubMed]
4. Bax, N.; Williamson, A.; Aguero, M.; Gonzalez, E.; Geeves, W. Marine invasive alien species: A threat to global biodiversity. *Mar. Policy* **2003**, *27*, 313–323. [CrossRef]
5. Bortolus, A.; Carlton, J.T.; Schwindt, E. Reimagining South American coasts: Unveiling the hidden invasion history of an iconic ecological engineer. *Divers. Distrib.* **2015**, *21*, 1267–1283. [CrossRef]
6. Blakeslee, A.M.H. Parasites and genetics in marine invertebrate introductions: Signatures of diversity declines across systems. In *Biological Invasions in Changing Ecosystems: Vectors, Ecological Impacts, Management and Predictions*; Canning-Clode, J., Ed.; Sciendo Migration: Warsaw, Poland, 2016; pp. 138–182.
7. Southgate, P.C.; Lucas, J.S. *The Pearl Oyster*; Elsevier Science: Amsterdam, The Netherlands, 2011.
8. Yu, D.H.; Zhu, J.H. Species status of Chinese *Pinctada fucata*, Japanese *P. fucata martensii* and Australian *P. imbricata* using ITS and AFLP markers. *South China Fish. Sci.* **2006**, *2*, 36–44.
9. Yu, D.H.; Chu, K.H.; Jia, X.P. Preliminary analysis on genetic relationship of common pearl oysters of *Pinctada* in China. *Oceanol. Limnol. Sin.* **2006**, *37*, 211–217.
10. Cunha, R.L.; Blanc, F.; Bonhomme, F.; Arnaud-Haond, S. Evolutionary patterns in pearl oysters of the genus *Pinctada* (Bivalvia: Pteriidae). *Mar. Biotechnol.* **2011**, *13*, 181–192. [CrossRef]
11. Choi, Y.H.; Chang, Y.J. Gametogenic cycle of the transplanted-cultured pearl oyster, *Pinctada fucata martensii* (Bivalvia: Pteriidae) in Korea. *Aquaculture* **2003**, *220*, 781–790. [CrossRef]
12. Tomaru, Y.; Udaka, N.; Kawabata, Z.; Nakano, S. Seasonal change of seston size distribution and phytoplankton composition in bivalve pearl oyster *Pinctada fucata martensii* culture farm. *Hydrobiologia* **2002**, *481*, 181–185. [CrossRef]
13. Wang, Z.R.; Chen, R.Q. The species of the Pterioda from the Nansha Islands waters. In *Studies on Marine Species of the Nansha Islands and Neighboring Waters*; Chen, Q.C., Ed.; China Ocean Press: Beijing, China, 1991; pp. 150–151.
14. Hynd, J.S. A Revision of the Australian Pearl-shells, Genus *Pinctada* (Lamelli-branchia). *Mar. Freshwater. Res.* **1955**, *6*, 98–138. [CrossRef]
15. Waller, T.R. Morphology, morphoclines and a new classification of the Pteriomorphia (Mollusca: Bivalvia). *Philos. Trans. R. Soc. B* **1978**, *284*, 345–365.
16. Masaoka, T.; Kobayashi, T. Natural hybridization between *Pinctada fucata* and *Pinctada maculata* inferred from internal transcribed spacer regions of nuclear ribosomal RNA genes. *Fish. Sci.* **2005**, *71*, 829–836. [CrossRef]
17. Wada, K.T.; Tëmkin, I. Taxonomy and Phylogeny. In *The Pearl Oyster*; Southgate, P.C., Lucas, J.S., Eds.; Elsevier: Amsterdam, The Netherlands, 2008.
18. Matsumoto, M. Phylogenetic analysis of the subclass Pteriomorphia (Bivalvia) from mtDNA COI sequences. *Mol. Phylogenet. Evol.* **2003**, *27*, 429–440. [CrossRef] [PubMed]
19. Wood, A.R.; Apte, S.; MacAvoy, E.S.; Gardner, J.P. A molecular phylogeny of the marine mussel genus Perna (Bivalvia: Mytilidae) based on nuclear (ITS1&2) and mitochondrial (COI) DNA sequences. *Mol. Phylogenet. Evol.* **2007**, *44*, 685–698. [PubMed]
20. Gavrilović, A.; Piria, M.; Guo, X.Z.; Jug-Dujaković, J.; Ljubučić, A.; Krkić, A.; Iveša, N.; Marshall, B.A.; Gardner, J.P. First evidence of establishment of the rayed pearl oyster, *Pinctada imbricata radiata* (Leach, 1814), in the eastern Adriatic Sea. *Mar. Pollut. Bull.* **2017**, *125*, 556–560. [CrossRef] [PubMed]
21. Somrup, S.; Sangsawang, A.; McMillan, N.; Winitchai, S.; Inthoncharoen, J.; Liu, S.; Muangmai, N. *Pinctada phuketensis* sp. nov. (Bivalvia, Ostreida, Margaritidae), a new pearl oyster species from Phuket, western coast of Thailand. *ZooKeys* **2022**, *1119*, 181–195. [CrossRef] [PubMed]

22. Muirhead, J.R.; Gray, D.K.; Kelly, D.W.; Ellis, S.M.; Heath, D.D.; Macisaac, H.J. Identifying the source of species invasions: Sampling intensity vs. genetic diversity. *Mol. Ecol.* **2008**, *17*, 1020–1035. [CrossRef]
23. Cristescu, M.E. Genetic reconstructions of invasion history. *Mol. Ecol.* **2015**, *24*, 2212–2225. [CrossRef]
24. Wu, L. *Studies of Genetic Diversity on* Pinctada martensii *and Sinonovacula Constricta*; China Ocean University: Qingdao, China, 2013.
25. Yu, D.H.; Chu, K.H. Species identity and phylogenetic relationship of the pearl oysters in *Pinctada* Röding, 1798 based on ITS sequence analysis. *Biochem. Syst. Ecol.* **2006**, *34*, 240–250. [CrossRef]
26. Tamura, K.; Peterson, D.; Peterson, N.; Stecher, G.; Nei, M.; Kumar, S. MEGA5: Molecular evolutionary genetics analysis using maximum likelihood, evolutionary distance, and maximum parsimony methods. *Mol. Biol. Evol.* **2011**, *28*, 2731–2739. [CrossRef]
27. Kimura, M. A simple method for estimating evolutionary rate base substitution through comparative studies of nucleotide sequences. *J. Mol. Evol.* **1980**, *16*, 111–120. [CrossRef] [PubMed]
28. Librado, P.; Rozas, J. DnaSP v5: A software for comprehensive analysis of DNA polymorphism data. *Bioinformatics* **2009**, *25*, 1451–1452. [CrossRef] [PubMed]
29. Excoffier, L.; Laval, G.; Schneider, S. Arlequin (version 3.0): An integrated software package for population genetics data analysis. *Evol. Bioinform.* **2005**, *1*, 47–50. [CrossRef]
30. Weir, B.S.; Cockerham, C.C. Estimating F-statistics for the analysis of population structure. *Evolution* **1984**, *38*, 1358–1370. [PubMed]
31. Nguyen, L.T.; Schmidt, H.A.; von Haeseler, A.; Minh, B.Q. IQ-TREE: A fast and effective stochastic algorithm for estimating maximum-likelihood phylogenies. *Mol. Biol. Evol.* **2015**, *32*, 268–274. [CrossRef]
32. Zhan, X.; Zhang, S.; Gu, Z.; Wang, A. Complete mitochondrial genomes of two pearl oyster species (Bivalvia: Pteriomorphia) reveal novel gene arrangements. *J. Shellfish Res.* **2018**, *37*, 1039–1050. [CrossRef]
33. Takakura, D.; Wada, K.T.; Kobayashi, T.; Masaoka, T.; Samata, T. Phylogenetic relationships of several pearl oyster species in *Pinctada* (Pteriidae; Bivalvia) inferred from mitochondrial 12S rDNA and cytochrome c oxidase subunit I genes sequences. *J. Foss. Res.* **2008**, *41*, 33–40.
34. Combosch, D.J.; Collins, T.M.; Glover, E.A.; Daniel, L.G.; Elizabeth, M.H.; John, M.H.; Gisele, Y.K.; Sarah, T.; Erin, M.; Ellen, E.S.; et al. A family-level tree of life for bivalves based on a Sanger-sequencing approach. *Mol. Phylogenet. Evol.* **2017**, *107*, 191–208. [CrossRef]
35. Sharif Ranjbar, M.; Zolgharnien, H.; Yavari, V.; Bita, A.; Mohammad, A.S.; Sophie, A.H.; Regina, L.C. Rising the Persian Gulf black-lip pearl oyster to the species level: Fragmented habitat and chaotic genetic patchiness in *Pinctada persica*. *Evol. Biol.* **2016**, *43*, 131–143. [CrossRef]
36. Meyer, J.B.; Cartier, L.E.; Pinto-Figueroa, E.A.; Michael, S.K.; Henry, A.H.; Bruce, A.M. DNA fingerprinting of pearls to determine their origins. *PLoS ONE* **2013**, *8*, e75606. [CrossRef]
37. Wu, X.; Li, X.; Li, L.; Yu, Z. A unique tRNA gene family and a novel, highly expressed ORF in the mitochondrial genome of the silver-lip pearl oyster, *Pinctada maxima* (Bivalvia: Pteriidae). *Gene* **2012**, *510*, 22–31. [CrossRef] [PubMed]
38. Hayami, I. Superfamily pteriacea. In *Marine Mollusks in Japan*; Okutani, T., Ed.; Tokai University: Tokyo, Japan, 2000; pp. 879–883.
39. Takemura, Y.; Okutani, T. On the identification of species of *Pinctada* found attached to *Pinctada maxima* (Jameson) in the Arafura Sea. *Bull. Tokai Reg. Fish. Res. Lab.* **1958**, *20*, 47–60.
40. Chen, Q.C. *Studies on Marine Biodiversity of the Nasha Islands and Neighboring Waters I.I.*; China Ocean Press: Beijing, China, 1996; p. 101.
41. Sousa, R.; Gutiérrez, J.; Aldridge, D. Non-indigenous invasive bivalves as ecosystem engineers. *Biol. Invas.* **2009**, *11*, 2367–2385. [CrossRef]
42. Terranova, M.S.; Brutto, S.L.; Arculeo, M.; Mitton, J.B. Population structure of *Brachidontes pharaonis* (P. Fisher, 1870) (Bivalvia: Mytilidae) in the Mediterranean Sea, and evolution of a novel mtDNA polymorphism. *Mar. Biol.* **2006**, *150*, 89–101. [CrossRef]
43. Fernández-Pérez, J.; Froufe, E.; Nantón, A.; Gaspar, M.B.; Méndez, J. Genetic diversity and population genetic analysis of *Donax vittatus* (Mollusca: Bivalvia) and phylogeny of the genus with mitochondrial and nuclear markers. *Estuar. Coast. Shelf Sci.* **2017**, *197*, 126–135. [CrossRef]
44. Takeuchi, T.; Masaoka, T.; Aoki, H.; Koyanagi, R.; Fujie, M.; Satoh, N. Divergent northern and southern populations and demographic history of the pearl oyster in the western Pacific revealed with genomic SNPs. *Evol. Appl.* **2019**, *13*, 837–853. [CrossRef]
45. Gerlach, G.; Jueterbock, A.; Kraemer, P.; Deppermann, J.; Harmand, P. Calculations of population differentiation based on G_{ST} and D: Forget G_{ST} but not all of statistics. *Mol. Ecol.* **2010**, *19*, 3845–3852. [CrossRef]
46. Chen, J.; Luo, H.; Zhai, Z.; Wang, H.; Liu, B.; Bai, L.; Yu, D. Estimates of additive and non-additive genetic effects on growth traits in a diallel cross of three strains of pearl oyster (*Pinctada fucata*). *Aquacult. Int.* **2021**, *29*, 1359–1371. [CrossRef]
47. Ruiz, G.M.; Fofonoff, P.W.; Steves, B.P.; Carlton, J.T. Invasion history and vector dynamics in coastal marine ecosystems: A North American perspective. *Aquat. Ecosyst. Health* **2015**, *18*, 299–311. [CrossRef]
48. Chang, Y.Q. *Aquaculture of Shellfish*; China Agriculture Press: Beijing, China, 2017.
49. Chen, M.; Chen, X.; Li, Y.; Wu, Q.; Yu, G.; Ma, Z.; Xing, K.; Wang, L. Cage culture of pearl oyster *Pinctada martensii* in Meiji Reef. *Fish. Sci.* **2014**, *35*, 379–383. (In Chinese)
50. Cheng, C. Meiji Reef grouper pelagic culture, a beautiful aquatic scenery line. *Curr. Fish.* **2014**, *39*, 62–63. (In Chinese)

51. Marques, R.C.; Breves, A. First record of *Pinctada imbricata* Röding, 1798 (Bivalvia: Pteroidea) attached to a rafting item: A potentially invasive species on the Uruguayan coast. *Mar. Biodivers.* **2015**, *45*, 333–337. [CrossRef]
52. Nackley, L.L.; West, A.G.; Skowno, A.L.; Bond, W.J. The nebulous ecology of native invasions. *Trends. Ecol. Evol.* **2017**, *32*, 814–824. [CrossRef] [PubMed]

Disclaimer/Publisher's Note: The statements, opinions and data contained in all publications are solely those of the individual author(s) and contributor(s) and not of MDPI and/or the editor(s). MDPI and/or the editor(s) disclaim responsibility for any injury to people or property resulting from any ideas, methods, instructions or products referred to in the content.

Article

Colonization in Artificial Seaweed Substrates: Two Locations, One Year

Diego Carreira-Flores [1,*], Regina Neto [1], Hugo R. S. Ferreira [2,3], Edna Cabecinha [4], Guillermo Díaz-Agras [5], Marcos Rubal [1] and Pedro T. Gomes [1]

[1] Centre of Molecular and Environmental Biology (CBMA)/Aquatic Research Network (ARNET), Department of Biology, University of Minho, 4704-553 Braga, Portugal; marcos.rubal@bio.uminho.pt (M.R.)
[2] Centro de Estudos do Ambiente e do Mar (CESAM), Departamento de Biologia, Universidade de Aveiro, Campus Universitário de Santiago, 3810-193 Aveiro, Portugal
[3] Tour du Valat, Research Institute for the Conservation of Mediterranean Wetlands, 13200 Le Sambuc, France
[4] Centre for the Research and Technology of Agro-Environmental and Biological Sciences (CITAB), Inov4Agro, Department of Biology and Environment, University of Trás-os Montes and Alto Douro, 5000-801 Vila Real, Portugal
[5] Estación de Bioloxía Mariña da Graña, University of Santiago de Compostela, Rede de Estacións Biolóxicas da USC (REBUSC),15590 Ferrol, Spain
* Correspondence: diego.carreira@bio.uminho.pt

Abstract: Artificial substrates have been implemented to overcome the problems associated with quantitative sampling of marine epifaunal assemblages. These substrates provide artificial habitats that mimic natural habitat features, thereby standardizing the sampling effort and enabling direct comparisons among different sites and studies. This paper explores the potential of the "Artificial Seaweed Monitoring System" (ASMS) sampling methodology to evaluate the natural variability of assemblages along a coastline of more than 200 km, by describing the succession of the ASMS' associated macrofauna at two Rías of the Galician Coast (NW Iberian Peninsula) after 3, 6, 9, and 12 months after deployment. The results show that macrofauna assemblages harbored by ASMS differ between locations for every type of data. The results also support the hypothesis that succession in benthic communities is not a linear process, but rather a mixture of different successional stages. The use of the ASMS is proved to be a successful standard monitoring methodology, as it is sensitive to scale-dependent patterns and captures the temporal variability of macrobenthic assemblages. Hence, the ASMS can serve as a replicable approach contributing to the "Good Environmental Status" assessment through non-destructive monitoring programs based on benthic marine macrofauna monitoring, capturing the variability in representative assemblages as long as sampling deployment periods are standard.

Keywords: epifaunal assemblages; artificial substrates; artificial seaweed monitoring system; succession

Citation: Carreira-Flores, D.; Neto, R.; Ferreira, H.R.S.; Cabecinha, E.; Díaz-Agras, G.; Rubal, M.; Gomes, P.T. Colonization in Artificial Seaweed Substrates: Two Locations, One Year. *Diversity* **2023**, *15*, 733. https://doi.org/10.3390/d15060733

Academic Editors: Renato Mamede and Bert W. Hoeksema

Received: 17 March 2023
Revised: 30 May 2023
Accepted: 31 May 2023
Published: 2 June 2023

Copyright: © 2023 by the authors. Licensee MDPI, Basel, Switzerland. This article is an open access article distributed under the terms and conditions of the Creative Commons Attribution (CC BY) license (https://creativecommons.org/licenses/by/4.0/).

1. Introduction

The evaluation of the natural variability of assemblages is essential when assessing and monitoring the natural succession processes in the marine environment, as well as when quantifying the scale of anthropogenic impacts [1–3]. Environmental and biological processes shape assemblages' structure at different spatial and temporal scales [4,5]. Consequently, studies on assemblages' temporal and spatial patterns are required in order to understand these succession patterns [6], constituting a fundamental key step when designing monitoring protocols. The monitoring of marine assemblages is an essential tool in environmental management, as envisaged by the European Marine Strategy Framework Directive (MSFD; 2008/56/EC), with species composition being a basic descriptor for evaluating 'Good Environmental Status' (GES). Still, scientific uncertainties regarding benthic processes and the difficulty of performing sampling and monitoring make assessing what constitutes GES regarding sea-floor integrity a challenging task [7,8]. This assessment of GES becomes even more complex when the study sites are located within protected marine areas, where the use of non-destructive sampling methods is mandatory.

Marine benthic monitoring is undertaken in three different scenarios: soft bottoms, hard bottoms, and mixtures of the two. Marine soft bottoms' highly diverse benthic macrofaunal communities [9] are relatively easy to sample using standard quantitative sampling devices such as grabs or corers [10–12]. For these well-studied ecosystems, indexes have been developed to assess the ecological quality of the surrounding marine environment [8,13]. Conversely, rocky bottoms present particular difficulties for standard surveys of biodiversity because of their spatial complexity and variability. The natural and physical heterogeneity of rocky reefs, such as the presence of sediments, large macroalgae settlements, rocky crevices, and the presence of biogenic structures such as mussels beds, conditions macrofauna assemblages [14–16]. Rocky reefs also support diverse assemblages of species from many phyla of invertebrates [17,18]. Heterogeneity directly influences the structure of the assemblages by creating new habitats, microenvironments, and resources. Additionally, it indirectly affects the intensity of biotic processes such as predation and competition by altering them [6]. Thus, sampling these habitats can be challenging due to their heterogeneity and complexity, which raises several questions about the most effective methods for capturing the spatial variability of these communities [11,19].

Natural assemblages are frequently complex and irregular along reduced spatial extension and/or time scales, often mainly driven by the availability of space [6] and the presence of disturbances [20,21]. The colonization process starts with the arrival of early colonizing species, with a given set of life-history characteristics. These early colonizing species have particular ecological features that influence the settlement of late-colonizing species, potentially leading to various succession processes based on the interactions between early and late colonizers [1,22]. Intense disturbances create new opportunities for settlement by providing additional uncolonized space, occasionally resulting in the colonization process being restarted [6,23]. The whole process of colonization and succession is continuous and traceable over different time scales (e.g., hours, days, months, or years). Hence, benthic communities can be understood to be a mixture of different successional stages, rather than as a linear process [24,25]. Nevertheless, the mechanisms of colonization and succession are not clearly understood, and the patterns of successional changes are not yet well documented [18], especially in NW Iberia.

Benthic invertebrates are reliable key biological elements for monitoring programs because they are able to reflect the state of environmental quality [26], as they are conspicuous, easily sampled, and respond quickly to changes [27,28]. Artificial substrates (AS) have been implemented to overcome problems with the quantitative sampling of benthic macrofauna in structurally complex environments [29–34] due to their success as collectors of macrofauna [35–37]. Artificial substrates are regarded as a viable non-destructive alternative, provided that the natural structural complexity is mimicked by the AS [21,29,31,35,38–41]. They have also been proved to be a valid tool for distinguishing macroalgae-like and crevice-like macrofauna assemblages from distinctive locations and over different time scales [18,38,41].

Artificial substrates provide uncolonized habitats that mimic the essential features of the natural habitat, standardizing the sampling effort and enabling direct comparisons between different sites and studies [19,21,29,42]. With known features and the ability to be developed with a deliberated target, AS are adequate for standardizing sampling devices for quantitative non-destructive sampling [19,21,32,34]. While standardizing deployment time and duration is crucial for obtaining comparable data in monitoring programs, there is variability in the literature regarding the periods of deployment for artificial substrates (AS). Some authors have utilized deployment periods ranging from a few days [34,39] to one month [29], of several months [18,39], or even lasting multiple years [19,42,43]. In certain cases, a standardized 3-month deployment period has been implemented to collect macrofaunal data using AS [44], but only a limited number of studies have examined the colonization process within this specific 3-month period [21].

A successful standard monitoring methodology based on AS must be sensitive to scale-dependent patterns and capture the variability of macrobenthic communities over time. This entails sampling the seasonal changes and fluctuations derived from the natural ecological variability of the deployment sites, especially if they have similar features. Thus, standard methodologies must guarantee their ability to represent the natural variability over small and large spatial scales, without simply being colonized by some opportunistic species that does not vary over time, leading the methods to not be representative of the surrounding natural assemblages. The process of validation of the "Artificial Seaweed Monitoring System" (ASMS) standard sampling methodology proposed by Carreira-Flores et al. [38,41] went through the standardization of the deployment (date and period) and the validation of the possibilities for capturing natural variability over time at different spatial scales.

Understanding succession patterns is a fundamental key step when designing monitoring protocols, and the knowledge of the mid-term (one-year) colonization process could complement or strengthen the effectiveness of any standard non-destructive methodology proposal for contributing to the assessment of GES. The objective of the present study is to evaluate the potential of ASMS for the assessment of the natural variability of assemblages on a kilometric spatial extent. This study also aims to complement the aforementioned methodology designed by Carreira-Flores et al. [38,41], by describing the succession of the ASMS-associated assemblages at two locations and over the course of one year. Two hypotheses were tested: (1) the captured variability of natural assemblages determined by ASMS will differ between locations; our experimental design is appropriate for capturing the variability in assemblages at centimetric (10s to 100s cm) and metric (10s to 100s m) scales, and we expect that the pattern will be the same on the kilometric scale (100s of km); and (2) that the captured variability of natural assemblages determined by ASMS will differ among dates.

2. Materials and Methods

2.1. Study Area

This study was carried out between June 2018 and June 2019 at two rocky reefs: Enseada de San Cristovo (ESC) (43°27′53.8″ N, 008°18′00.7″ W; 11 m in high tide) in the Ría de Ferrol and Bajo Tofiño (BT) (42°13 42.3 N, 008° 46 43.2 W; 11 m in high tide) in the Ría de Vigo (NW Iberian Peninsula) (Figure 1). The two sampling points are characterized by the presence of kelp forests of *Laminaria ochroleuca* Bachelot de la Pylaie, 1824 and *Laminaria hyperborea* (Gunnerus) Foslie, 1884, intercalated with sandbars within the rocky reef.

The Ría de Vigo is a very productive 26-km-long inlet due to the seasonal upwelling dynamics [45]. With its funnel-like morphology and SW-NE orientation, it is sheltered by the Illas Cíes against ocean waves [46]. The Ría de Ferrol exhibits a unique topography that regulates a complex current regime, resulting in a wide variety of sedimentary substrates [47]. It is also SW-NE oriented, and water exchange with the ocean is primarily controlled by tides through the narrow main channel in its middle part [48]. The outer part of the Ría de Ferrol is sheltered against ocean waves and storms by the "Porto Exterior de Ferrol". Both areas have significant human populations, with Ferrol, Narón, and Mugardos surrounding the Ría de Ferrol and totaling approximately 135,000 residents. Similarly, the Ría de Vigo is affected by high anthropic pressure, primarily from the nearly 340,000 inhabitants of Vigo, Cangas, and Moaña. These rías are impacted by human activities in the form of dockyards, commercial harbours, bivalve mollusc harvesting, sewage runoff pollution, and industrial discharge [48–50]. The Ría de Vigo also has an important area occupied by mussel raft cultures [50].

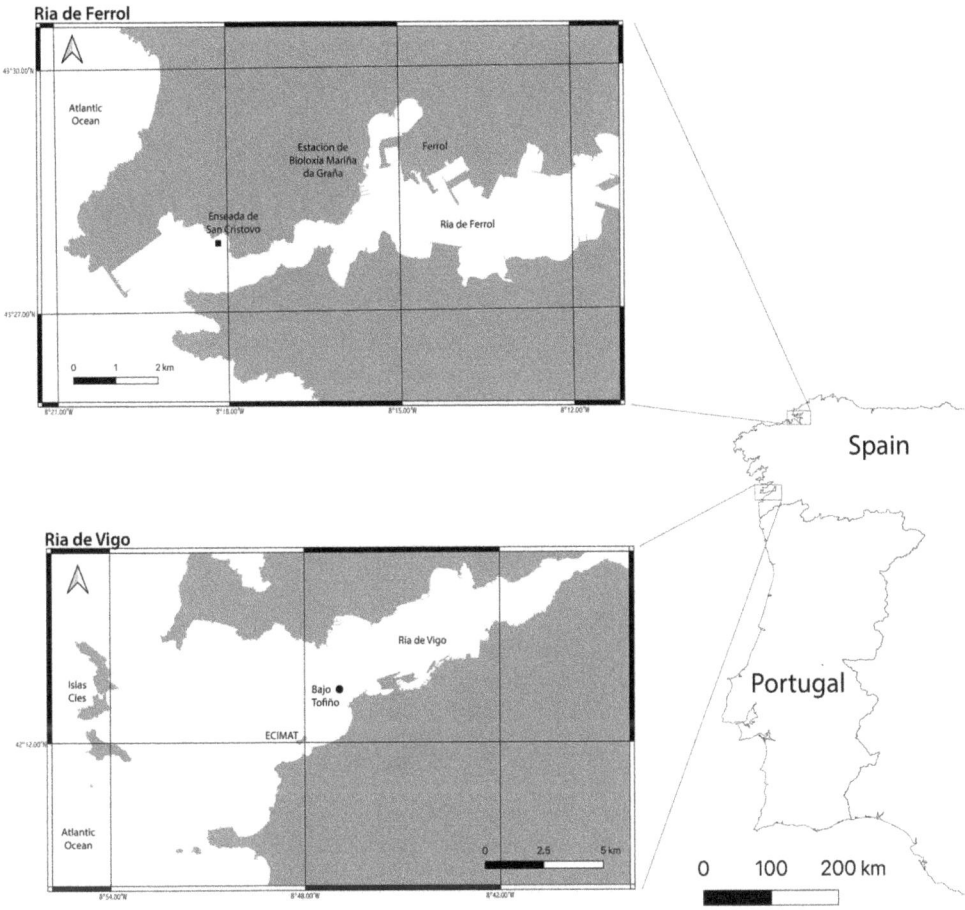

Figure 1. Location of study areas. ■ Enseada de San Cristovo, Ría de Ferrol; ● Bajo Tofiño, Ría de Vigo.

During the course of the study, temperature trends exhibited a similar pattern, with the lowest recorded temperatures occurring in February 2019 (12.79 ± 0.48 °C at Ría de Vigo; 12.87 ± 0.18 °C at Ría de Ferrol), while the highest temperatures were observed in July 2018 (19.46 ± 1.21 °C at Vigo; 15.54 ± 1.15 °C at Ferrol) (Figure 2). The variations in temperature during spring and summer can be attributed to differences in latitude and/or variations in the intensity of upwelling between the two locations.

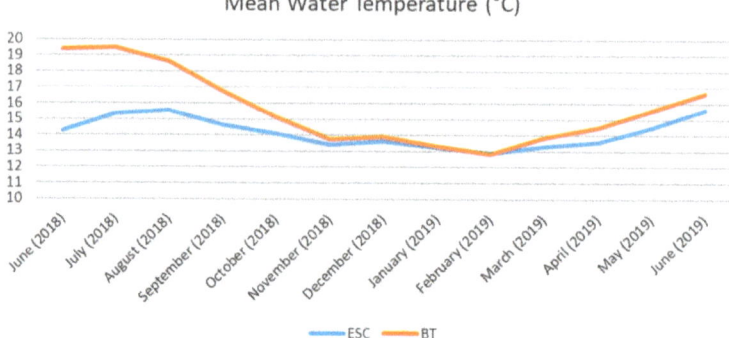

Figure 2. Mean temperature at Enseada de San Cristovo (ESC) and Bajo Tofiño (BT), obtained using Hobo pendant data loggers, installed at 11m in both locations.

2.2. Sampling Strategy

A total of forty ASMSs (Artificial Seaweed Monitoring Systems, ASMS_1 in Carreira-Flores et al. [38]) made of green polyethylene were deployed attached to 60 cm × 60 cm concrete plates within the natural settlements of macroalgae at the rocky reef (Figure 3). Specifically, twenty ASMS units were placed at ESC on 27 June 2018, and an additional twenty were placed at BT on 28 June 2018. To collect temperature data, water temperature loggers (TBI-32, Onset HOBO, Bourne, MA, USA) were attached to the concrete plates. These loggers recorded the water temperature at 5 min intervals from June 2018 to June 2019.

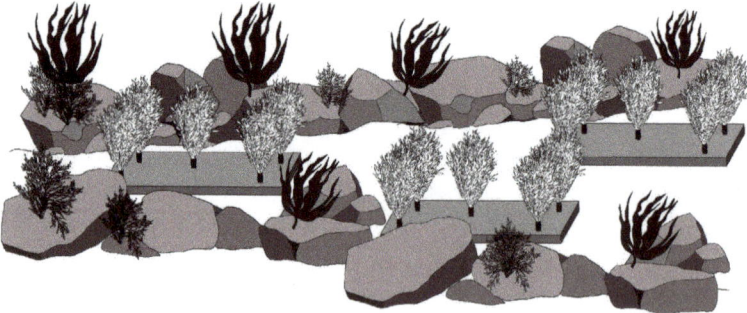

Figure 3. Experimental deployment of ASMS within the natural algae settlement of a rocky reef. Artificial substrates were attached to concrete plates (60 cm × 60 cm).

At both locations, a randomly selected set of five ASMS units were recovered by scuba diving after 3, 6, 9, and 12 months of deployment. The standard deployment period for ASMS units is three months, as described in previous studies [21,38,41,44]. Each substrate was carefully introduced into a 0.5 mm mesh bag and closed before being released from the base with a scraper to prevent small mobile organisms associated with the AS from escaping. Subsequently, the mesh bag was placed into a hermetic plastic bag. The associated macrofauna were washed off using filtered saltwater by shaking each AS vigorously through 0.5 mm sieves in the laboratory. The macrofauna were fixed in 99% ethanol before being quantified. Identification was performed to the species level in most cases, except when the condition of the specimen did not allow species-level identification. Taxonomic classification was performed following the World Register of Marine Species (WoRMS) [51].

2.3. Data Analysis

Data were analyzed using multivariate techniques to test the proposed hypotheses. The number of taxa (species richness), the total number of individuals (abundance), and the diversity (Simpson index) of the epifaunal assemblages were calculated and plotted in the R environment v 3.6.0 (Packages Vegan and Lattice) [52]. Non-parametric permutational multivariate analysis of variance (PERMANOVA; [53]) was used to test the hypotheses about differences in epifaunal assemblages. Analyses were performed based on Bray–Curtis dissimilarity matrixes from square-root-transformed density data to reduce the influence of the most abundant taxa [54]. Two hypotheses were tested: (1) the variability of natural assemblages captured using AS will differ between locations; and (2) the variability of natural assemblages captured using AS will differ among dates. For both hypotheses, the factors studied were Location (fixed, 2 levels, ESC vs. BT) and Time (random, four levels, time 1 vs. time 2 vs. time 3 vs. time 4). Although the Time factor is considered a random factor (i.e., a variance component in the model), comparisons between random factors can be allowed when there is a historical or biogeographical justification (sensu Anderson et al. [3]). These specific moments were taken into account to understand the role of time in the colonization process, assuming a 3-month period of colonization (starting at the beginning of each season), as is standard in AS colonization studies [21,38,41,44], as well as in succession studies [1,2]. When appropriate, multiple a posteriori comparisons were performed to test for differences between/within groups for pairs of levels of factors. The tests were based on 999 unrestricted random permutations of data. Additionally, non-metric multidimensional scaling (NMDS) (100 restarts) was used as the ordination method to explore differences in the assemblages' responses. Analyses of multivariate dispersion were also performed to test for the homogeneity of the dispersions between locations and dates (PERMDISP; [53]). The SIMPER procedure was used to identify each taxon's percentage contribution to the Bray–Curtis dissimilarity between the averages of groups. Taxa were considered important if their contribution to percentage dissimilarity was ≥5% and/or they contributed to explaining the first 40% (±2%) of the cumulative differences. Multivariate analyses were conducted using Primer v.6 [54] with PERMANOVA + add-on [55].

3. Results

Throughout this study, a total of 162 taxa and 122,822 individuals were collected (Supplementary Tables S1 and S2). At ESC, the abundance of individuals increased until the sixth month (1993.8 ± 472.5 ind/subst), declined after the ninth month (690 ± 184.8 ind/subst), and reached its maximum after the twelfth month (2724.8 ± 822.2 ind/subst) (Figure 4A). Conversely, at BT, the abundance increased at all times, reaching its peak after 12 months (8969.4 ± 470.83 ind/subst) (Figure 4A).

The Simpson index (species) was consistently higher at ESC compared to BT after 3, 6, and 9 months, with the highest value being observed after six months (0.91 ± 0.006). After 12 months, BT exhibited the greatest diversity value for temporal succession (0.89 ± 0.01), while ESC had the lowest value for this variable (0.8 ± 0.04) (Figure 4B).

Regarding species richness, at ESC, the pattern followed was similar to that of abundance, increasing and reaching its peak after six months (70.56 ± 11.36), decreasing after nine months (54 ± 8.7), and experiencing growth after twelve months (66.2 ± 7.85). At BT, species richness increased after six months (60.8 ± 10.32), remained stable after nine months (57 ± 12.3), and reached its maximum level after twelve months (73.4 ± 6.42) (Figure 4C).

The PERMANOVA results for the composition of the harbored assemblages showed a significant interaction between Location × Time (Table 1). PERMDISP analyses showed that these differences were due to the distance of the centroids rather than data dispersion (Table 1). Pairwise tests revealed significant differences in macrofaunal assemblages between locations at each time point (Table 2). These patterns were also evident in the NMDS plot (Figure 5), showing that the AS captured differences in environmental variability at the two locations over time.

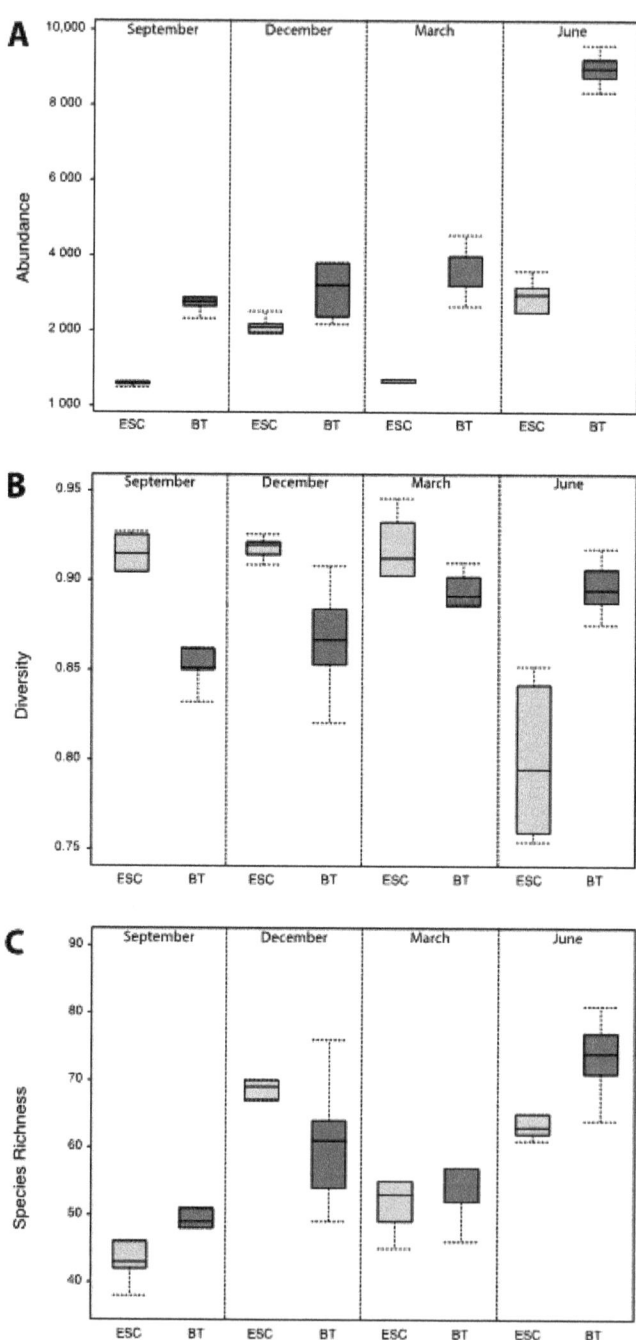

Figure 4. (**A**) Total abundance per substrate (number of individuals); (**B**) species (Simpson index), and (**C**) species richness per substrate (number of taxa) of macrofaunal assemblages associated with ASMS at Enseada de San Cristovo (ESC) and Bajo Tofiño (BT) at each data collection (3 months: September 2018; 6 months: December 2018; 9 months: March 2019; 12 months: June 2019).

Table 1. Summary of PERMANOVA results for total assemblages at ESC and BT.

Source	df	MS	Pseudo-F	P (perm)	Unique Perms
Location	1	21603	6.4063	0.002	579
Time	3	5373.7	21.164	0.001	997
Location × Time	3	3372.1	13.281	0.001	999
Residual	32	253.91			
Total	39				
Permdisp			P (perm):0.371		

Table 2. Results of pair-wise test between ASMS at Enseada de San Cristovo (ESC) and Bajo Tofiño (BT) at each data collection (3 months: September 2018; 6 months: December 2018; 9 months: March 2019; 12 months: June 2019). *, $p < 0.05$; **, $p < 0.01$. Number of unique perms = 126 for all taxonomic levels and combinations.

	3 Months	6 Months	9 Months	12 Months
	t	t	t	t
ESC vs. BT	5.0261 *	5.5261 *	5.3744 *	6.6794 **

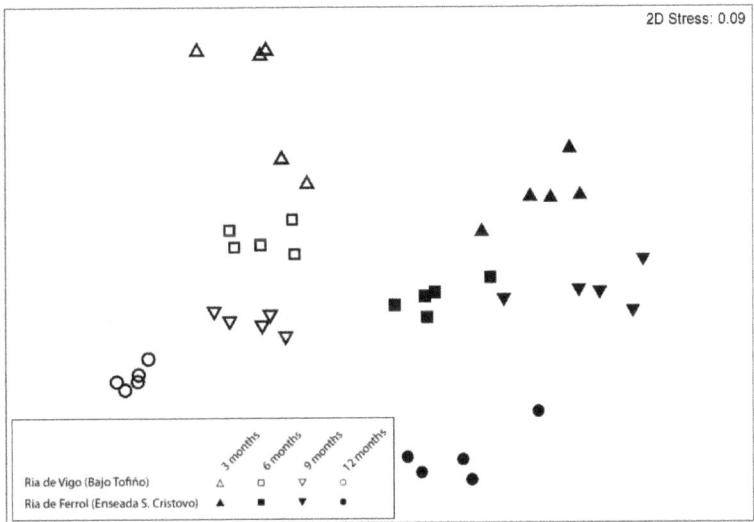

Figure 5. Non-metric multidimensional scaling (nMDS) ordination based on Bray–Curtis dissimilarity measures of macrofaunal assemblages density data of ASMS at Enseada de San Cristovo (ESC) and Bajo Tofiño (BT) at each data collection (3 months: September 2018; 6 months: December 2018; 9 months: March 2019; 12 months: June 2019).

After 3 months, both locations shared 43 taxa, but ESC supported 27 exclusive taxa, while BT had 26 exclusive taxa (Table 3). The dissimilarity between ES and BT was 59.37%, with six amphipod species being the taxa responsible for 40.49% of this dissimilarity (Table S3).

Table 3. Number of common and exclusive taxa of AS at Enseada de San Cristovo (ESC) and Bajo Tofiño (BT) at each data collection (3 months: September 2018; 6 months: December 2018; 9 months: March 2019; 12 months: June 2019).

	3 Months	6 Months	9 Months	12 Months	Total
Common	43	56	52	67	100
Only ESC	27	43	24	30	43
Only BT	26	27	35	26	19
Total ESC	70	99	76	95	143
Total BT	69	83	87	91	119

After 6 months, the two locations shared 56 taxa, but ESC supported 43 exclusive taxa versus the 27 exclusive taxa at BT (Table 3). The dissimilarity between ESC and BT was 55.47%, with eleven species of amphipod being the taxa responsible for 41.4% of the dissimilarity between locations (Table S3).

After 9 months, the two locations shared 52 taxa, but ESC supported 24 exclusive taxa versus the 35 exclusive taxa at BT (Table 3). The dissimilarity between ESC and BT was 61.82%, with eight amphipod species being the taxa responsible for the 40.28% dissimilarity between locations (Table S3).

After 12 months, the two locations shared 67 taxa, but ESC supported 30 exclusive taxa versus 26 exclusive taxa at BT (Table 3). The dissimilarity between locations was high (61.94%). The dissimilarity between locations was high (61.94%), with five amphipod species, one equinoderms species, and two annelids being the taxa responsible for 41.90% of the dissimilarity between the locations (Table S3).

Regarding changes in assemblages over time, the PERMANOVA results showed significant differences regarding the Time factor (Table 1). The previous pattern could be observed at the MDS (Figure 5).

At ESC, after 3 months, amphipods accounted for 52.44% of the total individuals, followed by decapods (11.82%) and sabellids (10.17%) (Figures S1–S3). After 6 months, the percentages of amphipods and sabellids remained stable, accounting for 51.74% and 9.74%, respectively (Figures S1–S3). After 9 months, the percentage of amphipods started to decrease, accounting for 42.50%, followed by sabellids (11.86%) (Figures S1–S3). Finally, after 12 months, sabellids accounted for 51.69% of the total individuals, followed by amphipods (19.78%) and decapods (15.13%) (Figures S1–S3). Between 3 and 6 months, seven species of amphipod and two species of polychaetes contributed to 40.50% of the dissimilarity at ESC (average dissimilarity = 47.27%) (Table S4). The dissimilarity between the 6-month and 9-month samples was 41.46%, with seven species of amphipod, two species of bivalve, one species of gastropod, and one species of annelid contributing to 41.68% of the dissimilarity (Table S4). The dissimilarity between the 9-month and 12-month samples was 48.93%, with three species of annelid, four species of amphipod, and one species of decapod comprising the taxa responsible for 41.16% of this dissimilarity (Table S4).

At BT, after 3 and 6 months, amphipods were the most abundant group, accounting for 84.92% and 80.91% of the total individuals, respectively (Figures S1–S3). After 9 months, the relative abundance of amphipods decreased to 74.93%, followed by sabellids (5.06%) and decapods (4.70%) (Figures S1–S3). After 12 months, the relative abundance of amphipods reached its minimum value, accounting for 49.24 %, followed by holoturoids (23.35%) and decapods (8.88%) (Figures S1–S3). At BT, six species of amphipod contributed to 39.47 % of the dissimilarity between the 3-month and 6-month samples (average dissimilarity: 39.50%) (Table S5). The taxa responsible for 39.34% of the dissimilarity between the 6-month and 9-month samples (average dissimilarity: 36.15) consisted of eight species of amphipod (Table S5). Between the 9-month and 12-month samples (average dissimilarity: 38.40), the taxa responsible for 40.91% of the dissimilarity consisted of two species of echinoderm, one species of bivalve, one species of decapod, one genus of annelid, and four species of amphipods (Table S5).

4. Discussion

Our results showed that the initial hypotheses were supported. The ASMS effectively captured the natural variability of the assemblages on a kilometric scale and over different deployment times, revealing differences in the epifaunal assemblages between locations and under similar conditions, as well as across different dates.

It can be assumed that a mostly local process is responsible for the availability of colonizing animals for ASMS (sensu Ref. [56]). Previous studies have demonstrated that in the early stages of deployment, the AS is colonized by most of the elements making up the mobile invertebrate fauna in their nearby area [31,56,57], which is sensitive to local variations and environmental conditions [32,57]. Our findings are in agreement with the work of Carreira-Flores et al. and Norderhaug et al. [34,38,41], who reported that in the early stages of colonization, complex artificial three-dimensional structures are predominantly colonized by peracarid amphipods, reflecting the horizontal dispersal patterns and mobility capabilities of these organisms. Conversely, in non-complex AS, such as the PVC plates commonly used in colonization studies in sessile epifauna [19], amphipods are not the most efficient colonizers during the early stages [1]. In non-complex structures, the development of complexity relies on sessile habitat-forming organisms to create suitable conditions for hosting high abundances of amphipods, which significantly prolongs the colonization process. This suggests that, despite the high mobility of amphipods, they exhibit a stronger affinity for higher structural complexity [58], making more complex AS more effective at capturing these early colonizers. Our results indicate that peracarid amphipods are the primary representatives responsible for the variability observed in the ASMS between different locations and dates. We observed an increasing trend in the number of colonizing amphipods, highlighting their dominant role in marine epifaunal assemblages and their successful colonization of complex AS [59,60]. However, while the number of amphipods increased over time, their relative abundance (considering the total number of individuals captured across all species) declined after 6 months of deployment, reaching its lowest value after 12 months at both locations. Conversely, the percentage of other taxa with pelagic larvae recruitment cycles, such as sessile polychaetes like *Filograna implexa* Berkeley, 1835, *Spirobranchus triqueter* (Linnaeus, 1758), and *Janua heterostropha* (Montagu, 1803), increased over time. This shift in the relative abundance from species with adult motile dispersal abilities ("early colonizing species") to species with pelagic larvae recruitment strategies ("late colonizing species") suggests an ongoing maturation process in the community [61]. Our findings also align with the observations of Underwood and Chapman [18], who reported variations and replacements of taxa over time. The use of complex ASMS in our methodology enabled the acquisition of accurate data regarding the status and composition of surrounding assemblages in the short term. ASMS are effectively able to capture a wide variety of amphipods, which are a dominant component of marine epifaunal assemblages, and also exhibit sensitivity to the arrival of recruited species in the medium term.

Comparing the abundance, species richness, and Simpson indexes with the results published by Carreira-Flores et al. [38,41], we did not observe a comparable trend on any of the examined dates. Specifically, for the ASMS deployed at ESC, there were no similarities in any of the aforementioned indexes when comparing the results after 3 months of deployment (March 2018 vs. September 2018) or 6 months (June 2018 vs. December 2018), or when comparing the same months in different years (March 2018 vs. March 2019; June 2018 vs. June 2019). This lack of similarity supports the findings of García-Sanz et al. [33], who suggested that even the same type of substrate can yield different results depending on the deployment time and the natural yearly variability. These differences can be attributed to variations in recruitment periods, the life cycle of the surrounding benthic fauna, or the presence of "early" and "late" colonizing fauna [1]. For instance, the NW Iberia region is characterized by pronounced seasonality and upwelling events that synchronize larval recruitment during spring and early summer [62,63]. Consequently, differences in deployment period could explain the variations observed when using the same standard ASMS structures. Our proposed non-destructive sampling methodology is sensitive to the natural variability of assemblages, regardless of their deployment date and period. This highlights the importance of standardizing the deployment and recovery periods for ASMS to avoid introducing biases in studies and to ensure the comparability of data. By establishing consistent timeframes, the potential biases arising from differences in deployment and the duration before recovery can be minimized.

The ASMS assemblage analysis revealed significant differences in the epifaunal assemblages across all sampling dates and locations, probably indicating the influence of inherent variability within the natural assemblages at each location. Our findings support the hypothesis that benthic assemblages do not exhibit linear dynamics, and must be understood to be a mixture of different successional stages [2,25] that change over time [64], driven by each site's unique characteristics and natural variability. At ESC, the abundance (total number of individuals) and species richness did not increase linearly. Instead, it exhibited two peaks: one in December after 6 months of deployment and another in June after 12 months. Similarly, at BT, species richness displayed a similar trend, while the abundance of individuals remained relatively stable after 3, 6, and 9 months, with a notable peak observed after 12 months of deployment. Satterthwaite et al. [65] reported variations in larval assemblage abundance and composition influenced by upwelling and relaxation dynamics, affecting recruitment cycles. Upwelling events, altering primary productivity by increasing nutrient availability, and influencing larval dynamics in the Galician Rías, could impact the colonization and succession processes of ASMS. Underwood and Chapman [18] suggested that succession did not stabilize within 6 months of deployment in their study. Consistent with their findings, our study did not identify a period in which the community stabilized, as evidenced by the significant differences observed between each dataset at both locations. This indicates that the climax stage may not have been reached after one year of deployment. Furthermore, Zupan et al. [66] found that a climax state was not achieved even after 11 years in large AS, highlighting the dynamic nature of marine ecosystems and the potential absence of an orthodox definition of climax. Consequently, to gain a comprehensive understanding of colonization dynamics in complex AS over the long term, extended deployment periods should be considered.

In light of there being some environmental factors that were common to both of the selected deployment locations (e.g., same depth, same surrounding macroalgae communities, comparable protection against waves, and comparable anthropic pressure), the differences in the assemblages observed between the deployment locations can be attributed to other drivers, with hydrodynamics and the latitudinal gradient being plausible explanatory factors. These drivers, individually or in combination, also influence the variability of the sampled assemblages captured by the ASMS, as evidenced by the distinct assemblages and numerous exclusive species at each site. Subtle changes in hydrodynamic conditions can regulate the epifaunal community, according to Conradi et al. [67], and the exclusive appearance of *Stenothoe monoculoides* (Montagu, 1813) and the greater abundance of *Erichtonius brasiliensis* (Dana, 1853) at BT suggests that, despite both locations "a priori" having the same features, BT may have been subjected to higher hydrodynamic conditions. Another plausible explanation for the differences in the assemblages is related to the latitudinal difference between the sites, with ESC being located at 41° and BT at 42° latitude N. These locations are subjected to different intensities of upwelling events, which, as mentioned earlier, can explain the differences in summer temperatures. Many species exhibit ecological variations across latitudes in response to large-scale environmental variability [68]. In the case of the Iberian Peninsula, latitudinal differences have been documented for oceanographic patterns (e.g., chlorophyll, water temperature, nutrient availability due to spatial variations in upwelling intensity and frequency), as well as macroalgae and fish assemblages [69]. Therefore, expecting a latitudinal gradient in the epifaunal assemblages captured by the ASMS is reasonable. Alternatively, the observed assemblage differences could be attributable merely to local inherent variability. Regardless, the assemblages captured by the ASMS were sensitive to the consequences of these different drivers, indicating that monitoring methodologies employing ASMS are responsive to large-scale variables.

Further studies encompassing a broader geographic and temporal range are necessary to elucidate the drivers underlying the observed assemblage differences and determine whether the latitudinal gradient, hydrodynamic conditions, or local intrinsic variability do indeed influence these differences.

5. Conclusions

Developing standardized and replicable monitoring protocols is essential for accurately describing species composition and assessing the seabed's GES. To achieve this, it is crucial to validate the effectiveness of candidate methodologies at capturing the natural variability of assemblages at different spatial and temporal scales. This study complements and reinforces the non-destructive standard methodology proposed by Carreira-Flores et al. for use in benthic monitoring studies, and can be applied irrespective of the geographic location. As highlighted in our previous works, the proposed methodology is suitable for capturing variability in assemblages at the centimeter (10s to 100s of centimeters) and meter (10s to 100s of meters) scales [38,41]. The results of this study further demonstrate its effectiveness at the kilometric scale (100s of km). Moreover, the results are in agreement with the observations made by Carreira-Flores et al. [38], confirming that this methodology can effectively capture the natural variability of assemblages and is able to distinguish different successional stages, enabling the tracking of the entire succession process. Finally, another key point is the requirement of standardizing deployment and recovery periods for the ASMS, ensuring total comparability of the data.

Supplementary Materials: The following supporting information can be downloaded at: https://www.mdpi.com/article/10.3390/d15060733/s1, Table S1. Number of individuals (average ± standard deviation) captured at Enseada de San Cristovo (ESC) and Bajo Tofiño (BT) in September 2018, 3 months (S) and December 2018, 6 months (D). Table S2. Number of individuals (average ± standard deviation) captured at Enseada de San Cristovo (ESC) and Bajo Tofiño (BT) in March 2019, 9 months (M) and June 2019, 12 months (J). Table S3. Contribution (%) of individual taxa and cumulative percentage (Cum %) from ASMS assemblages to the average Bray–Curtis dissimilarity between Enseada de San Cristovo (ESC) and Bajo Tofiño (BT) at each data collection (3 months: September 2018; 6 months: December 2018; 9 months: March 2019; 12 months: June 2019). Table S4. Contribution (%) of individual taxa and cumulative percentage (Cum %) from ASMS assemblages to the average Bray–Curtis dissimilarity at each data collection (3 months: September 2018; 6 months: December 2018; 9 months: March 2019; 12 months: June 2019) at Enseada de San Cristovo (ESC). Table S5. Contribution (%) of individual taxa and cumulative percentage (Cum %) from ASMS assemblages to the average Bray–Curtis dissimilarity at each data collection (3 months: September 2018; 6 months: December 2018; 9 months: March 2019; 12 months: June 2019) at Bajo Tofiño (BT). Figure S1. Total assemblage composition (%) of ASMS at Order-Level of Enseada de San Cristovo (ESC) and Bajo Tofiño (BT) at every time point (S: September 2018, 3 months; D: December 2018, 6 months; M: March 2019, 9 months; J: June 2019, 12 months). Figure S2. Amphipod assemblage composition (%) of ASMS s at family level of Enseada de San Cristovo (ESC) and Bajo Tofiño (BT) at every time point (S: September 2018, 3 months; D: December 2018, 6 months; M: March 2019, 9 months; J: June 2019, 12 months). Figure S3. Amphipod assemblage composition (%) of ASMS s at species level of Enseada de San Cristovo (ESC) and Bajo Tofiño (BT) at every time point (S: September 2018, 3 months; D: December 2018, 6 months; M: March 2019, 9 months; J: June 2019, 12 months).

Author Contributions: Conceptualization, E.C., G.D.-A. and P.T.G.; methodology, D.C.-F., R.N., H.R.S.F. and G.D.-A.; formal analysis, D.C.-F., G.D.-A. and M.R.; investigation, D.C.-F., R.N., H.R.S.F., G.D.-A., P.T.G.; resources, G.D.-A. and P.T.G.; data curation, D.C.-F., M.R. and E.C.; writing—original draft preparation, D.C.-F.; writing—review and editing, D.C.-F., M.R. and P.T.G.; supervision, E.C., G.D.-A. and P.T.G.; funding acquisition, P.T.G. All authors have read and agreed to the published version of the manuscript.

Funding: This study was supported by the project ATLANTIDA (ref. NORTE-01-0145-FEDER-000040), supported by the Norte Portugal Regional Operational Programme (NORTE 2020), under the PORTUGAL 2020 Partnership Agreement and through the European Regional Development Fund (ERDF). This work was also supported by the "Contrato-Programa" UIDB/04050/2020 funded by national funds through the FCT I.P.

Institutional Review Board Statement: Not applicable.

Data Availability Statement: The data presented in this study are available on request from the corresponding author. The data are not publicly available because this data set may be included as part of other ongoing studies.

Acknowledgments: We would like to thank all the members of the Marine Biology Station of A Graña (EBMG) and the Toralla Marine Science Station (ECIMAT) for providing the Stations' resources and for their assistance in the verification of the identification process. We would also like to acknowledge the two anonymous reviewers, as well as the editor, for their valuable comments.

Conflicts of Interest: The authors declare that they have no conflict of interest. The funders had no role in the design of the study; in the collection, analyses, or interpretation of data; in the writing of the manuscript, or in the decision to publish the results.

References

1. Antoniadou, C.; Voultsiadou, E.; Chintiroglou, C. Benthic colonization and succession on temperate sublittoral rocky cliffs. *J. Exp. Mar. Biol. Ecol.* **2010**, *382*, 145–153. [CrossRef]
2. Antoniadou, C.; Voultsiadou, E.; Chintiroglou, C. Seasonal patterns of colonization and early succession on sublittoral rocky cliffs. *J. Exp. Mar. Biol. Ecol.* **2011**, *403*, 21–30. [CrossRef]
3. Anderson, M.; Diebel, C.E.; Blom, W.M.; Landers, T.J. Consistency and variation in kelp holdfast assemblages: Spatial patterns of biodiversity for the major phyla at different taxonomic resolutions. *J. Exp. Mar. Biol. Ecol.* **2005**, *320*, 35–56. [CrossRef]
4. Benedetti-Cecchi, L. Variability in abundance of algae and invertebrates at different spatial scales on rocky sea shores. *Mar. Ecol. Prog. Ser.* **2001**, *215*, 79–92. [CrossRef]
5. Fraschetti, S.; Terlizzi, A.; Benedetti-Cecchi, L. From proof image to formal proof-a transformation. *Mar. Ecol. Prog. Ser.* **2005**, *296*, 13–29. [CrossRef]
6. Underwood, A.J.; Chapman, M.G. International Association for Ecology Scales of Spatial Patterns of Distribution of Intertidal Invertebrates. *Oecologia* **1996**, *107*, 212–224. [CrossRef] [PubMed]
7. Rice, J.; Arvanitidis, C.; Borja, A.; Frid, C.; Hiddink, J.G.; Krause, J.; Lorance, P.; Ragnarsson, S.Á.; Sköld, M.; Trabucco, B.; et al. Indicators for sea-floor integrity under the european marine strategy framework directive. *Ecol. Indic.* **2012**, *12*, 174–184. [CrossRef]
8. Danovaro, R.; Carugati, L.; Berzano, M.; Cahill, A.E.; Carvalho, S.; Chenuil, A.; Corinaldesi, C.; Cristina, S.; David, R.; Dell'Anno, A.; et al. Implementing and Innovating Marine Monitoring Approaches for Assessing Marine Environmental Status. *Front. Mar. Sci.* **2016**, *3*, 213. [CrossRef]
9. Gray, J.S. Marine biodiversity: Patterns, threats and conservation needs. *Biodivers. Conserv.* **1997**, *175*, 153–175. [CrossRef]
10. Aneiros, F.; Moreira, J.; Troncoso, J.S. A functional approach to the seasonal variation of benthic mollusc assemblages in an estuarine-like system. *J. Sea Res.* **2014**, *85*, 73–84. [CrossRef]
11. Beisiegel, K.; Darr, A.; Gogina, M.; Zettler, M.L. Benefits and shortcomings of non-destructive benthic imagery for monitoring hard-bottom habitats. *Mar. Pollut. Bull.* **2017**, *121*, 5–15. [CrossRef] [PubMed]
12. Veiga, P.; Redondo, W.; Sousa-Pinto, I.; Rubal, M. Relationship between structure of macrobenthic assemblages and environmental variables in shallow sublittoral soft bottoms. *Mar. Environ. Res.* **2017**, *129*, 396–407. [CrossRef] [PubMed]
13. Borja, A.; Franco, J.; Pérez, V. A marine Biotic Index to establish the ecological quality of soft-bottom benthos within European estuarine and coastal environments. *Mar. Pollut. Bull.* **2000**, *40*, 1100–1114. [CrossRef]
14. Veiga, P.; Torres, A.C.; Aneiros, F.; Sousa-Pinto, I.; Troncoso, J.S.; Rubal, M. Consistent patterns of variation in macrobenthic assemblages and environmental variables over multiple spatial scales using taxonomic and functional approaches. *Mar. Environ. Res.* **2016**, *120*, 191–201. [CrossRef]
15. Yakovis, E.L.; Artemieva, A.V.; Fokin, M.V.; Varfolomeeva, M.A.; Shunatova, N.N. Effect of habitat architecture on mobile benthic macrofauna associated with patches of barnacles and ascidians. *Mar. Ecol. Prog. Ser.* **2007**, *348*, 117–124. [CrossRef]
16. Commito, J.A.; Como, S.; Grupe, B.M.; Dow, W.E. Species diversity in the soft-bottom intertidal zone: Biogenic structure, sediment, and macrofauna across mussel bed spatial scales. *J. Exp. Mar. Biol. Ecol.* **2008**, *366*, 70–81. [CrossRef]
17. Terlizzi, A.; Anderson, M.J.; Fraschetti, S.; Benedetti-Cecchi, L. Scales of spatial variation in Mediterranean subtidal sessile assemblages at different depths. *Mar. Ecol. Prog. Ser.* **2007**, *332*, 25–39. [CrossRef]
18. Underwood, A.J.; Chapman, M.G. Early development of subtidal macrofaunal assemblages: Relationships to period and timing of colonization. *J. Exp. Mar. Biol. Ecol.* **2006**, *330*, 221–233. [CrossRef]

19. Kuklinski, P.; Balazy, P.; Porter, J.; Loxton, J.; Ronowicz, M.; Sokołowski, A. Experimental apparatus for investigating colonization, succession and related processes of rocky bottom epifauna. *Cont. Shelf Res.* **2022**, *233*, 104641. [CrossRef]
20. Wollgast, S.; Lenz, M.; Wahl, M.; Molis, M. Effects of regular and irregular temporal patterns of disturbance on biomass accrual and species composition of a subtidal hard-bottom assemblage. *Helgol. Mar. Res.* **2008**, *62*, 309–319. [CrossRef]
21. Baronio, M.D.A.; Bucher, D.J. Artificial crevice habitats to assess the biodiversity of vagile macro-cryptofauna of subtidal rocky reefs. *Mar. Freshw. Res.* **2008**, *59*, 661–670. [CrossRef]
22. Dean, R.L.; Connell, J.H. Marine invertebrates in an algal succession. I. Variations in abundance and diversity with succession. *J. Exp. Mar. Biol. Ecol.* **1987**, *109*, 195–215. [CrossRef]
23. Platt, W.J.; Connell, J.H. Natural disturbances and directional replacement of species. *Ecol. Monogr.* **2003**, *73*, 507–522. [CrossRef]
24. Chapman, M.G. Colonization of novel habitat: Tests of generality of patterns in a diverse invertebrate assemblage. *J. Exp. Mar. Biol. Ecol.* **2007**, *348*, 97–110. [CrossRef]
25. Olabarria, C. Role of colonization in spatio-temporal patchiness of microgastropods in coralline turf habitat. *J. Exp. Mar. Biol. Ecol.* **2002**, *274*, 121–140. [CrossRef]
26. Aguado-Giménez, F.; Gairín, J.I.; Martinez-Garcia, E.; Fernandez-Gonzalez, V.; Ballester Moltó, M.; Cerezo-Valverde, J.; Sanchez-Jerez, P. Application of "taxocene surrogation" and "taxonomic sufficiency" concepts to fish farming environmental monitoring. Comparison of BOPA index versus polychaete assemblage structure. *Mar. Environ. Res.* **2015**, *103*, 27–35. [CrossRef]
27. Van Der Linden, P.; Marchini, A.; Dolbeth, M.; Patrício, J.; Veríssimo, H.; Marques, J.C. The performance of trait-based indices in an estuarine environment. *Ecol. Indic.* **2015**, *61*, 378–389. [CrossRef]
28. Salas, F.; Marcos, C.; Neto, J.M.; Patrı, J. User-friendly guide for using benthic ecological indicators in coastal and marine quality assessment. *Ocean Coast. Manag.* **2006**, *49*, 308–331. [CrossRef]
29. Cacabelos, E.; Olabarria, C.; Incera, M.; Troncoso, J.S. Effects of habitat structure and tidal height on epifaunal assemblages associated with macroalgae. *Estuar. Coast. Shelf Sci.* **2010**, *89*, 43–52. [CrossRef]
30. Christie, H.; Norderhaug, K.M.; Fredriksen, S. Macrophytes as habitat for fauna. *Mar. Ecol. Prog. Ser.* **2009**, *396*, 221–233. [CrossRef]
31. Edgar, G.J. Distribution patterns of mobile epifauna associated with rope fibre habitats within the Bathurst Harbour estuary, south-western Tasmania. *Estuar. Coast. Shelf Sci.* **1991**, *33*, 589–604. [CrossRef]
32. Edgar, G.J. Artificial algae as habitats for mobile epifauna: Factors affecting colonization in a Japanese *Sargassum* bed. *Hydrobiologia* **1991**, *226*, 111–118. [CrossRef]
33. García-Sanz, S.; Navarro, P.G.; Tuya, F. Colonization of Prosobranch Gastropods onto Artificial Substrates: Seasonal Patterns between Habitat Patches. *Am. Malacol. Bull.* **2014**, *32*, 94–103. [CrossRef]
34. Norderhaug, K.M.; Christie, H.; Rinde, E. Colonisation of kelp imitations by epiphyte and holdfast fauna: A study of mobility patterns. *Mar. Biol.* **2002**, *141*, 965–973. [CrossRef]
35. Myers, A.A.; Southgate, T. Artificial substrates as a means of monitoring Rocky shore cryptofauna. *J. Mar. Biol. Assoc. United Kingd.* **1980**, *60*, 963–975. [CrossRef]
36. Hylkema, A.; Debrot, A.O.; Pistor, M.; Postma, E.; Williams, S.M.; Kitson-Walters, K. High peak settlement of *Diadema antillarum* on different artificial collectors in the Eastern Caribbean. *J. Exp. Mar. Biol. Ecol.* **2022**, *549*, 151693. [CrossRef]
37. García-Sanz, S.; Tuya, F.; Navarro, P.G.; Angulo-Preckler, C.; Haroun, R.J. Post larval, short-term, colonization patterns: The effect of substratum complexity across subtidal, adjacent, habitats. *Estuar. Coast. Shelf Sci.* **2012**, *112*, 183–191. [CrossRef]
38. Carreira-Flores, D.; Neto, R.; Ferreira, H.; Cabecinha, E.; Díaz-Agras, G.; Gomes, P.T. Artificial substrates as sampling devices for marine epibenthic fauna: A quest for standardization. *Reg. Stud. Mar. Sci.* **2020**, *37*, 101331. [CrossRef]
39. Zimmerman, T.L.; Martin, J.W. Artificial Reef Matrix Structures (Arms): An Inexpensive and Effective Method for Collecting Coral Reef-Associated Invertebrates. *Gulf Caribb. Res.* **2004**, *16*, 59–64. [CrossRef]
40. Plaisance, L.; Caley, M.J.; Brainard, R.E.; Knowlton, N. The diversity of coral reefs: What are we missing? *PLoS ONE* **2011**, *6*, e25026. [CrossRef]
41. Carreira-Flores, D.; Neto, R.; Ferreira, H.; Cabecinha, E.; Díaz-Agras, G.; Gomes, P.T. Two better than one: The complementary of different types of artificial substrates on benthic marine macrofauna studies. *Mar. Environ. Res.* **2021**, *171*, 105449. [CrossRef]
42. Ransome, E.; Geller, J.B.; Timmers, M.; Leray, M.; Mahardini, A.; Sembiring, A.; Collins, A.G.; Meyer, C.P. The importance of standardization for biodiversity comparisons: A case study using autonomous reef monitoring structures (ARMS) and metabarcoding to measure cryptic diversity on Mo'orea coral reefs, French Polynesia. *PLoS ONE* **2017**, *12*, e0175066. [CrossRef] [PubMed]
43. Pearman, J.K.; Leray, M.; Villalobos, R.; Machida, R.J.; Berumen, M.L.; Knowlton, N.; Carvalho, S. Cross-shelf investigation of coral reef cryptic benthic organisms reveals diversity patterns of the hidden majority. *Sci. Rep.* **2018**, *8*, 8090. [CrossRef] [PubMed]
44. Obst, M.; Exter, K.; Allcock, A.L.; Arvanitidis, C.; Axberg, A.; Bustamante, M.; Cancio, I.; Carreira-Flores, D.; Chatzinikolaou, E.; Chatzigeorgiou, G.; et al. A Marine Biodiversity Observation Network for Genetic Monitoring of Hard-Bottom Communities (ARMS-MBON). *Front. Mar. Sci.* **2020**, *7*, 1031. [CrossRef]
45. Prego, R.; Fraga, F. A simple model to calculate the residual flows in a Spanish Ría. Hydrographic consequences in the Ría of Vigo. *Estuar. Coast. Shelf Sci.* **1992**, *34*, 603–615. [CrossRef]
46. Torres, A.C.; Veiga, P.; Rubal, M.; Sousa-Pinto, I. The role of annual macroalgal morphology in driving its epifaunal assemblages. *J. Exp. Mar. Biol. Ecol.* **2015**, *464*, 96–106. [CrossRef]

47. DeCastro, M.; Gómez-Gesteira, M.; Prego, R.; Neves, R. Wind influence on water exchange between the ria of Ferrol (NW Spain) and the shelf. *Estuar. Coast. Shelf Sci.* **2003**, *56*, 1055–1064. [CrossRef]
48. Cunha, X. Macrofauna Bentónica da Enseada de Santa Lucía (Ría de Ferrol, Galicia): Diversidade e Dinámica Temporal a Longo Prazo. Ph.D. Thesis, University of Santiago de Compostela, A Coruña, Spain, 2017.
49. Díaz-Agras, G. Patrones de Distribución de Moluscos Gasterópodos en Sustratos Duros Intermareales Naturales y Artificiales en la Ría de Ferrol, Galicia. Ph.D. Thesis, University of Santiago de Compostela, A Coruña, Spain, 2015.
50. Prego, R.; Filgueiras, A.; Santos-Echeandía, J. Temporal and spatial changes of total and labile metal concentration in the surface sediments of the Vigo Ría (NW Iberian Peninsula): Influence of anthropogenic sources. *Mar. Pollut. Bull.* **2008**, *56*, 1022–1031. [CrossRef]
51. WoRMS Editorial Board. World Register of Marine Species. Available online: http://www.marinespecies.org (accessed on 29 September 2020).
52. R Core Team R. *A Language and Environment for Statistical Computing*; R Foundation for Statistical Computing: Vienna, Austria, 2019.
53. Anderson, M.J. Permutational Multivariate Analysis of Variance (PERMANOVA). *Wiley StatsRef Stat. Ref. Online* **2017**, 1–15. [CrossRef]
54. Clarke, K.R.; Gorley, R.N. *PRIMER v. 6: User Manual/Tutorial*; PRIMER-E: Plymouth, UK, 2001.
55. Anderson, M.; Gorley, R.N.; Clarke, K.R. *PERMANOVA+ for PRIMER: Guide to Software and Statistical Methods*; PRIMER-E: Plymouth, UK, 2008; Volume 1.
56. Tuya, F.; Wernberg, T.; Thomsen, M.S. Colonization of gastropods on subtidal reefs depends on density in adjacent habitats, not on disturbance regime. *J. Molluscan Stud.* **2008**, *75*, 27–33. [CrossRef]
57. Russo, A.R. The role of seaweed complexity in structuring Hawaiian epiphytal amphipod communities. *Hydrobiologia* **1990**, *194*, 1–12. [CrossRef]
58. Schreider, M.J.; Glasby, T.M.; Underwood, A.J. Effects of height on the shore and complexity of habitat on abundances of amphipods on rocky shores in New South Wales, Australia. *J. Exp. Mar. Biol. Ecol.* **2003**, *293*, 57–71. [CrossRef]
59. Ashton, G.V.; Burrows, M.T.; Willis, K.J.; Cook, E.J. Seasonal population dynamics of the non-native *Caprella mutica* (Crustacea, Amphipoda) on the west coast of Scotland. *Mar. Freshw. Res.* **2010**, *61*, 549–559. [CrossRef]
60. Hughes, L.E.; Ahyong, S.T. Collecting and processing amphipods. *J. Crustac. Biol.* **2016**, *36*, 584–588. [CrossRef]
61. Connell, J.H.; Slatyer, R. 0 Mechanisms of Succession in Natural Communities and Their Role in Community Stability and Organization. *Am. Nat.* **1977**, *111*, 1119–1144. [CrossRef]
62. Queiroga, H.; Cruz, T.; dos Santos, A.; Dubert, J.; González-Gordillo, J.I.; Paula, J.; Peliz, Á.; Santos, A.M.P. Oceanographic and behavioural processes affecting invertebrate larval dispersal and supply in the western Iberia upwelling ecosystem. *Prog. Oceanogr.* **2007**, *74*, 174–191. [CrossRef]
63. Wooster, W.S.; Bakun, A.; McLain, D.R. The seasonal upwelling cycle along the eastern boundary of the North Atlantic. *J. Mar. Res.* **1976**, *34*, 131–141.
64. Ponti, M.; Fava, F.; Perlini, R.A.; Giovanardi, O.; Abbiati, M. Benthic assemblages on artificial reefs in the northwestern Adriatic Sea: Does structure type and age matter? *Mar. Environ. Res.* **2015**, *104*, 10–19. [CrossRef]
65. Satterthwaite, E.V.; Morgan, S.G.; Ryan, J.P.; Harvey, J.B.J.; Vrijenhoek, R.C. Seasonal and synoptic oceanographic changes influence the larval biodiversity of a retentive upwelling shadow. *Prog. Oceanogr.* **2020**, *182*, 102261. [CrossRef]
66. Zupan, M.; Rumes, B.; Vaneverbeke, J.; Degraer, S.; Kerchof, F. Long-Term Succession on Offshore Wind Farms and the Role of Species Interactions. *Diversity* **2023**, *15*, 288. [CrossRef]
67. Conradi, M.; López-González, P.J.; García-Gomez, C. The Amphipod Community as a Bioindicator in Algeciras Bay (Southern Iberian Peninsula) Based on a Spatio-Temporal Distribution. *Mar. Ecol.* **1997**, *18*, 97–111. [CrossRef]
68. Brown, J. On the relationship between abundance and distribution of species. *Am. Nat.* **1984**, *124*, 255–279. [CrossRef]
69. Tuya, F.; Cacabelos, E.; Duarte, P.; Jacinto, D.; Castro, J.J.; Silva, T.; Bertocci, I.; Franco, J.N.; Arenas, F.; Coca, J.; et al. Patterns of landscape and assemblage structure along a latitudinal gradient in ocean climate. *Mar. Ecol. Prog. Ser.* **2012**, *466*, 9–19. [CrossRef]

Disclaimer/Publisher's Note: The statements, opinions and data contained in all publications are solely those of the individual author(s) and contributor(s) and not of MDPI and/or the editor(s). MDPI and/or the editor(s) disclaim responsibility for any injury to people or property resulting from any ideas, methods, instructions or products referred to in the content.

Article

Feeding and Reproductive Phenotypic Traits of the Sea Urchin *Tripneustes gratilla* in Seagrass Beds Impacted by Eutrophication

Helen Grace P. Bangi [1,2,*] and Marie Antonette Juinio-Meñez [1]

[1] The Marine Science Institute, University of the Philippines-Diliman, Quezon City 1101, Philippines; ajmenez@msi.upd.edu.ph

[2] College of Fisheries and Aquatic Sciences, Cagayan State University-Aparri Campus, Maura, Aparri 3515, northern, Philippines

* Correspondence: hgpbangi@gmail.com or helengracebangi@csu.edu.ph

Abstract: The sea urchin *Tripneustes gratilla* is a major grazer and is, hence, an excellent key model organism to study to gain a better understanding of responses to changes in its habitat. We investigated whether there are significant variations in the feeding and reproductive phenotypic traits of populations from three seagrass bed sites, with respect to their proximity to fish farms in Bolinao, northwestern Philippines. We established three stations in each of the three sites: the far, the intermediate, and those near the fish farms, and compared the sea urchins' phenotypic traits and determined whether these were related to seagrass productivity and water parameters. Regardless of the sampling period, adult sea urchins (66.92 ± 0.27 mm test diameter, TD, n = 157) from the areas intermediate and near to the fish farms had significantly lower indices of Aristotle's lantern, gut contents, gut and gonads, and lower gonad quality (high percentage of unusual black gonads), compared to those from the far stations. Multivariate analysis showed that the smaller feeding structures and gut, lower consumption rates and lower gonad indices and quality of sea urchins in the intermediate and near fish farms were positively related to lower shoot density, leaf production and species diversity, as well as lower water movement in those stations. The larger size of the Aristotle's lantern in the far stations was not related to food limitations. More importantly, the phenotypic variability in the feeding structures and gonads of sea urchins in the same seagrass bed provides new evidence regarding the sensitivity of this species to environmental factors that may affect variability in food quality.

Keywords: Aristotle's lantern; gut; gut contents; gonad; aquaculture; organic pollution

Citation: Bangi, H.G.P.; Juinio-Meñez, M.A. Feeding and Reproductive Phenotypic Traits of the Sea Urchin *Tripneustes gratilla* in Seagrass Beds Impacted by Eutrophication. *Diversity* 2023, 15, 843. https://doi.org/10.3390/d15070843

Academic Editors: Renato Mamede and Bert W. Hoeksema

Received: 5 May 2023
Revised: 1 July 2023
Accepted: 4 July 2023
Published: 11 July 2023

Copyright: © 2023 by the authors. Licensee MDPI, Basel, Switzerland. This article is an open access article distributed under the terms and conditions of the Creative Commons Attribution (CC BY) license (https://creativecommons.org/licenses/by/4.0/).

1. Introduction

Aquaculture has been expanding globally, accounting for nearly half of the global fish and seafood supply [1]. One of the main concerns with this expansion is the release of nutrients and organic matter into the environment as waste feeds or metabolic end-products [2,3]. Hence, these are sources of organic enrichment and eutrophication in coastal environments. While aquaculture wastes may increase food for wild fauna and primary production (e.g., [4–6]), they have negative direct effects on ecosystems. For example, excess nutrients increase the growth of filamentous turf algae and epiphytes, reducing the growth of important foundation species such as kelps, other macroalgae and seagrasses (e.g., [7,8]).

Seagrass beds are widely recognized as habitats for many epifauna, infauna and macrobenthic invertebrates, including sea urchins (e.g., [9]). However, they are highly susceptible to both natural and human influences, and there has been concern about the global decline in this important ecosystem (e.g., [10–13]). Studies related to seagrasses and associated herbivores are especially important in the tropical Indo-West Pacific region and southeast Asia, where the highest biodiversity in seagrass species is found [14–17]. Many seagrass beds, however, have been damaged. Increased nutrient loading related to eutrophication has been identified as a main driver of increasing rates of seagrass decline [16,18–21]. Moreover, fish farm-derived organic wastes has altered the natural isotopic composition of organic matter sources of primary producers and different trophic level consumers, thus modifying the natural food webs (e.g., [22,23]).

The sea urchin *Tripneustes gratilla* is widely distributed in the Indo-West Pacific region, and is usually found in shallow seagrass beds or reef habitats that have high potential primary productivity, but are exposed to natural and human-induced disturbances [24]. It is harvested commercially for its gonads or "roe" (e.g., [25–27]). In the Philippines, *T. gratilla* fisheries have been overexploited. For example, in our study site, the fishery collapsed and recovered after many years due to a restocking effort [26]. *T. gratilla* has high growth and respiration rates, as well as high feeding and reproductive capacities [28–30]. As a grazer, *T. gratilla* has a key role in the decomposition and recycling of nutrients, enhancing seagrass productivity [31–35]. It is also considered to be a keystone grazer in the control of the invasive macroalgae *Kappaphycus* in Kaneohe Bay, Hawaii [36] and a potential biological control agent for other invasive algae (e.g., [37–39]). This mechanism and the process of grazing of epiphytes increase the light and nutrient uptake by seagrasses (e.g., [40]), and open up more spaces for coral recruitment, survival, and growth, e.g., [41].

The composition and structure of the benthic community of associated epifauna, infauna and macrofauna have been shown to vary with changes in seagrass habitats, as influenced by the intensive milkfish (*Chanos chanos*) aquaculture in Santiago Island, Bolinao, northwestern Philippines (e.g., [42]). As a major grazer on seagrass beds [32], *T. gratilla* is an excellent model organism to be studied to gain insights into adaptations or responses to potential changes in this ecosystem. We hypothesize that the effect of eutrophication may also be evident among populations of *T. gratilla*, in relation to the differences in the quality of the seagrass habitat and the benthic environment. This could result in variability in phenotypic traits related to feeding and reproduction. In this study, we investigated whether there were significant variations in the feeding and reproductive phenotypic traits between populations of *T. gratilla* in three areas with different distances from the fish farms on Santiago Island in Bolinao. Specifically, we determined and compared different sea urchin parameters directly related to feeding and reproduction, such as the weight ratio of the Aristotle's lantern, gut contents, gut and gonads, to total body weight (expressed as body index), as well as the gonad quality of samples from the different stations in the three areas. Secondly, we determined whether seagrass productivity parameters (i.e., shoot density, estimated leaf production and above-ground fresh and dry biomass) and physical water parameters (i.e., relative water movement, sea surface temperature and mean depth) were related to the sea urchin parameters.

2. Materials and Methods

2.1. Study Stations and Sampling Design

Santiago Island (16°23′19.7″ N, 119°55′25.7″ E, Figure 1) in Bolinao is known to have the widest reef flat system in the northwestern Philippines [43]. The reef flat has an extensive seagrass bed with approximately 22,500 ha, and is an important source of fish and macroinvertebrate fishery resources, including the sea urchin *T. gratilla*. Fish pens and cages are located in the Guiguiwanen channel between the Bolinao mainland and the southern part of Santiago Island, and the fish aquaculture has been a major industry in the region. This, however, has contributed to the degradation in coastal waters, the sediment environment [44–48], and the seagrass ecosystem (e.g., [49,50]), specifically in the eastern area of the island, near the fish farms.

The study stations are dominated by *Thalassia hemprichii*, where adult *T. gratilla* are found. The stations are shallow, submerged even at the lowest tide, with depths ranging from about 0.8 m to 1.5 m. The study stations were established in three areas with respect to their proximity to the fish farms: the reference or the relatively unimpacted stations on the northwestern side being the farthest, named here as FAR, the eastern stations intermediate to the fish farms, named INT, and the eastern stations nearest the fish farms, referred to as NEAR. The FAR stations are approximately 13 to 15 km away from the fish farms, while the NEAR and INT stations are about 1 to 2 km from the fish farms, respectively. Sampling was conducted to coincide with the prevalence of the northeast (NE), southwest (SW), and easterly monsoon weather in the area. *T. gratilla* adult samples were found during the first sampling period in the NEAR stations; however, none was found during the subsequent sampling period. Hence, stations from a nearby location, i.e., INT, were then sampled. Sampling was conducted during the SW period in the FAR and NEAR stations, and during the NE and easterly periods in the FAR and INT stations.

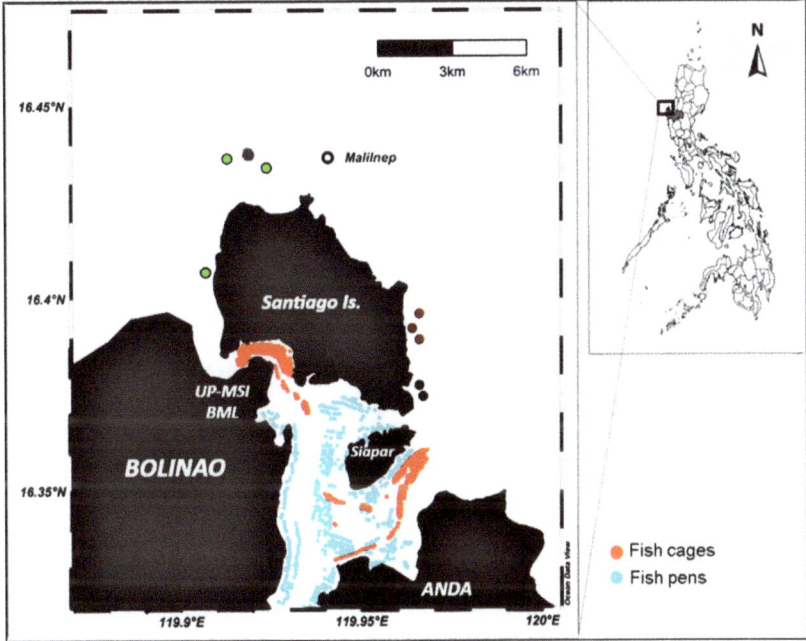

Figure 1. Map showing the study stations in Bolinao, northwestern Philippines, indicated by the filled circles (green: northwestern stations farthest to the fish farms (FAR), sampled during the southwest, SW, northeast, NE and easterly monsoon period; dark red: eastern side, intermediate stations (INT), sampled during NE and easterly period; black: eastern stations, nearest to the fish farms (NEAR), sampled during SW monsoon period). The location of fish cages and pens are shown within the Bolinao-Anda coastal waters based on Google Earth image of January 2014, modified from [51] using Ocean Data View software. Map of the Philippines on the right shows the location of Bolinao.

2.2. Physical Water Parameters

Physical parameters that usually vary in the sampling areas, such as sea surface temperature (SST), relative water movement and site depth, were monitored during the study period. Thermologgers (Onset HOBO Water Temp Prodata Logger, Onset Computer Corporation, MA) were installed in each location to record the SST throughout the duration of the study. Relative water movement was monitored at each of the FAR, INT and NEAR stations, with respect to the fish farms using the "diffusion factor (DF) technique" or clod card method [52]. Three replicate card sets (i.e., 3 cards per set) were deployed over two 24 h periods in each station during the middle of each sampling period.

2.3. Seagrass Parameters

The average above-ground biomass (fresh and dry weight, g·m^{-2}) of seagrass, and associated macroalgae, as well as the seagrass shoot density (number of individuals.m^{-2}), were determined in each station using a standard quadrat method [53,54]. Three replicate transects (50 m each) were used in each station and sampling period. Seagrass abundance was quantified by laying a quadrat (50 × 50 cm^2, divided into 25 squares, 10 × 10 cm^2 each) every 5 m (for non-uniform seagrass distribution) or 10 m interval (uniform distribution) along each transect. Representative samples of seagrasses (from 5 to 10 replicate quadrats per transect) were collected within the total quadrats assessed per station and brought to the laboratory for sorting and identification, including for associated macroalgae [55]. Samples were rinsed with fresh water, and epiphytes were gently scraped off prior to drying the seagrass samples at 60 °C to a constant weight. Seagrass leaf production (g DW·m^{-2}·day^{-1}) for each species per station was estimated and computed using the plastochrone method [56,57], considering the leaf weight (g), and the plastochrone index (PI, in days) for each species. Excessive turbidity precluded further replication within the first station NEAR the fish farms.

2.4. Sea Urchin Parameters

2.4.1. Feeding and Somatic Traits

The ability to obtain food is affected by body size [58], and the total body weight and components' weight (e.g., gonads) were found to be directly related to the test diameter (e.g., [26,59]); hence, adult *T. gratilla* samples in a similar size range (62 to 70 mm TD) were used in each station. To ensure that a minimum of 10 females in a similar size range could be analyzed in each station during each sampling period, around 30–35 adult *T. gratilla* (mixed males and females) were collected at a time, within 50 m × 2 m belt transects (using the same seagrass sampling area), for laboratory analyses. The analysis of gonad quality (or other phenotypic traits) was limited to females of the indicated size ranges, as females were more abundant than males. Each individual was measured and blotted dry with a cloth before being weighed (to the nearest 0.1 g) to obtain the total fresh body weight (BW). The gut of each individual was removed from the test, and the contents were carefully removed, set aside and weighed (to the nearest 0.1 g) prior to preservation in buffered 5% formalin and seawater, for further analysis. To correct for differences in body size, since the BW of the sea urchins varied between stations ($F_{2, 129}$ = 6.05, p = 0.003), the amount of food in the gut was evaluated using the repletion index (RI) or gut content index (GCI), modified from [60], which excluded the weight of the gut. This was determined using the following equation: *GCI* = (*FWdt content*/*BW*) × 100, where *FWdt content* is the fresh weight of the digestive tract contents in grams, and *BW* is the total fresh weight of the body in grams. Each individual sample was further dissected to determine the weight of the main body components: Aristotle's lantern (teeth and jaw pyramid), gut or digestive tract, and the gonad. The Aristotle's lantern and gut indices (*ALI* and *GI*, respectively) were determined as the ratio of the respective weights and *BW* of the sea urchin, multiplied by 100. The index, which is the ratio of the actual weight of each body component to the total BW, is a standard measure of the size of the body component, considering the individual differences in total BW, even with a uniform test diameter.

2.4.2. Reproductive Traits

The gonadosomatic index (*GSI*), which is the ratio of gonad fresh weight (*FWg*) to total body fresh weight (*BW*) in grams, multiplied by 100, was calculated as follows: *GSI* = (*FWg*/*BW*) × 100.

The same gonad samples (from 10 individuals per station and sampling period) that were used in for GSI determination were used for gonad quality evaluation. The relative quality (*GQlty*) of each gonad sample was based on color [61,62]. Fresh samples were scored by assessing the gonad color using a color table (standard colored cards based on

Munsell Color System, USA) and a reference guide (Pantone Matching System Color Chart, Pantone, Inc., Carlstadt, NJ, USA).

2.5. Data Analyses

A multivariate permutational analysis of variance (PERMANOVA, [63]), was used to test for significant differences in *T. gratilla* variables in the three areas, with respect to "proximity to fish farm" or "Pr". In the analyses, "Pr" was a fixed factor, with three levels (FAR, INT and NEAR the fish farms). Station (St) was a random factor, nested in "Pr", with two to three levels (station 1, 2 and 3). Season (Se) was an orthogonal factor, with two levels (SW and NE monsoon season and easterly season) per station. Data were computed using a resemblance matrix based on a Euclidean distance index on log-transformed data. Similar tests were performed to determine significant differences in the seagrass parameters (shoot density, leaf production, fresh and dry above-ground biomass), and physical water parameters with respect to the main factor, Pr. All p-values were obtained using 4999 random permutations of residuals under a reduced model or through Monte Carlo (MC), where appropriate. Posteriori pair-wise tests were conducted using PERMANOVA.

Multidimensional scaling or MDS [64,65] was used to visualize potential patterns in the response variables. A canonical analysis of principal coordinates, or CAP [66] was used to determine the distinctiveness of separation of variables in a multivariate space, with respect to the three sampling areas. To compute the correlational structure of the multiple sea urchin and seagrass variables, and their relationships with the physical water variables, vectors were overlaid, representing the Pearson's correlation of each variable with the ordination axes (principal coordinates analysis or PCO, and CAP coordinates). Data were computed on a resemblance matrix based on an Euclidean distance index on log- or square-root-transformed data. All p-values were obtained using 4999 random permutations of residuals under a reduced model [63,67]. The physical water variables were analyzed using principal component analysis (PCA) prior to PCO and CAP analyses. The significant correlational structure of the sea urchin variables and their relationships with the physical water and seagrass variables were further examined using a distance-based linear model, distLM [68]. Calculations were based on normalized Euclidean distance in log-transformed data.

All of the analyses were performed using the software Primer 6 [69] (Plymouth Marine Laboratory, Plymouth, UK) with the add-on package PERMANOVA (developed by Anderson, M.J.).

3. Results

3.1. Physical Water Parameters

Overall, the mean sea surface temperature (SST), mean diffusion rate (DR) or relative water movement and the mean water depth varied significantly between the FAR, INT and NEAR stations, during each sampling period.

Notably, the highest mean SST (36.2 °C) was recorded in the stations NEAR the fish farms, during the SW sampling period. The overall mean SST was 31.9 ± 0.1 °C in the stations INT to the fish farms and 29.43 ± 0.1 °C in the stations FAR from the fish farms, during two sampling periods: NE and easterly season.

On the other hand, the highest mean water movement (DR) was recorded in the FAR stations (24.46 ± 0.38 g·d^{-1}, during the NE sampling period), versus 11.18 ± 0.18 to 15.46 ± 0.32 g·d^{-1} only in the INT stations during the same period. The lowest mean value, however, was recorded in the INT stations (5.58 ± 0.15 g·d^{-1}) during the easterly sampling period. Slower water movement was recorded in the NEAR stations (7.25 ± 0.72 g·d^{-1}) compared to the FAR stations (13.11 ± 0.31 g·d^{-1}) during the SW sampling period.

The overall mean depth was 0.93 ± 0.02 m in the FAR stations, 0.92 ± 0.01 m in the INT stations and 0.79 ± 0.01 m in the NEAR stations.

Similar to the seagrass and sea urchin parameters, the physical water parameters were integrated in the multivariate analysis to look at their relationships, based on the objectives of the study.

3.2. Seagrass Parameters

The seagrass above-ground biomass, shoot density and leaf production are shown in Figure 2. The mean above-ground biomass was significantly ($t_{(35)}$ = 5.295, p = 0.004, p(perm)) higher by 25 to 33% in the INT stations compared to the mean values in the FAR stations, during the same NE and easterly sampling period. However, the biomass in the NEAR stations was not significantly different ($t_{(35)}$ = 1.036, p = 0.3558, p(perm)) from that in the FAR stations during the SW sampling period.

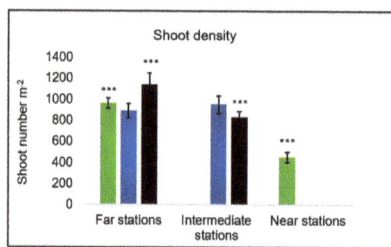
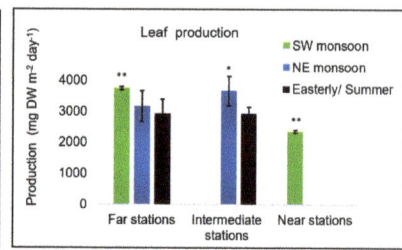
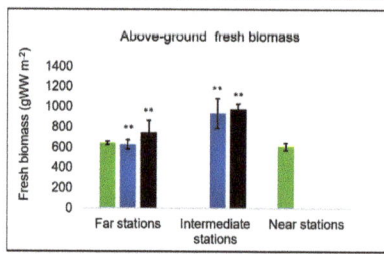
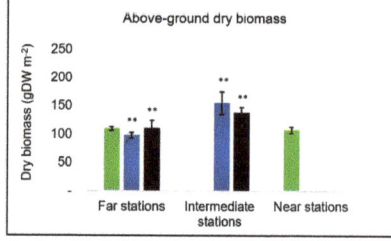

Figure 2. Overall mean (±SE) shoot density, estimated leaf production, above-ground fresh and dry biomass of seagrass resources measured in each station. Asterisks indicate statistical significance of samples between the stations, within each corresponding sampling period (* $p < 0.05$, ** $p < 0.01$, *** $p < 0.001$), obtained using 4999 permutations or through Monte Carlo (MC) where appropriate, based on PERMANOVA. Pair-wise tests were done as a posteriori check for significant effect.

On the other hand, the mean shoot density was significantly ($t_{(35)}$ = 4.063, p = 0.0002, p(perm)) higher by about 30% in the FAR stations compared to the INT and NEAR stations, during the SW and easterly sampling period. The mean leaf production was also significantly ($t_{(35)}$ = 3.5948, p = 0.0138, p(perm)) higher by about three orders of magnitude in the FAR stations compared to the NEAR stations during the SW sampling period. The mean values recorded in the NEAR stations were significantly lower by about 50% and 38% for shoot density and leaf production, respectively, than the mean values recorded in the FAR stations during the same SW sampling period. Despite spatial variations in shoot density, leaf production and biomass, *Thalassia hemprichii* was the most abundant seagrass in all of the stations, with relative abundance of up to 95% in the FAR stations and INT stations, and up to 84% in the NEAR stations (Figure S1). Notably, *Syringodium isoetifolium* was found only in the FAR stations.

Taken together, while the above-ground biomass of seagrass was high in all of the stations, the relative food abundance in terms of shoot densities, estimated leaf production and number of species was the lowest in the stations NEAR the fish farms.

3.3. Sea Urchin Parameters

3.3.1. Gut Contents

The mean gut content index (GCI) and gut content weights were significantly lower ($F_{2, 129} = 11.397$, $p = 0.0002$, $p(\text{perm})$) in samples from the INT and NEAR stations compared to those from the FAR stations during the NE and SW sampling periods (Figure 3). Despite this, there was no single sample observed to have an empty gut or with extremely low gut content. Preliminary gut content and Ivlev's electivity analyses (Figure S2a–c) indicated that *S. isoetifolium*, which was not found in the INT and NEAR stations, is a preferred species compared to other species. The gut contents consisted mostly of a mixture of seagrass and macroalgae in all of the stations. However, among the seagrass species, *T. hemprichii* formed the majority of the gut content, which was also the most abundant seagrass species in all of the stations based on shoot density, leaf production and above-ground biomass. Some sand, forams and sediments were also found in the gut of the sea urchin samples.

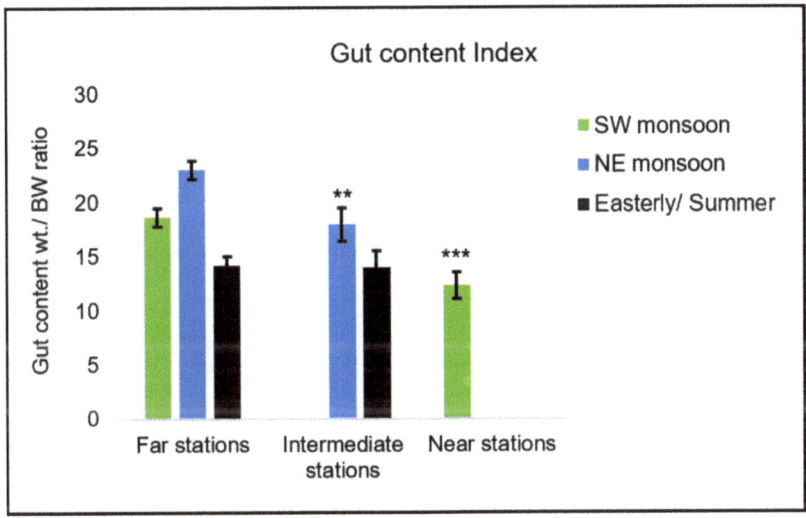

Figure 3. Overall mean (±SE) gut content percentage (in relation to total body weight) in similarly sized adult *T. gratilla* sampled from each study station. Asterisks indicate statistical significance of samples between stations, within each corresponding period (** $p < 0.01$, *** $p < 0.001$), obtained using 4999 permutations or through Monte Carlo (MC), where appropriate, based on PERMANOVA. Pair-wise tests were done as a posteriori check for significant effect.

3.3.2. Feeding Somatic Traits

Notably, significant differences were found in *T. gratilla* feeding somatic traits between the three sampling areas with respect to their proximity to the fish farms, despite the uniform test diameter size range of the samples used ($F_{2, 129} = 1.521$, $p = 0.222$).

The mean Aristotle's lantern index of *T. gratilla* was significantly ($t_{(129)} = 4.057$, $p = 0.0154$, $p(\text{perm})$) lower in the NEAR stations (8.30 ± 0.46) compared to those from the FAR stations (14.71 ± 0.72), during the same SW sampling period (Figure 4). The mean indices of samples in the INT stations (10.05 ± 0.71 and 9.87 ± 0.54) were likewise significantly lower ($t_{(129)} = 2.791$, $p = 0.0452$, $p(\text{MC})$) compared to those from the FAR stations (13.68 ± 0.65 and 11.22 ± 0.47) during the same NE and easterly sampling period, respectively. In terms of the gut index, a similar pattern in variations ($F_{2, 129} = 10.333$, $p = 0.0150$, $p(\text{perm})$, Figure 5) was observed as the Aristotle's lantern and gonads ($F_{2, 129} = 8.013$, $p = 0.0120$, $p(\text{perm})$, Figure 6), except during the easterly sampling period.

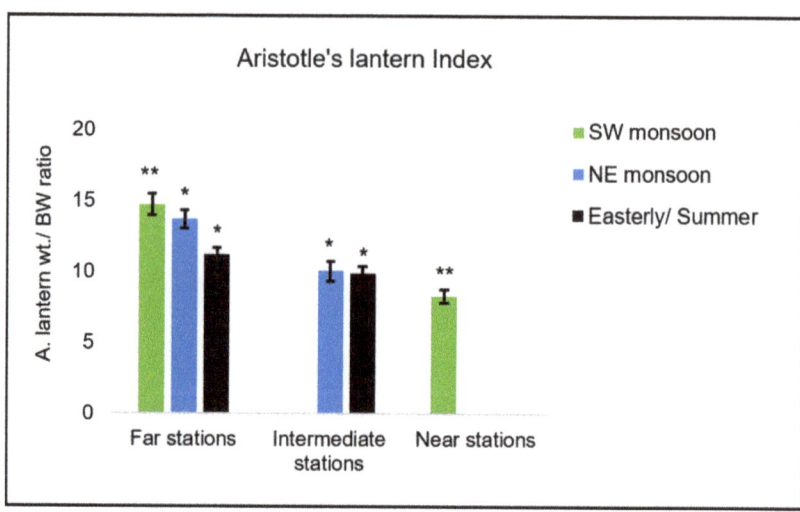

Figure 4. Overall mean (±SE) Aristotle's lantern weight percentage (in relation to total body weight) in similarly sized adult *T. gratilla* sampled from each station. Asterisks indicate statistical significance of samples between stations, within each corresponding sampling period (* $p < 0.05$, ** $p < 0.01$), obtained using 4999 permutations or through Monte Carlo (MC), where appropriate, based on PERMANOVA. Pair-wise tests were done as a posteriori check for significant effect.

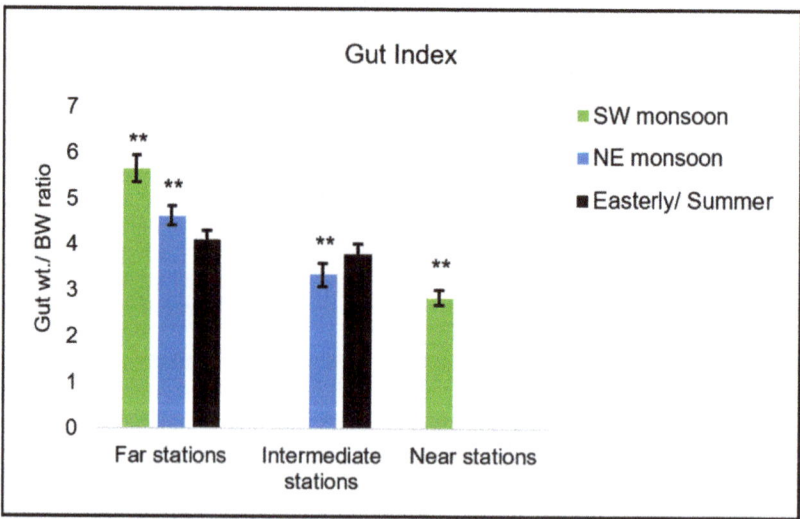

Figure 5. Overall mean (±SE) gut weight percentage (in relation to total body weight) in similarly sized adult *T. gratilla* sampled from each station. Asterisks indicate statistical significance of samples between stations, within each corresponding sampling period (** $p < 0.01$), obtained using 4999 permutations or through Monte Carlo (MC), where appropriate, based on PERMANOVA. Pair-wise tests were done as a posteriori check for significant effect.

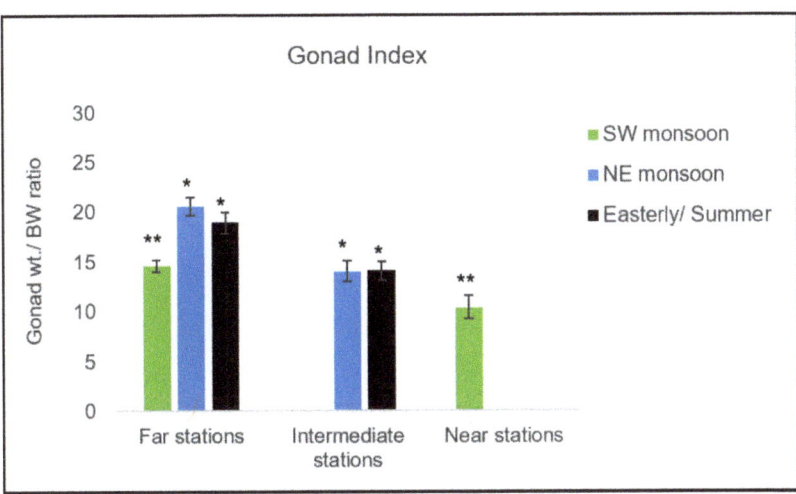

Figure 6. Overall mean (±SE) gonad weight percentage (in relation to total body weight) in similarly sized adult *T. gratilla* sampled from each station. Asterisks indicate statistical significance of samples between stations, within each corresponding sampling period (* $p < 0.05$, ** $p < 0.01$), obtained using 4999 permutations or through Monte Carlo (MC), where appropriate, based on PERMANOVA. Pair-wise tests were done as a posteriori check for significant effect.

3.3.3. Gonads

The overall mean gonadosomatic index (GSI) of *T. gratilla* showed similar patterns of variation ($F_{2, 129} = 8.013$, $p = 0.0120$, p(perm)) with the feeding somatic traits of Aristotle's lantern and gut. The mean GSI of the samples from the NEAR stations was also significantly lower (10.39 ± 1.15, Figure 6, $t_{(129)} = 3.155$, $p = 0.0432$, p(perm)) compared to those from the FAR stations (14.60 ± 0.64) during the same SW sampling period. Likewise, a significantly lower ($t_{(129)} = 2.635$, $p = 0.05$, p(MC)) mean GCI was recorded in the INT stations (14.04 ± 1.08 and 14.05 ± 0.96) compared to those sampled from the FAR stations (20.62 ± 0.92 and 18.91 ± 1.07) during the NE and easterly period, respectively. Similarly, gonad weight was significantly the lowest ($F_{2, 129} = 6.220$, $p = 0.0314$, p(perm)) in stations NEAR the fish farms.

Gonad quality (Figure 7) also varied significantly across the three areas ($F_{2, 129} = 15.99$, $p = 0.0046$, p(perm)). The quality of the gonads of samples from stations INT and NEAR to the fish farms was significantly lower (i.e., highest percentage of the "worst" category or Grade I, 77.3% and 35.0%, respectively), compared to those from the FAR stations (i.e., 9.6% only).

3.4. Relationship of Sea Urchin Phenotypic Traits, Seagrass and Water Parameters

The unconstrained (MDS) ordination of variables describing sea urchin and seagrass status showed fair separation between the impacted INT and NEAR stations close to the fish farms and those FAR from the fish farms (Figure S3), with impacted INT stations in a somehow intermediate position. A significant ($p = 0.0002$) maximized separation of multiple variables representing the groups, with respect to the three sampling areas, is evident. Moreover, CAP (Figure S4) correctly allocated every sample (100%) to the severely impacted NEAR or the FAR stations, with high but partial (83.33%) success occurring for the INT stations. Together with the high allocation success, the canonical correlation coefficient, δ^2, was high (0.8371, Table 1) for the first coordinate, explaining 71.0% or the majority of the total variations observed in the *T. gratilla* and seagrass parameters. For the first coordinate, the phenotypic traits significantly associated with the stations INT and NEAR the fish farms include the following: a lower gut content index and Aristotle's lantern and gut indices, as well as a lower gonad index and quality.

Figure 7. Gonad quality by overall mean % frequency of categories in similarly sized adult *T. gratilla* sampled from each station. Actual gonad colors are indicated. The five quality grades are as follows: Grade I (dark greenish to blackish brown), Grade II (greenish to yellowish brown), Grade III (light to vivid reddish yellow), Grade IV (bright reddish yellow) and Grade V (bright orange–yellow). Grade I is the "worst" quality and Grade V is the "best". Asterisks indicate statistical significance of samples between stations, within each corresponding sampling period (*** $p < 0.001$), obtained using 4999 permutations or through Monte Carlo (MC), where appropriate, based on PERMANOVA. Pair-wise tests were done as a posteriori check for significant effect.

Table 1. Canonical analyses of principal coordinates (CAP) examining the relationships of *T. gratilla* phenotypic traits (body components % weight, and gonad quality) and seagrass abundance measures, on the basis of Euclidean distance index on log- and square root-transformed data. "m" is the number of PCO or principal coordinate axes used in the CAP analysis, % Var is the percentage of the total variance explained by the first m PCO axis. Allocation % success is the percentage of points correctly allocated into each group, δ^2 = the squared canonical correlation. *p*-values given are the results of pair-wise comparisons of *T. gratilla* phenotypic traits and seagrass abundance measures in FAR vs. INT stations vs. NEAR stations, using permutational analysis of variance, with 999 permutations of individual.

Canonical Analysis of Principal Coordinates (CAP) Correlations				
Eigenvalue	Correlation	Correlation sq. (δ^2)		
1	0.9149	0.8371		
2	0.7861	0.618		
m	% Var (1st PCO)	Total Allocation success (%)	1st Correlation sq. (δ^2)	*p* value
3	71.04	92.424	0.83709	0.001
Cross Validation (Leave-one-out allocation of observation to groups)				
	Classified			Total	% correct
Original Groups	FAR stations	INTermediate stations	NEAR stations		
FAR stations	50	0	0	50	100
INTermediate stations	10	50	0	60	83.33
NEAR stations	0	0	22	22	100

Pearson correlation coefficient values which were significant for most of the traits ranged from −0.50 to −0.72 (Table 2). The opposite of these traits (higher quality, body component weights or indices) was true for the FAR stations.

Table 2. Result of CAP analyses examining the correlations of each *T. gratilla* phenotypic trait, density and seagrass abundance measure and environmental station variables. A positive correlation in the first axis indicates trait or variables associated with stations FAR from the fish farms, while a negative correlation indicates trait or variables associated with stations INT and NEAR to the fish farms.

	CAP Pearson correlatons (Response variables: *T. gratilla* and seagrass variables)								
	GSI	ALI	GI	RI	GQlty	Shoot den	Leaf prod	FW biomass	DW biomass
CAP1	−0.663	−0.599	−0.621	−0.504	−0.503	−0.849	−0.565	−0.132	−0.0152
CAP2	−0.182	−0.148	−0.140	−0.054	−0.175	0.145	0.464	0.922	0.875

	CAP Pearson correlation (Environmental variables)		
	SST	DF	Depth
CAP1	0.334	−0.707	−0.023
CAP2	0.079	−0.082	0.322

PERMUTATION TEST
Trace statistic = (tr(Q_m'HQ_m))
First squared canonical correlation = (delta_1^2) = δ_1^2
Trace statistic: tr (Q_m'HQ_m): 1.45507   P: 0.0002
δ_1^2: 0.83709   P: 0.0002
No. of permutations used: 4999

On the other hand, the seagrass and physical water variables associated with stations INT and NEAR the fish farms were lower shoot density, lower leaf production (NEAR stations), lower diffusion factor and higher SST, but higher above-ground biomass and higher mean depth (INT stations). The opposite was true for stations FAR from the fish farms. Pearson correlation coefficient values which were significant for most of the variables ranged from −0.55 to −0.85 (Table 2).

The results of the distance-based linear model analyses further showed significant ($p = 0.0002$) positive relationships between the respective average weights and indices for gut content, Aristotle's lantern and gonads. Likewise, gonad weight, and index and gonad quality were positively correlated (Table S1).

With respect to the response of the sea urchins, a lower gut content index and smaller Aristotle's lantern and gonads were significantly associated with the stations INT and NEAR the fish farms. The opposite was true for the stations FAR from the fish farms. In addition, the degrees of relationships of the *T. gratilla* phenotypic variables and seagrass and physical water parameters in the sampling areas, with respect to their proximity to the fish farms, are shown in Table S2. Sequential distance-based linear model tests showed that biological (seagrass shoot density and leaf production) and physical (DF, SST) factors contributed significantly (about 32.7%) to the variations observed in *T. gratilla* feeding and reproductive phenotypic traits.

4. Discussion

This study showed significant variability in multiple feeding and reproductive traits in *T. gratilla* populations impacted by eutrophication from intensive fish aquaculture. The results not only validate previous studies on the negative impacts of eutrophication on seagrasses, but also the potential negative impacts on the growth of a top grazer, as indicated by the feeding-related phenotypic traits and reproductive output and quality of populations. In addition, this study showed that the smaller feeding structures, lower consumption rates and lower reproductive outputs of sea urchins near the fish farms are positively related to the poor productivity and diversity in seagrasses in those areas. Variations among populations of *T. gratilla* sampled from stations that are only a few kilometers apart and with relative to proximity to fish farms have not been previously reported. This study supports an earlier study (e.g., [42]) that found that changes in the community structure of benthic macrofauna (e.g., *T. gratilla*) were associated with the seagrass beds along the eastern side of Santiago island due to mariculture-induced pollution. The species heterogeneity of the macrofauna was significantly reduced towards the polluted stations, near the fish farms that we sampled.

4.1. Seagrass Bed Condition

The lowest shoot densities, leaf production and number of seagrass species were found in the stations near the fish farms. While the above-ground biomass in these stations was high and comparable to those in stations far from the fish farms, only two species, *T. hemprichii* and *E. acoroides*, were present. The results of this study are consistent with previous studies, which showed a marked decrease in the number of seagrass species, and showed that the shoot density and cover of *T. hemprichii* were the lowest near the fish farms [44]. The cover and diversity in seagrass species decreased from 1995 [70] to 2012 [45,71] in the same eastern stations in Bolinao, near the fish farms. Moreover, four species (*C. rotundata, C. serrulata, Halodule uninervis, Halophila ovalis*) which were found in a station near the fish farm [45,70], very adjacent to our stations near the fish farms, were not observed in the present study.

The lower shoot densities and leaf production in the stations near the fish farms are unlikely caused by enhanced herbivory activities or overgrazing by sea urchins at high densities (1030 to 2000 individuals 100 m^{-2}), as reported in other studies (e.g., [38,72,73]). The mean density range of *T. gratilla* in our stations was very low (0.48 to 5.3 individuals 100 m^{-2}). The low seagrass productivity could be due to the low light extinction coefficient and low water velocity recorded in areas near the fish farm [74]. A decreasing pattern in leaf growth and an increase in shoot mortality in *T. hemprichii* was attributed to the anoxic sediment conditions and high sedimentation rates in areas close to the fish farms [75]. Under reduced sediment conditions, the high respiratory demand of the belowground parts alters the carbon balance of the *T. hemprichii* and *C. serrulata* shoots [76], resulting in shoot mortality. Hence, the driver of the poor health of the seagrass is likely the poor water and sediment quality, as this has already been shown to be heavily impacted by the milkfish aquaculture (e.g., [46,48,51]), even before and during the conduct of this study (Figure S5).

4.2. Condition of the Sea Urchin Populations

4.2.1. Diet

The absence of *E. acoroides* in the gut of *T. gratilla* indicates that it is not preferred by the sea urchin. *S. isoetifolium* was not found in the gut of *T. gratilla*; however, it primarily consumed *T. hemprichi*, the most abundant species in all of the stations. This was also shown in an earlier study of *T. gratilla* in Bolinao [32], and in Papua New Guinea [77]. Our data on proximate analysis on *T. hemprichii* showed higher concentrations of carbohydrates and proteins and higher calories compared to *S. isoethyfolium* (Table S3). The higher carbohydrates and proteins of the former species can promote better growth and reproduction, as shown for other sea urchins [78].

4.2.2. Feeding and Reproductive Phenotypic Traits

The feeding structure (Aristotle's lantern) of sea urchins from stations intermediate and near to the fish farms was significantly smaller compared to those from stations far from the fish farms. In addition, the gonad quality was higher in the latter stations. The smaller feeding structure of sea urchins in the stations intermediate and near to fish farms was positively related to lower consumption rates (gut content index), and a smaller gut. This suggests that feeding may be constrained in the impacted stations, considering the higher seagrass biomass in these stations. Conversely, the larger Aristotle's lantern of individuals in the stations far from the fish farms was related to higher consumption rates, a larger gut and consequently a larger gonad size. Our findings related to the size of the feeding structure concur with laboratory studies that have shown that sea urchins with a smaller feeding structure have a lower capacity to feed, [79] for *Diadema setosum*; [80] for *Echinometra mathaei*. However, the larger Aristotle's lantern in the stations far from the fish farms, where reproductive conditions and food quality were better, contrasts various studies which have reported that a larger Aristotle's lantern is a response to food limitations (e.g., [81–83]). The shoot density and leaf production in these stations were significantly higher compared to the eastern stations intermediate and near to the fish farms. None of the urchins sampled from any of the stations had empty gut. The positive relationship between the size of Aristotle's lantern to the gut size and content reflects feeding capacity rather than food availability.

The average gonad sizes of sea urchins from the stations far from the fish farms were 19 to 20% of the body weight, while those from the stations intermediate and near to the fish farms only ranged from 10.4% to 14%, regardless of season. The larger size and better quality of the gonads of sea urchins in the former stations may be due to the higher quality of the seagrass (i.e., higher shoot density, leaf production and diversity in species), together with the higher capacity of the sea urchins to feed (larger Aristotle's lantern) in these stations. On the other hand, the low gonad size of sea urchins in the stations intermediate and near to the fish farms may be related to the reduced feeding capacity of the sea urchins in these stations, as indicated by their significantly smaller feeding structures. In addition, there was a high incidence of black gonads in these stations. These black gonads were not previously encountered in our earlier studies in the area (e.g., [26,84]). Dark gonads have been reported to be related to old age and small gonad size in *Strongylocentrotus nudus* [61]. Our study, however, used adults of similar test diameter sizes, assumed to be of similar ages, based on the species short life span of about 2 years and fast gametogenic cycle of about 3 months [26]. The occurrence of black gonads could be an extreme indication of poor nutrition and the adverse physiological effects of polluted environmental conditions in the seagrass habitat. We observed that sea urchins with black gonads did not release gametes during induction in the laboratory (Bangi and Juinio-Meñez, unpublished data). Other samples that were induced to spawn in the laboratory from these stations had much lower fecundity (5021 ± 440 eggs per female, unpublished data) compared with the previous sea cage experiments of *T. gratilla* when fed with a high ration of *T. hemprichii* (503,261 ± 15,279 eggs per female) or unfed (12,770 ± 1286 eggs per female, [84], Bangi and Juinio-Meñez, unpublished data). These results further suggest that the fecundity and gamete quality of *T. gratilla* are also affected by the quality of the seagrass food and the environment.

Taken together, the difference in the traits of the sea urchins in the three study areas show that the degraded environmental conditions near the fish farms are affecting the feeding capacity and the quality of the food of these *T. gratilla* populations. Thus, growth rates may be affected, as well as the reproductive output. This study provides evidence on the effect of organic pollution on wild populations in their natural habitat, which has only been shown previously in laboratory studies. For example, studies have shown that chronic exposure to low (0.8 mgL^{-1} inorganic, 10 mgL^{-1} organic) and high (3.2 mgL^{-1} inorganic, 1000 mgL^{-1} organic) sublethal concentrations of phosphates inhibits feeding, fecal production, nutrient absorption and allocation, growth and righting behavior in the sea urchin *Lytechinus variegatus* [85], which has the same life history as *T. gratilla* [24]. Moreover, related studies have shown reduced fertilization success in *L. variegatus* exposed to all phosphate treatments, and arrested embryonic development in the highest concentrations [86]. In another study, the sea urchin *S. droebachiensis* exposed to chronic unionized ammonia up to 0.068 mg L^{-1} (high) caused 76% mortality and reduced gonad growth [87]. The exposure of *S. droebachiensis* to nitrite also caused a significant reduction of gonadal growth, starting from 0.5 mg N-NO$_2$ L^{-1} concentration [88]. In addition, one study [89] suggested that modern commercial aquafeeds may directly affect the reproductive fitness of sea urchins, based on their study on *Heriocidaris erythrogramma*. Hence, as seagrass habitats are exposed to organic pollution and eutrophication, the responses and the fine-scale phenotypic variability in top grazers such as *T. gratilla* provide important indicators on how this affects ecosystem health.

5. Conclusions

Overall, our results have shown significant variability in the feeding and reproductive traits of *T. gratilla* in different stations in three areas within the seagrass beds of Santiago Island in Bolinao. Sea urchins from the eastern stations intermediate and near to the fish farms had smaller feeding structures and gut, lower gut content and smaller gonads, compared to those from the western stations, located far from the fish farms. Moreover, those with smaller gonads also had a high percentage of unusual black gonads. These variabilities were related to the lower quality of seagrass parameters in those stations.

The lower quality and size of feeding and reproductive phenotypic traits in *T. gratila* possibly reflect the negative impacts of fish farming on the seagrass habitat, used as both food and refuge by the sea urchin. While *T. gratilla* is a resilient and stress-tolerant species with respect to hydrodynamically related disturbances, as shown in a previous study [90], this species is very vulnerable to other disturbances, i.e., eutrophication associated with intensive fish aquaculture. Furthermore, this has a negative effect on the reproductive potential of the sea urchin population in the region. Additionally, the decrease in gonad size and quality directly affects the production and economic value of the fishery of *T. gratilla*. These results highlight the need to consider the economic and social benefits of intensive fish farming against its related impact on other economic and environmental aspects of the surrounding environment.

Supplementary Materials: The following supporting information can be downloaded at: https://www.mdpi.com/article/10.3390/d15070843/s1, Figure S1: Percentage abundance by mean shoot density of the different seagrass species measured in each station in three locations; Figure S2a–c: Overall mean % abundance of food items by fresh biomass and the relative selectivity of similarly sized adult *T. gratilla* sampled from the FAR stations during the northeast NE monsoon season and easterly season (2a), from the FAR stations and NEAR stations during the southwest SW monsoon season (2b), and from the INTermediate stations during the northeast NE monsoon season and easterly season (2c); Figure S3: Unconstrained non-metric MDS plot done to compare the various biological variables (*T. gratilla* multiple phenotypic traits, repletion index, mean density and seagrass abundance measures) shown in CAP (Figure S3) with respect to the main factor "proximity to fish farm", based on Bray Curtis similarity measure on square root-transformed data; Figure S4: Constrained canonical analysis of principal coordinates (CAP) plots showing the correlation structures in various response variables measured in *T. gratilla* (multiple phenotypic traits, repletion index, mean density) as a response or proximity to the fish farm environment and related factors; Figure S5: Monthly concentrations of chlorophyll-*a* and major nutrients monitored in representative stations in the vicinity of the Milkfish cages and pens in Bolinao, with reference values based on ASEAN criteria, data from the project on "Monitoring of Water Parameters for a Mariculture Site in Bolinao" (San Diego-McGlone, personal communication); Table S1: Summary of results on the distance-based linear model (DistLM) examining the percentage of variance accounted for each biological variable (*T. gratilla* phenotypic traits, repletion index), and determining the relationships among the biological variables with respect to different stations in three areas and three seasons; Table S2: Distance-based linear model (DistLM) examining the percentage of variance accounted for each biological seagrass and environmental variables (diffusion factor (DF), sea surface temperature (SST), mean site depth, seagrass abundance measures such as shoot density (shootden), leaf production (leafprodn), and fresh or dry weight seagrass biomass (FW, DW biomass) in three sampling areas and three seasons; Table S3: Proximate analysis of macrophyte food for *T. gratilla* and its gonads.

Author Contributions: Conceptualization, H.G.P.B. and M.A.J.-M.; resources and supervision, H.G.P.B. and M.A.J.-M.; Field sampling and Lab preparation, H.G.P.B.; Lab sample processing and analysis, H.G.P.B.; Data analysis, H.G.P.B.; manuscript draft preparation and writing, H.G.P.B.; series of review and revision of the manuscript, H.G.P.B. and M.A.J.-M. All authors have read and agreed to the published version of the manuscript.

Funding: This study was mainly funded by the Department of Science and Technology (DOST)—Science Education Institute's Accelerated Science and Technology Human Resource Development Program (SEI-ASTHRDP, Fund SRSF, STSD-214). The Cagayan State University provided complete salary and one semester scholarship for H.G.P.B. Additional research funds, laboratory facilities and supplies were provided by the University of the Philippines Marine Science Institute's (UP MSI) Bolinao Marine Laboratory (PHD-16-02) and The Marine Invertebrate Laboratory of M.A.J.M.

Institutional Review Board Statement: Not applicable.

Data Availability Statement: The raw data and datasets collated and analyzed for the study may be available through a valid request from the corresponding author.

Acknowledgments: The authors are very grateful to the anonymous reviewers for providing significant comments and suggestions to improve the manuscript. They are also indebted to the following: Rene R. Rollon, for providing advice on seagrass sampling techniques; Symon Dworjanyn, for providing valuable inputs in the earlier version of this manuscript; Marilou San Diego-McGlone, for providing some water quality data in Bolinao; Charissa M. Ferrerra, for the assistance provided on the Ocean Data View mapping software; Ma Josefa R. Pante, for some statistical advice. The authors would like to thank Jay R Gorospe for reviewing and providing valuable suggestions on the revised version of the manuscript, likewise to Lambert Meñez, for critically editing the manuscript, and to Jerwin Baure for additional assistance in copy editing the manuscript. The authors are thankful to Larry Milan, Jack Rengel, Lawrence Ramoran, for assisting the authors in field sampling and laboratory processing of samples. L. Milan, Jan Noelle Rimando and Aphrodite Entoma assisted in laboratory analysis of samples, particularly in gut content analysis.

Conflicts of Interest: The authors declare no conflict of interest.

References

1. Campbell, B.; Pauly, D. Mariculture: A global analysis of production trends since 1950. *Mar. Policy* **2013**, *39*, 94–100. [CrossRef]
2. Tacon, A.G.J.; Forster, I.P. Aquafeeds and the environment: Policy implications. *Aquaculture* **2003**, *226*, 181–189. [CrossRef]
3. Primavera, J.H. Overcoming the impacts of aquaculture on the coastal zone. *Ocean Coast. Manag.* **2006**, *49*, 531–545. [CrossRef]
4. Tewfik, A.; Rasmussen, J.B.; McCann, K.S. Anthropogenic enrichment alters a marine benthic food web. *Ecology* **2005**, *86*, 2726–2736. [CrossRef]
5. Anton, A.; Cebrian, J.; Heck, K.L.; Duarte, C.M.; Sheehan, K.L.; Miller, M.-E.C.; Foster, C.D. Decoupled effects (positive to negative) of nutrient enrichment on ecosystem services. *Ecol. Appl.* **2011**, *1*, 991–1009. [CrossRef]
6. White, C.A.; Bannister, R.J.; Dworjanyn, S.A.; Husa, V.; Nichols, P.D.; Dempster, T. Aquaculture-derived trophic subsidy boosts populations of an ecosystem engineer. *Aquacult. Environ. Interact.* **2018**, *10*, 279–289. [CrossRef]
7. McGlathery, K.J. Macroalgal blooms contribute to the decline of seagrass in nutrient-enriched coastal waters. *J. Phycol.* **2001**, *37*, 453–456. [CrossRef]
8. Burkholder, J.M.; Tomasko, D.A.; Touchette, B.W. Seagrasses and eutrophication. *J. Exp. Mar. Biol. Ecol.* **2007**, *350*, 46–72. [CrossRef]
9. Williams, S.L.; Heck, K.L., Jr. Seagrass Community Ecology. In *Marine Community Ecology*; Bertness, M.D., Gaines, S.D., Hay, M.E., Eds.; Sinauer Associates: Sundergland, MA, USA, 2001; pp. 317–337.
10. Short, F.T.; Wyllie-Echeverria, S. Natural and human induced disturbance of seagrasses. *Environ. Conserv.* **1996**, *23*, 17–27. [CrossRef]
11. Duarte, C.M. The future of seagrass meadows. *Environ. Conserv.* **2002**, *29*, 192–206. [CrossRef]
12. Walker, D.I.; Kendrick, G.A.; McComb, A.J. Decline and Recovery of Seagrass Ecosystems-The Dynamics of Change. In *Seagrasses: Biology, Ecology and Conservation*; Larkum, A.W.D., Orth, E.J., Duarte, C.M., Eds.; Springer: Dordrecht, The Netherlands, 2007; pp. 551–565.
13. Coles, R.; Grech, A.; Rasheed, M.; McKenzie, L.; Unsworth, R.; Short, F. Seagrass Ecology and Threats in the Tropical Indo-Pacific Bioregion. In *Seagrass: Ecology, Uses and Threats*; Pirog, R.S., Ed.; Nova Science Publishers, Inc.: New York, NY, USA, 2011.
14. Short, F.; Carruthers, T.; Dennison, W.; Waycott, M. Global seagrass distribution and diversity: A bioregional model. *J. Exp. Mar. Biol. Ecol.* **2007**, *350*, 3–20. [CrossRef]
15. Short, F.T.; Polidoro, B.; Livingstone, S.R.; Carpenter, K.E.; Bandeira, S.; Bujang, J.S.; Calumpong, H.P.; Carruthers, T.J.; Coles, R.G.; Dennison, W.C.; et al. Extinction risk assessment of the world's seagrass species. *Biol. Conserv.* **2011**, *144*, 1961–1971. [CrossRef]

16. Short, F.T.; Coles, R.; Fortes, M.D.; Victor, S.; Salik, M.; Isnain, I.; Andrew, J.; Seno, A. Monitoring in the Western Pacific region shows evidence of seagrass decline in line with global trends. *Mar. Pollut. Bull.* **2014**, *83*, 408–416. [CrossRef]
17. Fortes, M.D.; Ooi, J.L.S.; Tan, Y.M.; Prathep, A.; Bujang, J.S.; Yaakub, M.S. Seagrass in Southeast Asia: A review of status and knowledge gaps, and a road map for conservation. *Bot. Mar.* **2018**, *61*, 269–288. [CrossRef]
18. Orth, R.J.; Carruthers, T.J.; Dennison, W.C.; Duarte, C.M.; Fourqurean, J.W.; Heck, K.L.; Hughes, A.R.; Kendrick, G.A.; Kenworthy, W.J.; Olyarnik, S.; et al. A global crisis for seagrass ecosystems. *Bioscience* **2006**, *56*, 987–996. [CrossRef]
19. Ralph, P.J.; Tomasko, D.; Moore, K.; Seddon, S.; Macinnis, O.C.M. Human impacts on seagrasses: Eutrophication, sedimentation, and contamination. In *Seagrasses: Biology, Ecology, and Conservation*; Larkum, A.W.D., Orth, E.J., Duarte, C.M., Eds.; Springer: Dordrecht, The Netherlands, 2007; pp. 567–593.
20. Waycott, M.; Duarte, C.M.; Carruthers, T.J.B.; Orth, R.J.; Dennison, W.C.; Olyarnik, S.; Calladine, A.; Fourqurean, J.W.; Heck, K.L.; Hughes, A.R.; et al. Accelerating loss of seagrasses across the globe threatens coastal ecosystems. *Proc. Natl. Acad. Sci. USA* **2009**, *106*, 12377–12381. [CrossRef]
21. Jiang, Z.; Cui, L.; Liu, S.; Zhao, C.; Wu, Y.; Chen, Q.; Yu, S.; Li, J.; He, J.; Fang, Y.; et al. Historical changes in seagrass beds in a rapidly urbanizing area of Guangdong Province: Implications for conservation and management. *Glob. Ecol. Conserv.* **2020**, *22*, e01035. [CrossRef]
22. Vizzini, S.; Mazzola, A. Stable isotope evidence for the environmental impact of a land-based fish farm in the western Mediterranean. *Mar. Pollut. Bull.* **2004**, *49*, 61–70. [CrossRef]
23. Vizzini, S.; Mazzola, A. The effects of anthropogenic organic matter inputs on stable carbon and nitrogen isotopes in organisms from different trophic levels in a southern Mediterranean coastal area. *Sci. Total Environ.* **2006**, *368*, 723–731. [CrossRef]
24. Lawrence, J.M. Sea urchin life history strategies. In *Sea Urchins: Biology and Ecology*, 4th ed.; Lawrence, J.M., Ed.; Elsevier: London, UK, 2020; pp. 19–28.
25. Talaue-McManus, L.T.; Kesner, K.P. Valuation of a Philippine municipal sea urchin fishery and implications of its collapse. In *Philippine Coastal Resources under Stress*; Juinio-Meñez, M.A.R., Newkirk, G.F., Eds.; Selected papers from the Fourth Annual Common Property Conference; University of the Philippines: Quezon City, Philippines, 1995; pp. 229–239.
26. Juinio-Meñez, M.A.; Bangi, H.G.P.; Malay, M.C.D.; Pastor, D. Enhancing the recovery of depleted *Tripneustes gratilla* stocks through grow-out culture and restocking. *Rev. Fish. Sci.* **2008**, *16*, 35–43. [CrossRef]
27. Andrew, N.L.; Agatsuma, Y.; Ballesteros, E.; Bazhin, A.G.; Creaser, E.P.; Barnes, D.K.A.; Botsford, L.W.; Bradbury, A.; Campbell, A.; Dixon, J.D.; et al. Status and management of the world sea urchin fisheries. *Oceanogr. Mar. Biol. Annu. Rev.* **2002**, *40*, 343–425.
28. Lawrence, J.M.; Agatsuma, Y. Tripneustes. In *Sea Urchins: Biology and Ecology*, 4th ed.; Lawrence, J.M., Ed.; Academic Press: London, UK; Elsevier: London, UK, 2020; pp. 681–703.
29. Shimabukuro, S. *Tripneustes gratilla* (sea urchin). In *Aquaculture in Tropical Areas*; English ed.; Shokita, S., Kakazu, K., Tomori, A., Toma, T., Eds.; Prepared by M. Yamaguchi; Midoro Shobo: Tokyo, Japan, 1991; pp. 313–328.
30. Lawrence, J.M.; Bazhin, A. Life-history strategies and the potential of sea urchins for aquaculture. *J. Shellfish Res.* **1998**, *17*, 1515–1522.
31. Koike, I.; Mukai, H.; Nojima, S. The role of the sea urchin, *Tripneustes gratilla* (Linnaeus), in decomposition and nutrient cycling in a tropical seagrass bed. *Ecol. Res.* **1987**, *2*, 19–29. [CrossRef]
32. Klumpp, D.W.; Salita-Espinosa, J.T.; Fortes, M.D. Feeding ecology and trophic role of sea urchins in a tropical seagrass community. *Aquat. Bot.* **1993**, *45*, 205–229. [CrossRef]
33. Valentine, J.F.; Heck, K.L., Jr.; Busby, J.; Webb, D. Experimental evidence that herbivory can increase shoot density in a subtropical turtlegrass (*Thalassia testudinum*) meadow. *Oecologia* **1997**, *112*, 193–200. [CrossRef]
34. Vonk, J.A.; Pijnappels, M.H.J.; Stapel, J. In situ quantification of *Tripneustes gratilla* grazing and its effects on three co-occurring tropical seagrass species. *Mar. Ecol. Prog. Ser.* **2008**, *360*, 107–114. [CrossRef]
35. Chiu, S.-H.; Huang, Y.-H.; Lin, H.-J. Carbon budget of leaves of the tropical intertidal seagrass *Thalassia hemprichii*. *Estuar. Coast. Shelf Sci.* **2013**, *125*, 27–35. [CrossRef]
36. Conklin, E.J.; Smith, J.E. Abundance and spread of the invasive red algae, *Kappaphycus* spp. in Kane'ohe Bay, Hawai'i and an experimental assessment of management options. *Biol. Invasions* **2005**, *7*, 1029–1039. [CrossRef]
37. Stimson, J.; Cunha, T.; Philippoff, J. Food preferences and related behavior of the browsing sea urchin *Tripneustes gratilla* (Linnaeus) and its potential for use as a biological control agent. *Mar. Biol.* **2007**, *151*, 1761–1772. [CrossRef]
38. Valentine, J.P.; Edgar, G.J. Impacts of a population outbreak of the urchin *Tripneustes gratilla* amongst Lord Howe Island coral communities. *Coral Reefs* **2010**, *29*, 399–410. [CrossRef]
39. Westbrook, C.E.; Ringang, R.R.; Cantero, S.M.A.; HDAR, TNC Urchin Team; Toonen, R.J. Survivorship and feeding preferences among size classes of outplanted sea urchins, *Tripneustes gratilla*, and possible use as biocontrol for invasive alien algae. *PeerJ* **2015**, *3*, e1235. [CrossRef]
40. Cebrian, J.; Enriquez, S.; Fortes, M.; Agawin, N.; Vermaat, J.E.; Duarte, C.M. Epiphyte accrual on *Posidonia oceanica*. L Delile leaves: Implications for light absorption. *Bot. Mar.* **1999**, *42*, 123–128. [CrossRef]
41. Furman, B.; Heck, K.I., Jr. Differential impacts of echinoid grazers on coral recruitment. *Bull. Mar. Sci.* **2009**, *85*, 121–132.
42. Leopardas, V.; Honda, K.; Go, G.A.; Bolisay, K.; Pantallano, A.D.; Uy, W.; Fortes, M.D.; Nakaoka, M. Variation in macrofaunal communities of sea grass beds along a pollution gradient in Bolinao, northwestern Philippines. *Mar. Pollut. Bull.* **2016**, *105*, 310–318. [CrossRef]

43. McManus, J.W.; Nañola, C.L.; Reyes, R.B.; Kesner, K.N. *Resource Ecology of the Bolinao Coral Reef System*; ICLARM Studies Review, 22; International Center for Living Aquatic Resources Management: Manila, Philippines, 1992; p. 117.
44. Holmer, M.; Marba, N.; Terrados, J.; Duarte, C.M.; Fortes, M.D. Impacts of milkfish (*Chanos chanos*) aquaculture on carbon and nutrient fluxes in the Bolinao area, Philippines. *Mar. Pollut. Bull.* **2002**, *44*, 685–696. [CrossRef]
45. Holmer, M.; Duarte, C.M.; Heilskov, A.; Olesen, B.; Terrados, J. Biogeochemical conditions in sediments enriched by organic matter from net-pen fish farms in the Bolinao area, Philippines. *Mar. Pollut. Bull.* **2003**, *46*, 1470–1479. [CrossRef]
46. David, C.P.; Sta. Maria, Y.Y.; Siringan, F.P.; Reotita, J.M.; Zamora, P.B.; Villanoy, C.L.; Sombrito, E.Z.; Azanza, R.V. Coastal pollution due to increasing nutrient flux in aquaculture sites. *Environ. Geol.* **2009**, *58*, 447–454. [CrossRef]
47. Geček, S.; Legović, T. Towards carrying capacity assessment for aquaculture in the Bolinao Bay, Philippines: A numerical study of tidal circulation. *Ecol. Modell.* **2010**, *221*, 1394–1412. [CrossRef]
48. San Diego-McGlone, M.L.; Azanza, R.V.; Villanoy, C.L.; Jacinto, G.S. Eutrophic waters, algal bloom and fish kill in fish farming areas in Bolinao, Pangasinan, Philippines. *Mar. Poll. Bull.* **2008**, *57*, 295–301. [CrossRef]
49. Fortes, M.D.; Go, G.A.; Bolisay, K.; Nakaoka, M.; Uy, W.H.; Lopez, M.R.; Leopardas, V.; Leriorato, J.; Pantallano, A.; Paciencia, F.; et al. Seagrass response to mariculture-induced physico-chemical gradients in Bolinao, northwestern Philippines. In Proceedings of the 12th International Coral Reef Symposium, Cairns, Australia, 9–13 July 2012.
50. Tanaka, T.; Go, G.A.; Watanabe, A.; Miyajima, T.; Nakaoka, M.; Uy, W.H.; Nadaoka, K.; Watanabe, S.; Fortes, M.D. 17-year change in species composition of mixed seagrass beds around Santiago Island, Bolinao, the northwestern Philippines. *Mar. Pollut. Bull.* **2014**, *88*, 81–85. [CrossRef]
51. Ferrera, C.M.; Watanabe, A.; Miyajima, T.; San Diego-McGlone, M.L.; Morimoto, N.; Umezawa, Y.; Herrera, E.; Tsuchiya, T.; Yoshikai, M.; Nadaoka, K. Phosphorus as a driver of nitrogen limitation and sustained eutrophic conditions in Bolinao and Anda, Philippines, a mariculture-impacted tropical coastal area. *Mar. Pollut. Bull.* **2016**, *105*, 237–248. [CrossRef]
52. Doty, S.M. Measurement of water movement in reference to benthic algal growth. *Bot. Mar.* **1971**, *14*, 32–35. [CrossRef]
53. English, S.; Wilkinson, C.; Baker, V. Seagrass Communities. Chapter 5. In *Survey Manual for Tropical Marine Resources*; ASEAN-Australia Marine Science Project; Australian Institute of Marine Science: Townsville, Australia, 1997; pp. 241–252.
54. Burdick, D.M.; Kendrick, G.A. Standards for seagrass collection, identification, and sample design. In *Global Seagrass Research Methods*; Short, F.T., Coles, R.G., Eds.; Elsevier Science B.V.: Amsterdam, The Netherlands, 2001; pp. 79–100.
55. Duarte, C.M.; Kirkman, H. Methods for the measurement of seagrass abundance and depth distribution. In *Global Seagrass Research Methods*; Short, F.T., Coles, R.G., Eds.; Elsevier Science B.V.: Amsterdam, The Netherlands, 2001; pp. 141–154.
56. Vermaat, J.E.; Fortes, M.D.; Agawin, N.R.; Duarte, C.M.; Marbà, N.; Uri, J.S. Meadow maintenance, growth and productivity in a mixed Philippine seagrass bed. *Mar. Ecol. Prog. Ser.* **1995**, *124*, 215–255. [CrossRef]
57. Short, F.T.; Duarte, C.M. Methods for the measurement of seagrass growth and production. In *Global Seagrass Research Methods*; Short, F.T., Coles, R.G., Eds.; Elsevier Science B.V.: Amsterdam, The Netherlands, 2001; pp. 155–182.
58. Lawrence, J.M. The effect of stress and disturbance in Echinoderms. *Zool. Sci.* **1990**, *7*, 17–28.
59. Tuason, A.Y.; Gomez, E.D. The reproductive biology of *Tripneustes gratilla* Linnaeus (Echinodermata: Echinoidea) with some notes on *Diadema setosum* Leske. *Proc. Int. Symp. Mar. Biogeogr. Evol. South. Hemisph.* **1979**, *2*, 707–716.
60. Vaïtilingon, D.; Rasolofonirina, R.; Jangoux, M. Feeding preferences, seasonal gut repletion indices, and diel feeding patterns of the sea urchin *Tripneustes gratilla* (Echinodermata: Echinoidea) on a coastal habitat off Toliara (Madagascar). *Mar. Biol.* **2003**, *143*, 451–458. [CrossRef]
61. Agatsuma, Y.; Sato, M.; Taniguchi, K. Factor causing brown-colored gonads of the sea urchin, *Strongylocentrotus nudus*, in northern Honshu, Japan. *Aquaculture* **2005**, *249*, 449–458. [CrossRef]
62. Agnette, T.; Zairin, M.; Mokoginta; Suprayudi, M.A.; Yulianda, F. Protein level and protein energy ratio that produce the best gonad quality of sea urchin *Tripneustes gratilla*. *J. Biol. Life Sci.* **2014**, *5*, 95–105.
63. Anderson, M.J. Permutation tests for univariate or multivariate analysis of variance and regression. *Can. J. Fish. Aquat. Sci.* **2001**, *58*, 626–639. [CrossRef]
64. Shepard, R.N. The analysis of proximities: Multidimensional scaling with an unknown distance function. II. *Psychometrika* **1962**, *27*, 219–246. [CrossRef]
65. Kruskal, J.B. Multidimensional Scaling by Optimizing Goodness of Fit to a Nonmetric Hypothesis. *Psychometrika* **1964**, *29*, 1–27. [CrossRef]
66. Anderson, M.J.; Willis, T.J. Canonical analysis of principal coordinates: A useful method of constrained ordination for ecology. *Ecology* **2003**, *84*, 511–525. [CrossRef]
67. Anderson, M.J.; ter Braak, C.J.F. Permutation tests for multi-factorial analysis of variance. *J. Stat. Comput. Simul.* **2003**, *73*, 85–113. [CrossRef]
68. McArdle, B.H.; Anderson, M.J. Fitting multivariate models to community data: A comment on distance-based redundancy analysis. *Ecology* **2001**, *82*, 290–297. [CrossRef]
69. Clarke, K.R.; Warwick, R.M. *Change in Marine Communities: An Approach to Statistical Analysis and Interpretation*, 2nd ed.; Primer-e Ltd., Plymouth Marine Laboratory: Plymouth, UK, 2001.
70. Bach, S.S.; Borum, J.; Fortes, M.D.; Duarte, C.M. Species composition and plant performance of mixed seagrass beds along a siltation gradient at Cape Bolinao, the Philippines. *Mar. Ecol. Prog. Ser.* **1998**, *174*, 247–256. [CrossRef]

71. Watai, M.; Nakamura, Y.; Honda, K.; Bolisay, K.O.; Miyajima, T.; Nakaoka, M.; Fortes, M.D. Diet, growth, and abundance of two seagrass bed fishes along a pollution gradient caused by milkfish farming in Bolinao, northwestern Philippines. *Fish Sci.* **2015**, *81*, 43–51. [CrossRef]
72. Eklöf, J.S.; de la Torre-Castro, M.; Gullström, M.; Uku, J.; Muthiga, N.; Lyimo, T.; Bandeira, S.O. Sea urchin overgrazing of seagrasses: A review of current knowledge on causes, consequences, and management. *Estuar. Coast. Shelf Sci.* **2008**, *79*, 569–580. [CrossRef]
73. Fernandez, C.; Ferrat, L.; Pergent, G.; Pasqualini, V. Sea urchin–seagrasses interactions: Trophic links in a benthic ecosystem from a coastal lagoon. *Hydrobiologia* **2012**, *699*, 21–33. [CrossRef]
74. Rollon, R.N.; Van Steveninck, E.D.D.R.; Van Vierssen, W.; Fortes, M.D. Contrasting recolonization strategies in multi-species seagrass meadows. *Mar. Pollut. Bull.* **1998**, *37*, 450–459. [CrossRef]
75. Terrados, J.; Duarte, C.M.; Kamp-Nielsen, L.; Borum, J.; Agawin, N.S.R.; Fortes, M.D.; Gacia, E.; Lacap, D.; Lubanski, M.; Greve, T. Are seagrass growth and survival affected by reducing conditions in the sediment? *Aquat. Bot.* **1999**, *65*, 175–197. [CrossRef]
76. Uy, W.H.; Vermaat, J.E.; Hemminga, M.A. The Interactive Effect of Shading and Sediment Conditions on Growth and Photosynthesis of Two Seagrass Species, *Thalassia hemprichii* and *Halodule uninervis*. Functioning of Philippine Seagrass Species under Deteriorating Light Conditions. Ph.D. Dissertation, Wageningen University-International Institute for Infrastructural, Hydraulic and Environmental Engineering, Swets and Zeitlinger B.V., Lisse, The Netherlands, 2001; pp. 49–72.
77. Mukai, H.; Nojima, S. A preliminary study on grazing and defecation rates of a seagrass grazer, *Tripneustes gratilla* (L.) (Echinidermata: Echinoidea), in Papua New Guinean seagrass beds. *Spec. Publ. Makaishishima Mar. Biol. Stn.* **1985**, 84–191.
78. Watts, S.A.; Lawrence, A.L.; Lawrence, J.M. Nutrition. In *Sea Urchins: Biology and Ecology*, 4th ed.; Lawrence, J.M., Ed.; Elsevier: London, UK, 2020; pp. 191–208.
79. Ebert, T.A. Relative growth of sea urchin jaws: An example of plastic resource allocation. *Bull. Mar. Sci.* **1980**, *30*, 467–474.
80. Black, R.; Codd, C.; Hebbert, D.; Vink, S.; Burt, J. The functional significance of the relative size of Aristotle's lantern in the sea urchin *Echinometra mathaei* (de Blainville). *J. Exp. Mar. Biol. Ecol.* **1984**, *77*, 81–97. [CrossRef]
81. Levitan, D.R. Skeletal changes in the test and jaws of the sea urchin *Diadema antillarum* in response to food limitation. *Mar. Biol.* **1991**, *111*, 431–435. [CrossRef]
82. Edwards, P.B.; Ebert, T.A. Plastic responses to limited food availability and spine damage in the sea urchin *Strongylocentrotus purpuratus*. *J. Exp. Mar. Biol. Ecol.* **1991**, *145*, 205–220. [CrossRef]
83. Smith, J.G.; Garcia, S.C. Variation in purple sea urchin (*Strongylocentrotus purpuratus*) morphological traits in relation to resource availability. *PeerJ* **2021**, *9*, e11352. [CrossRef]
84. Bangi, H.G.P. The Effect of Adult Nutrition on Somatic and Gonadal Growth, Egg Quality and Larval Development of the Sea Urchin *Tripneustes gratilla* Linnaeus 1758 (Echinodermata: Echinoidea). Master's Thesis, Marine Science Institute, University of the Philippines Diliman, Quezon City, Philippines, 2001.
85. Böttger, S.A.; McClintock, J.B.; Klinger, T.S. Effects of organic and inorganic phosphates on feeding, absorption, nutrient allocation, growth, righting responses of the sea urchin *Lytechinus variegatus*. *Mar. Biol.* **2001**, *138*, 741–751. [CrossRef]
86. Böttger, S.A.; McClintock, J.B. The effects of organic and inorganic phosphates on fertilization and early development in the sea urchin *Lytechinus variegatus* (Echinodermata: Echinoidea). *Comp. Biochem. Physiol. Part C* **2001**, *129*, 307–315. [CrossRef]
87. Siikavuopio, S.I.; Dale, T.; Foss, A.; Mortensen, A. Effects of chronic ammonia exposure on gonad growth and survival in green sea urchin *Strongylocentrotus droebachiensis*. *Aquaculture* **2004**, *242*, 313–320. [CrossRef]
88. Siikavuopio, S.I.; Dale, T.; Christiansen, J.S.; Nevermo, I. Effects of chronic nitrite exposure on gonad growth in green sea urchin *Strongylocentrotus droebachiensis*. *Aquaculture* **2004**, *242*, 357–363. [CrossRef]
89. White, C.A.; Dworjanyn, S.A.; Nichols, P.D.; Mos, B.; Dempster, T. Future aquafeeds may compromise reproductive fitness in a marine invertebrate. *Mar. Environ. Res.* **2016**, *122*, 67–75. [CrossRef]
90. Bangi, H.G.P.; Juinio-Meñez, M.A. Resource allocation trade-offs in the sea urchin *Tripneustes gratilla* under relative storminess and wave exposure. *Mar. Ecol. Prog. Ser.* **2019**, *608*, 165–182. [CrossRef]

Disclaimer/Publisher's Note: The statements, opinions and data contained in all publications are solely those of the individual author(s) and contributor(s) and not of MDPI and/or the editor(s). MDPI and/or the editor(s) disclaim responsibility for any injury to people or property resulting from any ideas, methods, instructions or products referred to in the content.

Article

Echinoids and Crinoids from Terra Nova Bay (Ross Sea) Based on a Reverse Taxonomy Approach

Alice Guzzi [1,2,*], Maria Chiara Alvaro [2], Matteo Cecchetto [2] and Stefano Schiaparelli [2]

[1] Department of Earth, Environmental and Life Sciences (DISTAV), University of Genoa, Corso Europa 26, 16132 Genoa, Italy

[2] Italian National Antarctic Museum (MNA, Section of Genoa), University of Genoa, Viale Benedetto XV No. 5, 16132 Genoa, Italy; chiara.alvaro@yahoo.it (M.C.A.); matteocecchetto@gmail.com (M.C.); stefano.schiaparelli@unige.it (S.S.)

* Correspondence: aliceguzzi@libero.it

Abstract: The identification of species present in an ecosystem and the assessment of a faunistic inventory is the first step in any ecological survey and conservation effort. Thanks to technological progress, DNA barcoding has sped up species identification and is a great support to morphological taxonomy. In this work, we used a "Reverse Taxonomy" approach, where molecular (DNA barcoding) analyses were followed by morphological (skeletal features) ones to determine the specific status of 70 echinoid and 22 crinoid specimens, collected during eight different expeditions in the Ross and Weddell Seas. Of a total of 13 species of sea urchins, 6 were from the Terra Nova Bay area (TNB, Ross Sea) and 4 crinoids were identified. Previous scientific literature reported only four species of sea urchins from TNB to which we added the first records of *Abatus cordatus* (Verrill, 1876), *Abatus curvidens* Mortensen, 1936 and *Abatus ingens* Koehler, 1926. Moreover, we found a previous misidentification of *Abatus koehleri* (Thiéry, 1909), erroneously reported as *A. elongatus* in a scientific publication for the area. All the crinoid records are new for the area as there was no previous faunistic inventory available for TNB.

Keywords: Southern Ocean; COI; morphology; DNA barcoding; Echinoidea; Crinoidea; Antarctica

1. Introduction

The increasing application of integrated taxonomy, coupled with new modelling approaches, requires data to be findable, accessible, interoperable and reusable in the long term [1].

The most common challenges facing studies or the construction of biodiversity inventories are accurate species identification and the absence of detailed information on the distribution of taxa throughout the different geographical regions of the planet [2]. Morphology-based identification represents the classic approach to taxonomy but is strongly dependent on the level of experience and expertise of the identifier. This method is, thus, largely prone to mistakes whenever intraspecific variability has not been previously tested. However, the increase in molecular advances has made it evident that this approach comes with some inherent limitations [3]. Taxonomic discrepancies, such as synonymous or cryptic species, are extremely common when a traditional taxonomic approach is used. Neither molecular nor morphological taxonomic methods are sufficient on their own [4] and the number of examples where this integrated approach is applied to identify species is rapidly increasing (sea stars (e.g., [5–11]), brittle stars (e.g., [12]), holothurians (e.g., [13]), fish (e.g., [14]) and many more.

With the rapid accumulation of samples in museums and the co-occurring decline in taxonomic expertise in recent years [15], molecular tools, phylogenetics and coalescent-based analyses have become the practices used for species identification or discrimination. Among all these, DNA barcoding in particular has provided a useful method for fast, efficient and reliable species identification and discovery [5,16,17]. It is based on the concept of the "barcode gap" routine occurrence [16], where significantly higher interspecific divergence for the cytochrome c oxidase subunit I (COI) gene is lower compared to an intraspecific one. DNA barcoding exhibits remarkable effectiveness in taxonomic assignments, overcoming challenges commonly associated with morphology-based identification methods. It not only circumvents the difficulties encountered when diagnostic traits are lacking due to damaged specimens, but also establishes connections between various stages of animal development [18]. A 658-bp region of the COI gene is, thus, largely used as an effective marker to pinpoint species delimitation boundaries in different groups of marine organisms [5,12,19–24].

Several studies have demonstrated the efficiency of COI sequencing and that integrative taxonomy brings added value to address some species complexes or identify cryptic species within Echinodermata phylum [5,11,25,26].

Crinoidea, an important class of echinoderms, include a stalked form commonly referred to as sea lily and a more mobile form known as feather star. Sea lilies, which account for 80 crinoid species, live mostly at depths greater than several hundred meters. Sea lilies' calyx and arms are supported above the substrate by a stalk composed of disc-shaped plates called columnals. By contrast, feather stars are more successful ecologically, with about 570 species occupying diverse habitats from the intertidal to the deep sea, and from tropic to polar sea. They lack a stem in the adult stage and are usually anchored much closer to the substrate by claw-like cirri that radiate from the centrodorsal plate [27]. Although pentacrinoid larvae of feather stars do have stalks, they abandon them during development. For this class, there is no specific diagnostic tool and information is scattered in a variety of scientific papers.

Among the five echinoderm extant classes, echinoids represent a conspicuous and important element of many marine benthic communities and are always reported in local checklists. They exploit a wide array of marine habitats, from the poles to the equator and from the intertidal zone to the deep sea, although they achieve their greatest levels of diversity and abundance in shallow shelf areas [28,29]. Their typically large size coupled with their purportedly fairly easily identifiable characters have contributed to this success (e.g., [30–32]). As a result, a comprehensive database and identification guides have been produced, comprising the Southern Ocean [33]. For this area, in fact, there exists the Southern Ocean Antarctic Echinoidea database assembled by David et al. in 2005 [33,34], which is a powerful interactive database synthesizing the results of more than 130 years of Antarctic expeditions and providing the main morphological characteristics for species identification. It represents one of the most complete collections of information for any Antarctic taxa but still reveals major gaps in the geographic and bathymetric distributions of many species.

Since the establishment of the Italian research station "Mario Zucchelli", the benthic fauna of Terra Nova Bay (TNB) has been widely investigated. Many of these ecological studies were conducted on some of the most conspicuous and easy to find species of echinoderms, mainly asteroids ([35,36]). With the introduction in 2016 of the Ross Sea region Marine Protected Area (RSMPA) through the RSMPA Monitoring Plan (CCAMLR Conservation Measure 91-05: Ross Sea region Marine Protected Area. 2016 [37]) and the inclusion of the TNB area in the marine protected area (MPA), the accurate description of the benthic communities has become a new priority, with the aim of implementing monitoring and conservation efforts for species and communities.

However, to date, a complete faunistic inventory for all echinoderm classes is still lacking, despite the fact that continuous research has been undertaken in the area. Specifically, the TNB echinoid species inventory was assessed for the first time by Chiantore et al., 2006 [38], based on morphological identifications leading to a list of four species in the area. For crinoids, there is a general assessment for the Ross Sea [39], but a comprehensive list for this taxon in the TNB area is still lacking.

The objectives of the study are: (i) to update the checklist for echinoids of the TNB area; and (ii) to evaluate the first comprehensive inventory for crinoids from the same location. To achieve this goal, we used an integrated approach using both DNA barcoding and morphological characters. The results will serve as a baseline for future works in ecology, monitoring and management of the study area. The current paper represents a further contribution of the Italian National Antarctic Museum (MNA), Genoa section, as the custodian of biodiversity data for the Ross Sea area. Many contributions to the Antarctic Biodiversity Portal have been published by the MNA over the years, with the aim of increasing the knowledge of the area [11,40–48] (http://www.biodiversity.aq, accessed on 25 November 2022).

2. Materials and Methods

The samples available at Italian National Antarctic Museum (MNA), Genoa section, analysed in this study derive from the Antarctic Peninsula (Figure 1a) and the Ross Sea sector, specifically the TNB area, which is part of both the marine protected area and the Antarctic Special Protected Area (n.161) (CIT 62) (Figure 1b).

Specimens were collected in the framework of several recent scientific expeditions performed in the Southern Ocean and which are now permanently stored and curated at MNA. The Italian National Antarctic Program (PNRA) expedition "XVII" (2001/2002), "XIX" (2003/2004), "XXV" (2009/2010), "XXVII" (2011/2012), "XXVIII" (2012/2013), "XXIX" (2013/2014), "XXXII" (2016/2017) were all from the Ross Sea, and additional samples collected from the Antarctic Peninsula were obtained from the Alfred Wegner Institute (AWI) ANT-XXIX/3, PS81 expedition (2013).

2.1. Sampling and DNA Extraction

A total of 92 samples, 70 belonging to echinoids and 22 to crinoids, were analysed and the distributional data considered here originated from 31 different sampling stations, ranging from 15 and 750 m in depth (Table 1). Sampling was performed through deployments of a variety of sampling gear comprising SCUBA diving (see Table 1 for details). Six pentacrinoid larvae (MNA-03760, MNA-03766, MNA-03795, MNA-03855 and MNA-07967) were included in the analyses and were obtained by examining biological materials (e.g., polychaetes tubes) on which they settled; one of these (MNA-09159) was found on the metal structures of Mooring L.

Whenever possible, following the collection and sorting phase, the live specimens were photographed by one of us (SS) to avoid the loss of potential diagnostic characteristics such as colours that would fade or disappear once the organism was fixed in ethanol.

Samples were fixed in ethanol (95% Et-OH) or frozen ($-20\,°C$) in order to preserve them for further genetic analysis. Sorting and classification on a morphological basis were performed at the MNA using the available literature and keys from Koehler (1926) [49], Clark (1967) [50], Moore (1983) [51] and Speel at al. (1983) [52]. All the samples were acquired as permanent vouchers at the MNA (https://steu.shinyapps.io/MNA-generale/, accessed on 21 November 2022). The clipped material from each sample was sent to the Canadian Center for DNA Barcoding using microplates (University of Guelph, Guelph, ON, Canada), which performed extraction, amplification and sequencing. Primers used for amplification were LCOech1aF1 (5'-TTTTTTCTACTAAACACAAGGATATTGG-3') or EchinoF1 (5'-TTTCAACTAATCATAAGGACATTGG-3') and HCO2198 (5'-TAAACTTCAGGGT GACCAAAAAATCA-3'). Sequences were uploaded to the BOLD platform (Barcode Of Life Data systems, http://www.boldsystems.org, accessed on 21 November 2022).

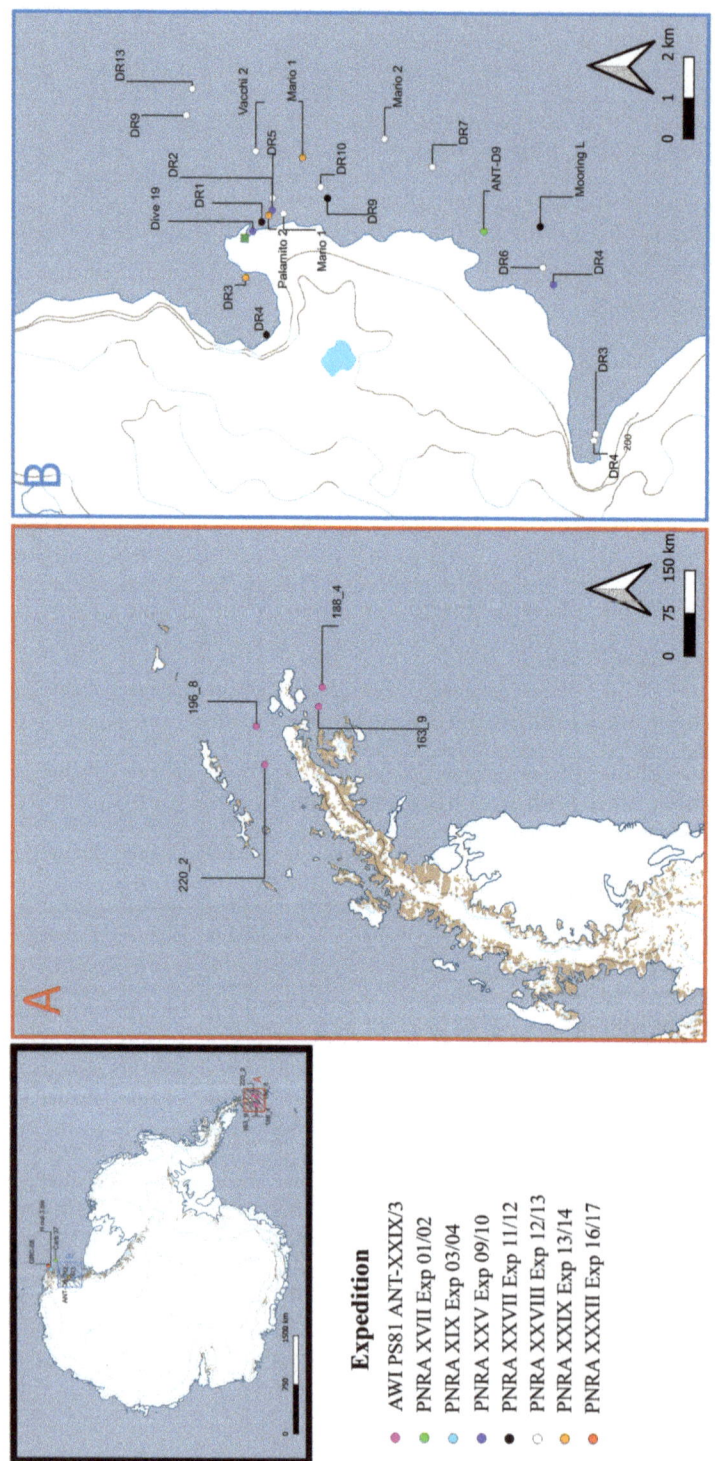

Figure 1. Antarctica (high left); highlighted in red is the Antarctic Peninsula and in blue the Terra Nova Bay (Ross Sea) sector. (**A**) Sampling station of the Antarctic Peninsula and (**B**) sampling sites in Terra Nova Bay with Mario Zucchelli Station (Italy) highlighted in green square. Legend is colour coded for expedition.

Table 1. Sampling stations and data. Abbreviations: Mario Zucchelli Station (MZS); number of specimens (N). Additional information on individual specimens can be found in Supplementary Materials S1.

Expedition	Station	Location	Year	Latitude	Longitude	Depth (m)	Sampling Methods	N
AWI PS81 ANT-XXIX/3	163_9	Weddell Sea	2013	−63.79600	−56.31000	550.9	AGT (Agassiz Trawl)	3
	188_4	Weddell Sea	2013	−63.83933	−55.62367	427	AGT (Agassiz Trawl)	1
	196_8	Bransfield Strait	2013	−62.79667	−57.08917	580	AGT (Agassiz Trawl)	3
	220_2	Bransfield Strait	2013	−62.94533	−58.39383	792	AGT (Agassiz Trawl)	1
PNRA XVII Exp 01/02	ANT-D9	Tethys Bay	2002	−74.74860	164.12467	113	Dredge	1
	Carb 37	Mawson Bank	2002	−73.15133	174.29467	309	Dredge	1
	H out 3 bis	Cape Hallett	2004	−72.29000	170.44000	258	AGT (Agassiz Trawl)	1
PNRA XIX Exp 03/04	R3	Cape Russell	2004	−74.82167	164.19167	330	AGT (Agassiz Trawl)	1
	R2	Cape Russell	2004	−74.81667	164.30167	364	AGT (Agassiz Trawl)	1
PNRA XXV Exp 09/10	Dive 19	Road Bay	2010	−74.69647	164.12007	15	Dive	2
	DR2	Road Bay	2010	−74.70082	164.13762	148	Dredge	11
	DR4	Adelie Cove	2010	−74.76450	164.08202	100	Dredge	8
PNRA XXVII Exp 11/12	DR1	Road Bay	2012	−74.69848	164.12812	100	Dredge	12
	DR4	Tethys Bay	2012	−74.70010	164.03502	198	Dredge	10
	DR9	Faraglione	2012	−74.71337	164.14903	150	Dredge	4
	Mooring L	Terra Nova Bay	2012	−74.76130	164.13032	149	Mooring	1

Table 1. Cont.

Expedition	Station	Location	Year	Latitude	Longitude	Depth (m)	Sampling Methods	N
PNRA XXVIII Exp 12/13	DR3	Adelie Cove	2013	−74.77468	163.95948	77	Dredge	2
	DR4	Adelie Cove	2013	−74.77430	163.95400	78	Dredge	1
	DR5	Road Bay	2013	−74.70087	164.14793	150	Dredge	3
	DR6	Caletta	2013	−74.76207	164.09623	146	Dredge	1
	DR7	Terra Nova Bay	2013	−74.73675	164.17702	240	Dredge	1
	DR9	MZS	2013	−74.68090	164.21433	522	Dredge	2
	DR10	Faraglione	2013	−74.71178	164.15802	250	Dredge	2
	DR13	MZS	2013	−74.68210	164.23640	525	Dredge	2
	Mario 1	Terra Nova Bay	2013	−74.70348	164.13550	137	GN (Gill Net)	3
	Mario 2	Terra Nova Bay	2013	−74.72597	164.19908	319	LL (Long Line)	1
	Vacchi 2	Tethys Bay	2013	−74.69677	164.18622	460	GN (Gill Net)	1
PNRA XXIX Exp 13/14	DR3	Tethys Bay	2014	−74.69508	164.08137	60	Dredge	3
	Mario 1	Terra Nova Bay	2014	−74.70750	164.18167	242	TN (Trammel Net)	3
	Palamito 2	Terra Nova Bay	2014	−74.70000	164.13333	100	LL (Long Line)	3
PNRA XXXII Exp 16/17	GRC-08	Cape Hallett	2017	−71.98111	172.19383	750	Dredge	1

Manual taxonomic assignation was performed in BOLD (Accessed 16 November 2022) of which sequences are available for 5147 echinoid and 4291 crinoid specimens, representing, respectively, 307 and 203 species. Comparison was also performed in the National Center for Biotechnology Information (NCBI) database with BLAST (https://blast.ncbi.nlm.nih.gov/Blast.cgi, accessed on 21 November 2022) for definitive assignment. A correct identification was defined as a sequence match that exceeds 98% similarity to the reference database [53]. The taxonomic names and classification used in this study were obtained from the World Register of Marine Species (WoRMS) website (www.marinespecies.org, last accessed on 21 November 2022). Sequences were edited and corrected in CodonCode Aligner v9.0.1, developed by CodonCode Corporation in Centerville, MA, USA (http://www.codoncode.com/aligner/, accessed on 21 November 2022). The MUSCLE algorithm was used to align sequences, which is available within CodonCode Aligner, and result was visually inspected for accuracy. *Odontaster validus* Koehler, 1906 (GenBank accession number: ON103477) was selected as outgroup. The substitution pattern was determined by analysing the Bayesian information criterion (BIC) scores in MEGA X [43], and the T92 + G (Tamura 3-parameter + Gamma distribution [44]) model was found to have the lowest scores, indicating the best fit. Phylogenies were inferred using Bayesian, maximum likelihood (ML) and maximum parsimony (MP) approaches. Bayesian estimation of phylogeny was carried out using Mr Bayes [54,55]. Additionally, a generalized time reversible (GTR) model with gamma(G)-correction was used to avoid risk of obtaining unsupported results with under parametrization in Bayesian inference. Markov chain Monte Carlo (MCMC) algorithm with two simultaneous independent runs was performed starting from different random trees. Each run comprised four chains (one cold and three heated), which were sampled every 100 generations for a total of 2×10^8 generations. To ensure appropriate effective sampling size (ESS all > 100), Tracer v.1.6 was utilized. The final result trees for comparison were performed using FigTree v1.4.4 (http://tree.bio.ed.ac.uk/software/figtree/, accessed on 21 November 2022) for graphical representation. Sequences obtained in this study have been deposited in GenBank: accession numbers OR157781-OR157864.

2.2. Species Delimitation Methods

The assumption of using species delimitation methods dictates that two or more species are distinct by exhibiting a "barcode gap" [56]; that is, genetic variation between species (interspecific) greater than genetic variation within species (intraspecific) [57]. Four methods were conducted for primary species hypotheses to identify the number of molecular operational taxonomic units (MOTUs) within our dataset. The Barcode Index Number System (BIN) [58] and Automatic Barcode Gap Discovery (ABGD) [16] rely on pairwise sequence distances between specimens to determine the number of OTUs within a dataset. Standard BIN assignments are available on BOLD (http://www.boldsystems.org, accessed on 13 December 2022), but they are generated through the analysis of all barcode sequences on BOLD, meaning that the results are not strictly comparable with those obtained with other methods (because they are based on a more inclusive dataset). ABGD analysis was performed on the web interface (http://wwwabi.snv.jussieu.fr/public/abgd/, accessed on accessed on 15 December 2022), Kimura (K80) was used as the genetic distance with default range of 0.001 to 0.1 and was examined for intraspecific distances, while gap values from 1 to 1.5 were employed. The Generalized Mixed Yule Coalescent (GMYC) [59] differs strongly from the other methods because it is a model-based approach, aiming to discover the maximum likelihood solution for the threshold between the branching rates of speciation and coalescent processes on a tree. The tree-based methods employ a coalescent framework to independently identify evolving lineages without gene flow, each representing a putative species [60]. They can be performed using a single marker and are used to establish a threshold that identifies the separation of intraspecific population substructure from interspecific divergence. It therefore identifies those groups that may be candidate species [61]. The last species delimitation approach was implemented using a PTP process [62]. Here, we used the Bayesian implementation of the Poisson tree processes model (bPTP) [62], the ML tree was used as input. The bPTP analysis (species.h-its.org/ptp) runs parameters that were 500,000 generations of MCMC, a thinning of 100 and a 25% burn-in. In all the species partition methods used, the outgroup (*Odontaster validus*) was removed.

2.3. Morphological Identification

Following the "Reverse Taxonomy" approach [63,64], morphological analyses were conducted for a re-examination of our molecular results on available specimens. Observations were carried out under a stereoscopic microscope. For determination to species level, each sea urchin individual was identified according to the morphological features indicated in the taxonomic keys for Antarctic Echinoidea by Thomas Saucède (http://echinoidea-so.identificationkey.org/mkey.html, accessed on 13 January 2023). Crinoids were identified with available literature from Clark (1967) [50], Moore (1983) [51] and Speel at al. (1983) [52]. For echinoids, our morphological analysis focused on morphological skeletal features, such as accessory structures and spines. We particularly focused our attention on pedicellariae, which are defensive structures consisting of a head composed of two or more valves hinged to one another, a stem and sometimes a neck. The four main types of pedicellariae analysed were: globiferous, dentate, triphyllous and ophicephalous.

Given the taxonomic relevance of pedicellariae shape morphology for species identification, the small mandibular appendage that articulates on the test was removed from selected samples corresponding to putative species partition highlighted by the molecular analysis. The tissue portion was treated with sodium hypochlorite (NaClO) to remove organic matter. Subsequently, the skeletal elements obtained were washed with deionized water then after with water and ethanol (Et-OH). Proportions were increased until the skeletal elements were completely washed with 100% Et-OH. This made it possible to observe skeletal characteristics in detail under the stereomicroscope in order to obtain the correct identification of the species.

For crinoids, we compared the external morphological features. All diagnostic characters were analysed in detail, including the cirri, oral pinnules, genital pinnules, arm number, and segments of the cirri and arms under a stereomicroscope. Specimens identified in this study showed morphological characteristics corresponding to those described in the literature [39–41], and molecular species identification was cross-referred with the morphological result.

3. Results

A total of 92 specimens were analysed in the current study, and all were correctly sequenced to obtain a final COI sequence length of 628 bp. The COI dataset employed for analyses is reported as Supplementary Material (M1). Of the 92 sequences generated in this study, 70 belonged to echinoids and 22 to crinoids. All sequences were barcode-compliant (Supplementary File S2) and received a BIN, which aided species delimitation [58]. The other species delimitation methods recovered a different number of secondary species hypotheses (SSH) for sea urchin, but were all in agreement regarding the crinoid's investigation (Supplementary File S2). The most problematic method was bPTP because in the echinoids' SSH investigation, it showed an overestimation in species partition. The maximum likelihood and Bayesian analysis results were consistent and revealed 13 putative species of echinoids and 4 of crinoids (Figure 2 and Supplementary File S2).

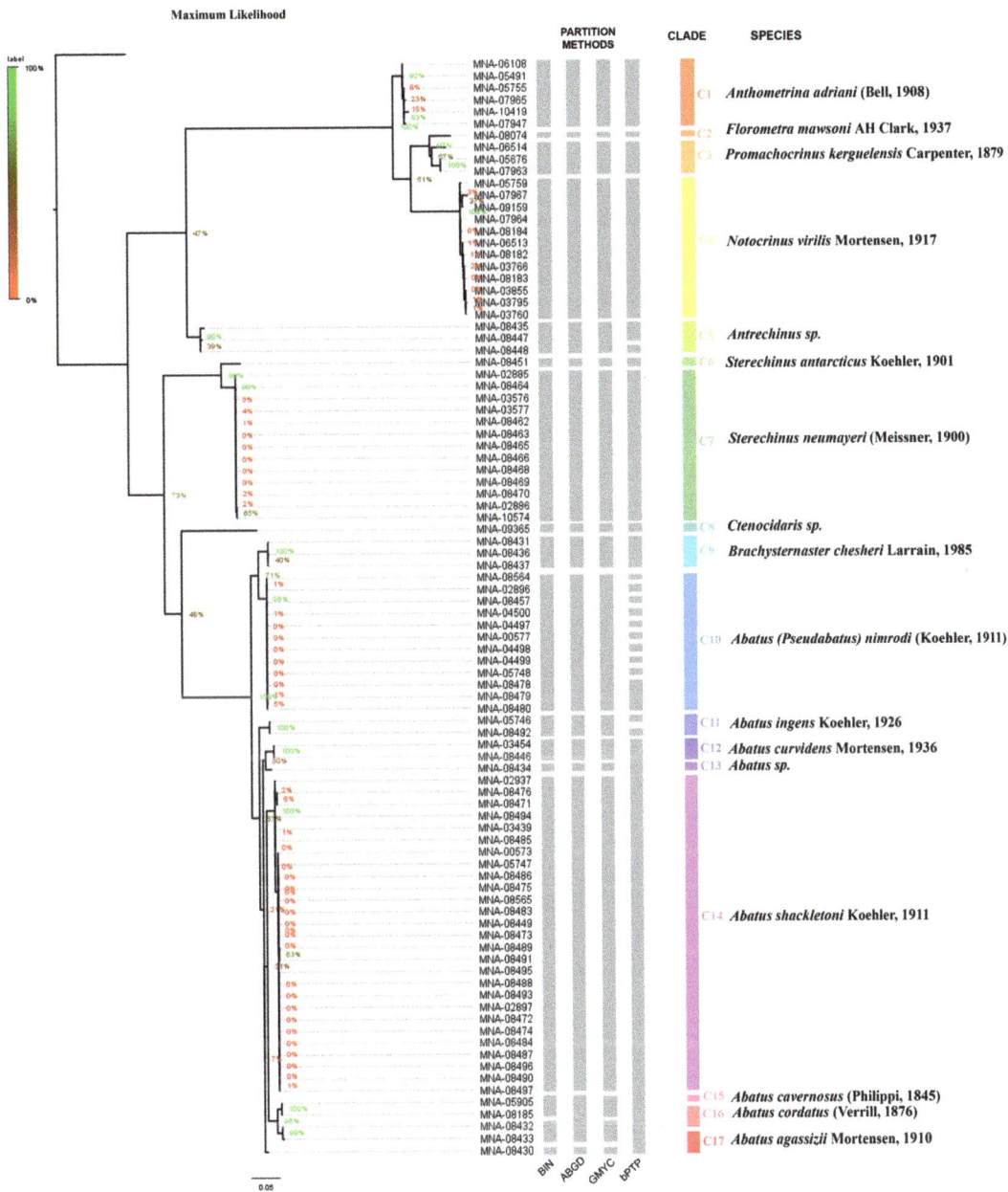

Figure 2. Tree topology comparison of maximum likelihood. Posterior probability node values are shown on the tree with corresponding legend for each analysis. BIN: barcode index number; BOLD: automatic species delimitation [58]; ABGD: results from automatic barcode gap discovery method [16]; GMYC: species delimitation from generalized mixed Yule coalescent method [59]; bPTP: species delimitation using Bayesian Poisson tree processes method [62].

3.1. Molecular Results

Identification through barcoding requires specimens from the same species to cluster together using the barcode markers. Detailed and high-resolution trees' comparison (ML and Bayesian interference) with species partition method results are available in Supplementary File S3.

3.1.1. Crinoidea

The 22 crinoids analysed here were assigned to four morphospecies, all of them corresponding to described and well-known species. Our crinoid specimens were correctly grouped into four putative species by the species delimitation methods, showing consistency between the analyses. Clade 1 (posterior probability 92% ML and value of 0.97 in Bayesian interference) corresponded to individuals of *Anthometrina adriani* (Bell, 1908), and Clade 2 (posterior probability 97% ML and value of 1.00 in Bayesian) to *Florometra mawsoni* AH Clark, 1937individuals. *Promachocrinus kerguelensis* Carpenter, 1879 individuals were included in Clade 3 (posterior probability 97% ML and value of 1.00 in Bayesian). *Notocrinus virilis* Mortensen, 1917 individual were included in Clade 4 (posterior probability 100% ML and value of 1.00 in Bayesian). All pentacrinoids stage larva feltt in the *N. virilis* clade.

3.1.2. Echinoidea

The 70 echinoids studied represented 13 morphospecies, five of which were given provisional identifications based on molecular taxonomy: Clades 12, 13 and 16 (*Abatus* sp.), Clade 5 (*Antrechinus* sp.) and Clade 8 (*Ctenocidaris* sp.). This was due to a lack of matching sequences in the online database (cross-check on BOLD and GenBank. Accessed 22 November 2022). COI-based species delimitation methods identified 13 (BIN and GMYC), 12 (ABGD) and 19 putative species (bPTP). The results are consistent between the species delimitation methods for *Sterechinus antarcticus* Koehler, 1901 (Clade 6), *Sterechinus neumayeri* (Meissner, 1900) (Clade 7) and *Brachysternaster chesheri* Larrain, 1985 (Clade 9).

Sequences belonging to Clades 16 and 17 were grouped together by ABGD. In bPTP, those sequences are similarly grouped with Clades 12, 13, 14 and 15. However, bPTP seems to overpartition putative species of *Abatus ingens* Koehler, 1926 (Clade 11), *Antrechinus* sp. (Clade 5) and *Abatus (Pseudabatus) nimrodi* (Koehler, 1911) (Clade 10).

Posterior probability node values, which are shown on the tree (Supplementary File S3), range from 47% to 100% for ML tree reconstruction and a value included from 0.56 to 1 in Bayesian interference. In our samples, no corresponding sequence matched *Abatus koehleri* (Thiéry, 1909) (previously reported from the TNB area as *A. elongatus*), a species previously reported from Terra Nova Bay water [55].

3.2. Morphological Analysis

A total of 70 echinoid and 22 crinoid individuals were morphologically examined after primary species partition based on molecular screening following the "Reverse Taxonomy" approach. Clades 12 and 16 were assigned on a sole morphological base, as they did not match any sequence in the online databases.

The main descriptors to distinguish the species are given below for crinoids and echinoids, respectively, in Tables 2 and 3; the more in-depth descriptions of the species are reported in Supplementary Materials S4. The results of the molecular analysis were combined with the morphological results following the "Reverse Taxonomy" approach, and the species partition was consistent. To optimize the visualization and understanding of the results, the tree in Figure 2 was subdivided, highlighting the class of crinoids, in Figure 3, and echinoids, in Figure 4, with the available representative photos of selected specimens.

Table 2. Main descriptors to distinguish crinoid species identified in this work.

Species	Arms	Cirri	Lateral Perisome
A. adriani	10 with protuberance on the dorsal face of most of the arm joints	50–60	
F. mawsoni		40–77	
P. kerguelensis	20 each ray divided at primibrachial		
N. virilis		up to 40	pinnules contains large plates that are usually triangular in shape with rounded angles

Table 3. Main descriptors to distinguish echinoid species identified in this work.

Species	Globiferous Pedicellariae Shape	Diameter	Apical System Position	Labrum Size	Frontal Sinus	Apical System Plating
S. antarcticus	valves with 1 to 3 lateral teeth	20–50 mm				
S. neumayeri	valves with 1 to 3 lateral teeth	10–20 mm				
A. (Pseuabatus) nimrodi	valves terminating in a series of small teeth					
A. agassizii	valves terminating in two long teeth		apical system subcentral	short, reaching no farther than the 1st adjacent ambulacral plates		
A. cavernosus	valves terminating in 4 teeth					
A. cordatus	globiferous pedicellariae absent					
A. curvidens	valves terminating in two long teeth		apical system anterior		present	
A. elongatus	valves terminating in two long teeth		apical system anterior		absent	
A. ingens	valves terminating in two long teeth		apical system subcentral	long, extending between the 3rd and the 4th adjacent ambulacral plates		apical system not disjunct separate genitals 2 and 3
A. shackletoni	valves terminating in two long teeth		apical system subcentral	long, extending between the 3rd and the 4th adjacent ambulacral plates		

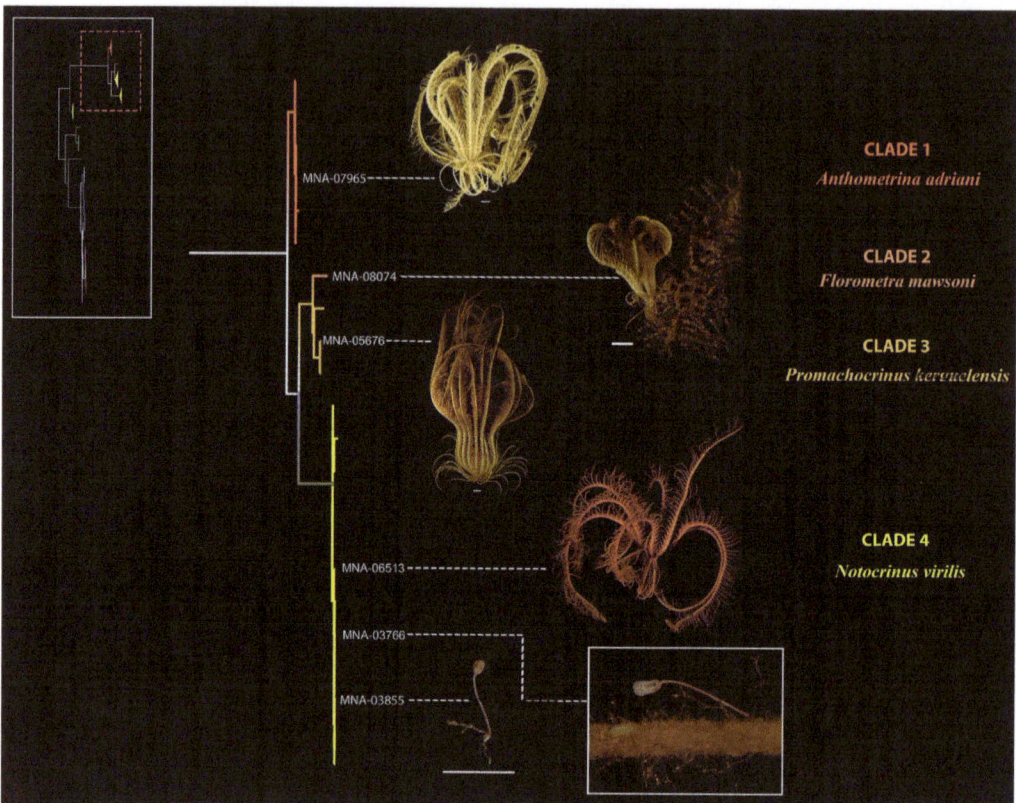

Figure 3. Representative photos of selected specimens. In the tree, the different species identified are highlighted by different colours. The species present in the Terra Nova Bay area are listed on the right. Bottom left is the schematic view of the tree in Figure 2, the portion analysed in detail in the image is highlighted in red. Scale bar: 1 cm in grey.

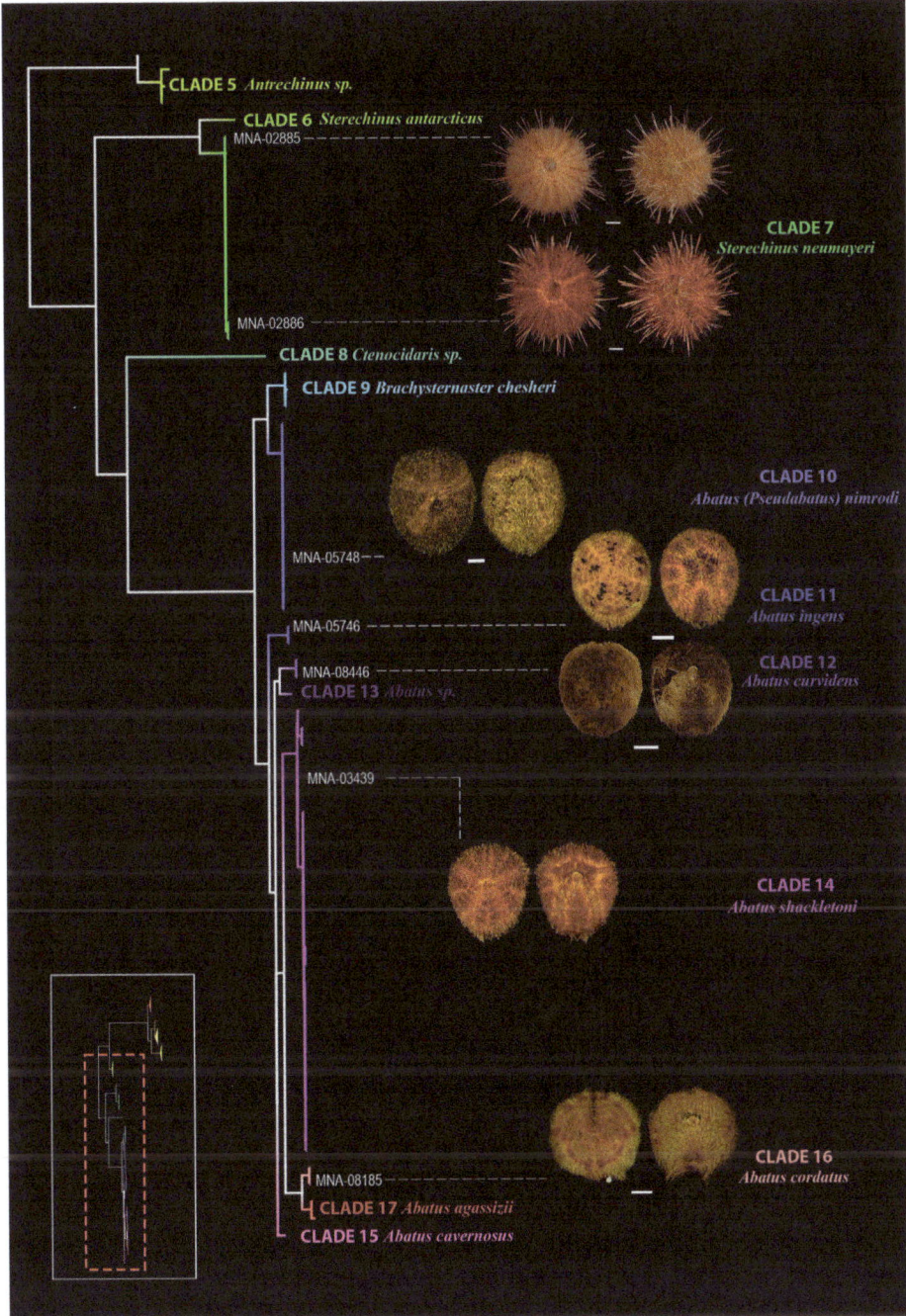

Figure 4. Representative photos of selected specimens (left—aboral view, right—oral view) from Terra Nova Bay. In the tree, the different species identified are highlighted by different colours (only the species from TNB are figured). Bottom left is the schematic view of the tree in Figure 2, the portion analysed in detail in the image is highlighted in red. Scale bar: 1 cm in grey.

3.3. Faunistic Inventory Revision

In our analysis no samples corresponding, morphologically or molecularly, to *A. koehleri*, a species identified with classical morphology by Chiantore et al., 2006 and reported in that publication with the old name of *A. elongatus* [38] were found.

This was unexpected and we thus cross-checked all the available materials present in the Italian National Antarctic Museum (MNA, Section of the Genoa) collections. Unfortunately, only a small amount of previously studied and published material has been later given to the museum, preventing a general in-depth re-evaluation. However, sample MNA-00573 was found to belong to the bulk of specimens published by Chiantore et al., 2006, and still reported an original identification label indicating it as *A. elongatus*. During our study, this same sample was successfully sequenced, morphologically reviewed bringing to an undoubted identification of *A. shackletoni* Koehler, 1911.

In the light of this result, we believe that the presence of *A. koehleri*, previously reported as *A. elongates* by Chiantore et al., 2006, in the Terra Nova Bay area has to be considered questionable. Hence this is the same for the published in the Southern Ocean Echinoid database (e.g., [33,34]) that are based on the same publication. This modifies the number of previous identified species from TNB area from four to three.

Overall, by combining molecular and morphological identifications, we found three more echinoids species, i.e., *A. cordatus*, *A. curvidens* and *A. ingens*, not previously reported in TNB, which bring the total number of echinoids present here to 6 species. The revised check list is given in Table 4, together with an updated depth range for the considered species (Figure 5).

Table 4. Faunistic inventory for echinoids and crinoids of Terra Nova Bay with updated information based on the present study.

Class	Family	Species	Depth Range (m)	Chaintore et al., 2006	This Work
Echinoidea	Echinidae	*S. neumayeri*	15–380	x	x
	Schizasteridae	*A. (Pseudabatus) nimrodi*	60–150	x	x
		A. cordatus	78–146		x
		A. curvidens	100		x
		A. ingens	148–150		x
		A. shackletoni	36–380	x	x
Crinoidea	Antedonidae	*A. adriani*	250–522		x
		F. mawsoni	522		x
		P. kerguelensis	137–525		x
	Notocrinidae	*N. virilis*	137–525		x

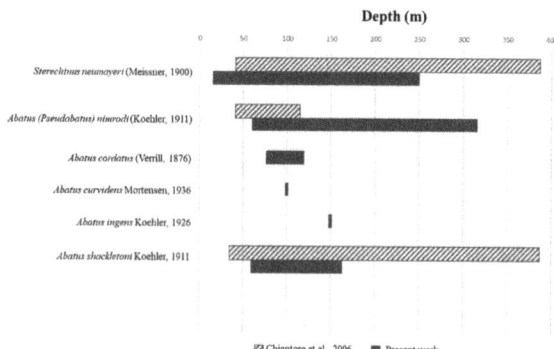

Figure 5. List of echinoids species found in Terra Nova Bay and updated depth range [38].

4. Discussion

The identification of species in an ecosystem is the first step in any ecological and conservation study, but even to provide a list of species for a given area is not an easy task. In recent years, one of the main hurdles in this kind of activity has become the chronic lack of experienced taxonomists. This fact, coupled to the generally time-consuming nature of morphological investigation, had the general effect of a significant slowdown of the duration of a given study.

The possible presence of cryptic species complexes represents another challenging aspect of biodiversity studies. These organisms are remarkably similar in appearance to other closely related species, resulting in them being virtually indistinguishable from the latter based on traditional morphological characters alone. Consequently, cryptic species are often overlooked, thus introducing serious biases in species richness estimates and in conservation efforts. Moreover, it is always possible that some of the unrecognized species in a sample collected in a newly studied area are new to science. This is a not a remote risk if it is considered that only 25% out of the 0.7 to 1.0 million marine species seems to have been described to date [65]. In terms of conservation efforts, this means that species could face extinction before they can be described [66].

A solution proposed to overcome all of the above issues relies on the use of molecular tools, such as DNA barcoding [3,19,57]. This method gained more attention in the last decade due to the increase in the speed of laboratory procedures, the availability of ad hoc software to better define species hypotheses and the high reproducibility of all these analyses.

The efficiency of identification through barcoding, however, depends on the quality and completeness of a reference database of sequences [67,68]. To this aim, projects such as the "The Barcode of Life" (BOLD) [69] directly and indirectly encourage large-scale molecular studies with a higher focus on quality. One of the main foundations of BOLD is the attention to voucher specimens that intrinsically provide opportunities for morphological and molecular studies using the same specimens as well as for subsequent cross-checking of identifications whenever disagreement is found. Another key point of a good reference library is the geographical coverage. In fact, data from few locations may only lead to an incomplete understanding of intraspecific genetic diversity, and a comprehensive DNA barcode library should include a broad range of species from as many locations as possible.

However, neither molecular nor morphological methods are sufficient per se for accurate taxonomy, and only the combined use of several methods provides a robust way to obtain a more precise estimation of species boundaries.

An ideal approach is, thus, that of "reverse taxonomy", where morphological analyses are performed after an initial molecular assessment [63,64].

In the case of echinoderm research, the application of "reverse taxonomy" has provided numerous benefits. Echinoderms, in fact, are known for their complex morphological features and challenging taxonomy, resulting in subjective and prone-to-misidentification species recognition [5]. This is especially true in cases where external morphology alone may not provide sufficient differentiation or when it is applied to large-scale sampling activities or to museum collections where thousands of samples have to be processed.

Application of this integrated approach also allows accurate identification of damaged specimens or larval stages, which can be challenging or simply impossible to determine at the specific level by using traditional taxonomy alone.

The Southern Ocean is not an exception to all these problems, and whenever this combined approach was applied, unexpected outcomes emerged and even apparently complicated taxonomical situations were resolved (e.g., [11,25]).

In this contribution, we applied a "classic" reverse taxonomy approach on two classes of echinoderms, i.e., echinoids and crinoids, in order to verify the number of species present in our samples and to test its usefulness in enhancing our understanding of echinoderm biodiversity.

Molecular data were fully resolved by current available algorithmic methods for species delineation applied to DNA barcodes. Although this test involved low sample sizes, it provided an estimate of the relative efficacies of OTU designation via DNA-based methods and external morphology, speeding up the process of species identification.

Since in this case limited or no previous knowledge was available for the considered group, if the initial phase of taxonomic work would have been based on a pre-sorting based on external morphological characters alone, this step would only have slowed down the whole process and several species would not have been recognized.

Overall, in this study, we contributed nearly 100 new sequences, including sequences from two species not yet available in public databases (i.e., *A. curvidens* and *A. cordatus*), enhancing our understanding of echinoid and crinoid biodiversity not only for the TNB area, but also for the Southern Ocean in general. This is a significant result when considering that echinoderms represent a considerable biomass in marine habitats and play a major role in the Antarctic marine ecosystems [35,70–76].

Species recognition was also possible for the six pentacrinoids larvae that were identified by comparing COI sequences against the reference sequences held in GenBank, confirming the importance of DNA barcodes in order to identify juvenile organisms and larval stages where morphological identification could be challenging.

Two putative new species (*Antrechinus* sp. and *Ctenocidaris* sp.), defined here based on COI sequences alone, need to be better characterized with integrated taxonomy to resolve their status. The sharing of this information may speed up comparisons with museum materials from other institutions, allowing, in the end, to formally assign a species name to these COI-based putative species.

The application of a reverse taxonomy approach has proven to be an efficient tool, even for checking the identity of old, already published, museum materials, highlighting the necessity of maintaining permanent repositories of scientific samples for future generations and comparisons.

This reinforces the pivotal role that museums play, not only as conservation centres for biological collections but also as hubs for information sharing. It is desirable that in the coming years, all available museum collections undergo molecular identifications to accurately assess species determinations and occurrences, and that all data are at the foundation of any monitoring activity.

Additionally, this work lays the foundations for future research on the diversity within the TNB area, now part of the Ross Sea Marine Protected Area, where a variety of monitoring activities are requested by the conservation measures of Annex 91-05/C [59].

Supplementary Materials: The following supporting information can be downloaded at: https://www.mdpi.com/article/10.3390/d15070875/s1. Supplementary Materials M1: COI sequence dataset produced in this study. Supplementary File S1: List of specimens analysed in this study and corresponding GenBank accession numbers for COI sequences; File S2: Species partition methods; File S3: Tree topology of the 92 samples analysed and species partition methods; File S4: Morphological description of identified species. The data presented in this study are openly available in GenBank (accession numbers: OR157781-OR157864).

Author Contributions: Data curation, A.G.; formal analysis, A.G.; funding acquisition, S.S.; investigation, A.G., M.C.A., M.C. and S.S.; resources, A.G., M.C.A., M.C. and S.S.; supervision, S.S.; validation, A.G.; visualization, A.G.; writing—original draft, A.G.; writing—review and editing, A.G., M.C.A., M.C. and S.S. All authors have read and agreed to the published version of the manuscript.

Funding: This research was funded by project "BAMBi" (Barcoding of Antarctic Marine Diversity; PNRA 2010/A1.10; PI Stefano Schiaparelli) and "TNB-CODE" (Terra Nova Bay barCODing and mEtabarcoding of Antarctic organisms from marine and limno-terrestrial environments; PNRA 16_00120; PI Stefano Schiaparelli). The Italian National Antarctic Museum (MNA), Genoa section, played an essential cooperation role with the projects funded for biological specimen repository and outreach activity. The authors are grateful to the Italian National Antarctic Museum (MNA) for the financial support.

Institutional Review Board Statement: All Italian sampling activities in Antarctica were authorized by the Italian National Antarctic Program (PNRA).

Data Availability Statement: In accordance with FAIR principles, the COI sequence dataset produced in this study (Supplementary Materials M1) can be found in the Supplementary Materials and is openly available in GenBank (https://www.ncbi.nlm.nih.gov/genbank/, accession numbers: OR157781-OR157864).

Acknowledgments: We are grateful to Julian Gutt (AWI) for the donation to MNA of the materials from the 2013 "Polarstern" ANT-XXIX/3 expedition. We also want to thank Santorio Mario and Paschini Elio who contributed to the collection of the material used for this study. This paper is an Italian contribution to the CCAMLR CONSERVATION MEASURE 91-05 (2016) for the Ross Sea region Marine Protected Area, specifically addressing the priorities of Annex 91-05/C. Many thanks are due to Josh Murray and Daniel Harvey who contributed to significant linguistic improvements. The authors are grateful to the two anonymous reviewers for their useful comments and suggestions on the manuscript.

Conflicts of Interest: The authors declare no conflict of interest. The funders had no role in the design of the study; in the collection, analyses, or interpretation of data; in the writing of the manuscript, or in the decision to publish the results.

References

1. Wilkinson, M.D.; Dumontier, M.; Aalbersberg, I.J.; Appleton, G.; Axton, M.; Baak, A.; Blomberg, N.; Boiten, J.-W.; da Silva Santos, L.B.; Bourne, P.E. The FAIR Guiding Principles for Scientific Data Management and Stewardship. *Sci. Data* **2016**, *3*, 160018. [CrossRef] [PubMed]
2. Rosenfeld, S.; Maturana, C.S.; Spencer, H.G.; Convey, P.; Saucède, T.; Brickle, P.; Bahamonde, F.; Jossart, Q.; Poulin, E.; Gonzalez-Wevar, C. Complete Distribution of the Genus Laevilitorina (Littorinimorpha: Littorinidae) in the Southern Hemisphere: Remarks and Natural History. *ZooKeys* **2022**, *1127*, 61–77. [CrossRef]
3. Hebert, P.D.; Cywinska, A.; Ball, S.L.; DeWaard, J.R. Biological Identifications through DNA Barcodes. *Proc. R. Soc. London. Ser. B Biol. Sci.* **2003**, *270*, 313–321. [CrossRef] [PubMed]
4. Carstens, B.C.; Pelletier, T.A.; Reid, N.M.; Satler, J.D. How to Fail at Species Delimitation. *Mol. Ecol.* **2013**, *22*, 4369–4383. [CrossRef] [PubMed]
5. Layton, K.K.; Corstorphine, E.A.; Hebert, P.D. Exploring Canadian Echinoderm Diversity through DNA Barcodes. *PLoS ONE* **2016**, *11*, e0166118. [CrossRef] [PubMed]
6. Wright, A.G.; Pérez-Portela, R.; Griffiths, C.L. Determining the Correct Identity of South African Marthasterias (Echinodermata: Asteroidea). *Afr. J. Mar. Sci.* **2016**, *38*, 443–455. [CrossRef]
7. Knott, K.E.; Ringvold, H.; Blicher, M.E. Morphological and Molecular Analysis of Henricia Gray, 1840 (Asteroidea: Echinodermata) from the Northern Atlantic Ocean. *Zool. J. Linn. Soc.* **2018**, *182*, 791–807. [CrossRef]
8. Peck, L.S.; Clark, M.S.; Dunn, N.I. Morphological Variation in Taxonomic Characters of the Antarctic Starfish Odontaster Validus. *Polar Biol.* **2018**, *41*, 2159–2165. [CrossRef]
9. Ringvold, H.; Moum, T. On the Genus Crossaster (Echinodermata: Asteroidea) and Its Distribution. *PLoS ONE* **2020**, *15*, e0227223.
10. Jossart, Q.; Kochzius, M.; Danis, B.; Saucède, T.; Moreau, C.V. Diversity of the Pterasteridae (Asteroidea) in the Southern Ocean: A Molecular and Morphological Approach. *Zool. J. Linn. Soc.* **2021**, *192*, 105–116. [CrossRef]
11. Guzzi, A.; Alvaro, M.C.; Danis, B.; Moreau, C.; Schiaparelli, S. Not All That Glitters Is Gold: Barcoding Effort Reveals Taxonomic Incongruences in Iconic Ross Sea Sea Stars. *Diversity* **2022**, *14*, 457. [CrossRef]
12. Jossart, Q.; Sands, C.J.; Sewell, M.A. Dwarf Brooder versus Giant Broadcaster: Combining Genetic and Reproductive Data to Unravel Cryptic Diversity in an Antarctic Brittle Star. *Heredity* **2019**, *123*, 622–633. [CrossRef] [PubMed]
13. Uthicke, S.; Byrne, M.; Conand, C. Genetic Barcoding of Commercial Bêche-de-Mer Species (Echinodermata: Holothuroidea). *Mol. Ecol. Resour.* **2010**, *10*, 634–646. [CrossRef] [PubMed]
14. Christiansen, H.; Dettai, A.; Heindler, F.M.; Collins, M.A.; Duhamel, G.; Hautecoeur, M.; Steinke, D.; Volckaert, F.A.; Van de Putte, A.P. Diversity of Mesopelagic Fishes in the Southern Ocean-A Phylogeographic Perspective Using DNA Barcoding. *Front. Ecol. Evol.* **2018**, *6*, 120. [CrossRef]
15. Hopkins, G.W.; Freckleton, R.P. Declines in the Numbers of Amateur and Professional Taxonomists: Implications for Conservation. *Anim. Conserv. Forum* **2002**, *5*, 245–249. [CrossRef]
16. Puillandre, N.; Lambert, A.; Brouillet, S.; Achaz, G. ABGD, Automatic Barcode Gap Discovery for Primary Species Delimitation. *Mol. Ecol.* **2012**, *21*, 1864–1877. [CrossRef]
17. Puillandre, N.; Modica, M.V.; Zhang, Y.; Sirovich, L.; Boisselier, M.-C.; Cruaud, C.; Holford, M.; Samadi, S. Large-Scale Species Delimitation Method for Hyperdiverse Groups. *Mol. Ecol.* **2012**, *21*, 2671–2691. [CrossRef]
18. Meiklejohn, K.A.; Wallman, J.F.; Dowton, M. DNA Barcoding Identifies All Immature Life Stages of a Forensically Important Flesh Fly (Diptera: Sarcophagidae). *J. Forensic Sci.* **2013**, *58*, 184–187. [CrossRef]

19. Ward, R.D.; Holmes, B.H.; O'HARA, T.D. DNA Barcoding Discriminates Echinoderm Species. *Mol. Ecol. Resour.* **2008**, *8*, 1202–1211. [CrossRef]
20. Bribiesca-Contreras, G.; Solís-Marín, F.A.; Laguarda-Figueras, A.; Zaldívar-Riverón, A. Identification of Echinoderms (Echinodermata) from an Anchialine Cave in C Ozumel I Sland, M Exico, Using DNA Barcodes. *Mol. Ecol. Resour.* **2013**, *13*, 1137–1145.
21. Moreau, C.; Danis, B.; Jossart, Q.; Eléaume, M.; Sands, C.; Achaz, G.; Agüera, A.; Saucède, T. Is Reproductive Strategy a Key Factor in Understanding the Evolutionary History of Southern Ocean Asteroidea (Echinodermata)? *Ecol. Evol.* **2019**, *9*, 8465–8478. [CrossRef]
22. Arbizu, P.M.; Khodami, S.; Stöhr, S.; Laakmann, S. Molecular Species Delimitation of Icelandic Brittle Stars (Ophiuroidea). *Pol. Polar Res.* **2014**, *35*, 243–260.
23. Boissin, E.; Hoareau, T.B.; Paulay, G.; Bruggemann, J.H. DNA Barcoding of Reef Brittle Stars (Ophiuroidea, Echinodermata) from the Southwestern Indian Ocean Evolutionary Hot Spot of Biodiversity. *Ecol. Evol.* **2017**, *7*, 11197–11203. [CrossRef] [PubMed]
24. Christodoulou, M.; O'Hara, T.; Hugall, A.F.; Khodami, S.; Rodrigues, C.F.; Hilario, A.; Vink, A.; Martinez Arbizu, P. Unexpected High Abyssal Ophiuroid Diversity in Polymetallic Nodule Fields of the Northeast Pacific Ocean and Implications for Conservation. *Biogeosciences* **2020**, *17*, 1845–1876. [CrossRef]
25. Janosik, A.M.; Mahon, A.R.; Halanych, K.M. Evolutionary History of Southern Ocean Odontaster Sea Star Species (Odontasteridae; Asteroidea). *Polar Biol.* **2011**, *34*, 575–586. [CrossRef]
26. Janosik, A.M.; Halanych, K.M. Unrecognized Antarctic Biodiversity: A Case Study of the Genus Odontaster (Odontasteridae; Asteroidea). *Integr. Comp. Biol.* **2010**, *50*, 981–992. [CrossRef]
27. Meyer, D.L.; Macurda, D.B. Adaptive Radiation of the Comatulid Crinoids. *Paleobiology* **1977**, *3*, 74–82. [CrossRef]
28. Smith, A.B.; Gale, A.S.; Monks, N.E. Sea-Level Change and Rock-Record Bias in the Cretaceous: A Problem for Extinction and Biodiversity Studies. *Paleobiology* **2001**, *27*, 241–253. [CrossRef]
29. Linse, K.; Walker, L.J.; Barnes, D.K. Biodiversity of Echinoids and Their Epibionts around the Scotia Arc, Antarctica. *Antarct. Sci.* **2008**, *20*, 227–244. [CrossRef]
30. Kier, P.M.; Grant, R.E. Echinoid Distribution and Habits, Key Largo Coral Reef Preserve, Florida. *Smithson. Misc. Collect.* **1965**, *149*, 1–68.
31. Nebelsick, J.H. Biodiversity of Shallow-Water Red Sea Echinoids: Implications for the Fossil Record. *J. Mar. Biol. Assoc. United Kingd.* **1996**, *76*, 185–194. [CrossRef]
32. Barnes, D.K.; Brockington, S. Zoobenthic Biodiversity, Biomass and Abundance at Adelaide Island, Antarctica. *Mar. Ecol. Prog. Ser.* **2003**, *249*, 145–155. [CrossRef]
33. David, B.; Choné, T.; Festeau, A.; Mooi, R.; De Ridder, C. Biodiversity of Antarctic Echinoids: A Comprehensive and Interactive Database. *Sci. Mar.* **2005**, *69*, 201–203. [CrossRef]
34. David, B.; Choné, T.; Mooi, R.; de Ridder, C. Antarctic Echinoidea. In *Synopsis of the Antarctic Benthos, Volume 10. Theses Zoologicae*; Wägele, J.W., Sieg, J., Eds.; Koeltz Scientific Books: Königstein, Germany, 2005; Volume 35, pp. 1–275.
35. Cerrano, C.; Bavestrello, G.; Calcinai, B.; Cattaneo-Vietti, R.; Sarà, A. Asteroids Eating Sponges from Tethys Bay, East Antarctica. *Antarct. Sci.* **2000**, *12*, 425–426. [CrossRef]
36. Chiantore, M.; Cattaneo-Vietti, R.; Elia, L.; Guidetti, M.; Antonini, M. Reproduction and Condition of the Scallop Adamussium Colbecki (Smith 1902), the Sea-Urchin Sterechinus Neumayeri (Meissner 1900) and the Sea-Star Odontaster Validus Koehler 1911 at Terra Nova Bay (Ross Sea): Different Strategies Related to Inter-Annual Variations in Food Availability. *Polar Biol.* **2002**, *25*, 251–255.
37. C-CAMLR-XXXV, Report of the Thirty-Fifth Meeting of the Scientific Committee, Hobart, Australia, 17–21 October, Annex 6, 3.2, 3.7-3.9. 2016. CCAMLR CONSERVATION MEASURE 91-05 (2016) for the Ross Sea Region Marine Protected Area, Specifically, Addressing the Priorities of Annex 91-05/C. 2016. Available online: Https://Www.Ccamlr.Org/En/System/Files/e-Sc-Xxxv.Pdf (accessed on 30 May 2023).
38. Chiantore, M.; Guidetti, M.; Cavallero, M.; De Domenico, F.; Albertelli, G.; Cattaneo-Vietti, R. Sea Urchins, Sea Stars and Brittle Stars from Terra Nova Bay (Ross Sea, Antarctica). *Polar Biol.* **2006**, *29*, 467–475. [CrossRef]
39. Eléaume, M.; Hemery, L.G.; Améziane, N.; Roux, M. Southern Ocean Crinoids. In *Biogeographic Atlas of the Southern Ocean*; The Scientific Committee on Antarctic Research: Cambridge, UK, 2014.
40. Ghiglione, C.; Alvaro, M.C.; Griffiths, H.J.; Linse, K.; Schiaparelli, S. Ross Sea Mollusca from the Latitudinal Gradient Program: R/V Italica 2004 Rauschert Dredge Samples. *ZooKeys* **2013**, 37–48.
41. Ghiglione, C.; Alvaro, M.C.; Cecchetto, M.; Canese, S.; Downey, R.; Guzzi, A.; Mazzoli, C.; Piazza, P.; Rapp, H.T.; Sarà, A. Porifera Collection of the Italian National Antarctic Museum (MNA), with an Updated Checklist from Terra Nova Bay (Ross Sea). *ZooKeys* **2018**, 137–156. [CrossRef]
42. Piazza, P.; Błażewicz-Paszkowycz, M.; Ghiglione, C.; Alvaro, M.C.; Schnabel, K.; Schiaparelli, S. Distributional Records of Ross Sea (Antarctica) Tanaidacea from Museum Samples Stored in the Collections of the Italian National Antarctic Museum (MNA) and the New Zealand National Institute of Water and Atmospheric Research (NIWA). *ZooKeys* **2014**, 49–60.
43. Selbmann, L.; Onofri, S.; Zucconi, L.; Isola, D.; Rottigni, M.; Ghiglione, C.; Piazza, P.; Alvaro, M.C.; Schiaparelli, S. Distributional Records of Antarctic Fungi Based on Strains Preserved in the Culture Collection of Fungi from Extreme Environments (CCFEE) Mycological Section Associated with the Italian National Antarctic Museum (MNA). *MycoKeys* **2015**, *10*, 57.

44. Cecchetto, M.; Alvaro, M.C.; Ghiglione, C.; Guzzi, A.; Mazzoli, C.; Piazza, P.; Schiaparelli, S. Distributional Records of Antarctic and Sub-Antarctic Ophiuroidea from Samples Curated at the Italian National Antarctic Museum (MNA): Check-List Update of the Group in the Terra Nova Bay Area (Ross Sea) and Launch of the MNA 3D Model 'Virtual Gallery'. *ZooKeys* **2017**, 61–79. [CrossRef] [PubMed]
45. Cecchetto, M.; Lombardi, C.; Canese, S.; Cocito, S.; Kuklinski, P.; Mazzoli, C.; Schiaparelli, S. The Bryozoa Collection of the Italian National Antarctic Museum, with an Updated Checklist from Terra Nova Bay, Ross Sea. *ZooKeys* **2019**, 1–22. [CrossRef]
46. Garlasché, G.; Karimullah, K.; Iakovenko, N.; Velasco-Castrillón, A.; Janko, K.; Guidetti, R.; Rebecchi, L.; Cecchetto, M.; Schiaparelli, S.; Jersabek, C.D. A Data Set on the Distribution of Rotifera in Antarctica. *Biogeogr./Ital. Biogeogr. Soc.* **2020**, *35*, 17–25. [CrossRef]
47. Bonello, G.; Grillo, M.; Cecchetto, M.; Giallain, M.; Granata, A.; Guglielmo, L.; Pane, L.; Schiaparelli, S. Distributional Records of Ross Sea (Antarctica) Planktic Copepoda from Bibliographic Data and Samples Curated at the Italian National Antarctic Museum (MNA): Checklist of Species Collected in the Ross Sea Sector from 1987 to 1995. *ZooKeys* **2020**, *969*, 1–22. [CrossRef] [PubMed]
48. Grillo, M.; Huettmann, F.; Guglielmo, L.; Schiaparelli, S. Three-Dimensional Quantification of Copepods Predictive Distributions in the Ross Sea: First Data Based on a Machine Learning Model Approach and Open Access (FAIR) Data. *Diversity* **2022**, *14*, 355. [CrossRef]
49. Koehler, R. Echinodemata Echinoidea Australasian Antarctic Expedition 1911–1914. *Sci. Rep. Ser. C Zool. Bot.* **1926**, *3*, 1–134.
50. Clark, A.H. *A Monograph of the Existing Crinoids*; US Government Printing Office: Washington, DC, USA, 1967.
51. Moore, R.C. *Treatise on Invertebrate Paleontology,(T) Echinodermata 2 (1-3),(T) Echinodermata*; Paleontological Institute, University of Kansas: Lawrence, KS, USA, 1978.
52. Speel, J.A.; Dearborn, J.H. Comatulid Crinoids from R/V Eltanin Cruises in the Southern Ocean. *Biol. Antarct. Seas XIII* **1983**, *38*, 1–60.
53. Leray, M.; Knowlton, N. Censusing Marine Eukaryotic Diversity in the Twenty-First Century. *Philos. Trans. R. Soc. B Biol. Sci.* **2016**, *371*, 20150331. [CrossRef] [PubMed]
54. Ronquist, F.; Huelsenbeck, J.P. MrBayes 3: Bayesian Phylogenetic Inference under Mixed Models. *Bioinformatics* **2003**, *19*, 1572–1574. [CrossRef] [PubMed]
55. Huelsenbeck, J.P.; Ronquist, F. MRBAYES: Bayesian Inference of Phylogenetic Trees. *Bioinformatics* **2001**, *17*, 754–755. [CrossRef] [PubMed]
56. Meyer, C.P.; Paulay, G. DNA Barcoding: Error Rates Based on Comprehensive Sampling. *PLoS Biol.* **2005**, *3*, e422. [CrossRef] [PubMed]
57. Hebert, P.D.; Ratnasingham, S.; De Waard, J.R. Barcoding Animal Life: Cytochrome c Oxidase Subunit 1 Divergences among Closely Related Species. *Proc. R. Soc. London. Ser. B Biol. Sci.* **2003**, *270*, S96–S99. [CrossRef] [PubMed]
58. Ratnasingham, S.; Hebert, P.D. A DNA-Based Registry for All Animal Species: The Barcode Index Number (BIN) System. *PLoS ONE* **2013**, *8*, e66213. [CrossRef] [PubMed]
59. Fujisawa, T.; Barraclough, T.G. Delimiting Species Using Single-Locus Data and the Generalized Mixed Yule Coalescent Approach: A Revised Method and Evaluation on Simulated Data Sets. *Syst. Biol.* **2013**, *62*, 707–724. [CrossRef] [PubMed]
60. Fujita, M.K.; Leaché, A.D.; Burbrink, F.T.; McGuire, J.A.; Moritz, C. Coalescent-Based Species Delimitation in an Integrative Taxonomy. *Trends Ecol. Evol.* **2012**, *27*, 480–488. [CrossRef] [PubMed]
61. Leavitt, S.D.; Moreau, C.S.; Thorsten Lumbsch, H. The Dynamic Discipline of Species Delimitation: Progress toward Effectively Recognizing Species Boundaries in Natural Populations. In *Recent Advances in Lichenology*; Springer: Berlin/Heidelberg, Germany, 2015; pp. 11–44.
62. Zhang, J.; Kapli, P.; Pavlidis, P.; Stamatakis, A. A General Species Delimitation Method with Applications to Phylogenetic Placements. *Bioinformatics* **2013**, *29*, 2869–2876. [CrossRef]
63. Markmann, M.; Tautz, D. Reverse Taxonomy: An Approach towards Determining the Diversity of Meiobenthic Organisms Based on Ribosomal RNA Signature Sequences. *Philos. Trans. R. Soc. B Biol. Sci.* **2005**, *360*, 1917–1924. [CrossRef]
64. Michaloudi, E.; Papakostas, S.; Stamou, G.; Neděla, V.; Tihlaříková, E.; Zhang, W.; Declerck, S.A. Reverse Taxonomy Applied to the Brachionus Calyciflorus Cryptic Species Complex: Morphometric Analysis Confirms Species Delimitations Revealed by Molecular Phylogenetic Analysis and Allows the (Re) Description of Four Species. *PLoS ONE* **2018**, *13*, e0203168. [CrossRef]
65. Appeltans, W.; Ahyong, S.T.; Anderson, G.; Angel, M.V.; Artois, T.; Bailly, N.; Bamber, R.; Barber, A.; Bartsch, I.; Berta, A. The Magnitude of Global Marine Species Diversity. *Curr. Biol.* **2012**, *22*, 2189–2202. [CrossRef]
66. Scheffers, B.R.; Joppa, L.N.; Pimm, S.L.; Laurance, W.F. What We Know and Don't Know about Earth's Missing Biodiversity. *Trends Ecol. Evol.* **2012**, *27*, 501–510. [CrossRef]
67. Ekrem, T.; Willassen, E.; Stur, E. A Comprehensive DNA Sequence Library Is Essential for Identification with DNA Barcodes. *Mol. Phylogenetics Evol.* **2007**, *43*, 530–542. [CrossRef]
68. Puillandre, N.; Strong, E.E.; Bouchet, P.; Boisselier, M.C.; Couloux, A.; Samadi, S. Identifying Gastropod Spawn from DNA Barcodes: Possible but Not yet Practicable. *Mol. Ecol. Resour.* **2009**, *9*, 1311–1321. [CrossRef]
69. Hajibabaei, M.; deWaard, J.R.; Ivanova, N.V.; Ratnasingham, S.; Dooh, R.T.; Kirk, S.L.; Mackie, P.M.; Hebert, P.D. Critical Factors for Assembling a High Volume of DNA Barcodes. *Philos. Trans. R. Soc. B Biol. Sci.* **2005**, *360*, 1959–1967. [CrossRef]

70. Dayton, P.K. Toward an Understanding of Community Resilience and the Potential Effects of Enrichments to the Benthos at McMurdo Sound, Antarctica. In Proceedings of the Colloquium on Conservation Problems in Antarctica, Blacksberg, VA, USA, 10–12 September 1971; pp. 81–96.
71. Dayton, P.K. Observations of growth, dispersal and population dynamics of some sponges in mcmurds sound, antarctica. *Colloq. Int. Du CNRS* **1979**, *291*, 271–282.
72. Dayton, P.K. Interdecadal Variation in an Antarctic Sponge and Its Predators from Oceanographic Climate Shifts. *Science* **1989**, *245*, 1484–1486. [CrossRef]
73. Dayton, P.K.; Robilliard, G.A.; Paine, R.T.; Dayton, L.B. Biological Accommodation in the Benthic Community at McMurdo Sound, Antarctica. *Ecol. Monogr.* **1974**, *44*, 105–128. [CrossRef]
74. Dearborn, J.H. Foods and Feeding Characteristics of Antarctic Asteroids and Ophiuroids. *Adapt. Within Antarct. Ecosyst.* **1977**, 293–326.
75. Dearborn, J.H.; Edwards, K.C. Analysis of Data on the Feeding Biology of Antarctic Sea Stars and Brittle Stars. *Antarct. J. United States* **1985**, *19*, 138–139.
76. McClintock, J.B. Investigation of the Relationship between Invertebrate Predation and Biochemical Composition, Energy Content, Spicule Armament and Toxicity of Benthic Sponges at McMurdo Sound, Antarctica. *Mar. Biol.* **1987**, *94*, 479–487. [CrossRef]

Disclaimer/Publisher's Note: The statements, opinions and data contained in all publications are solely those of the individual author(s) and contributor(s) and not of MDPI and/or the editor(s). MDPI and/or the editor(s) disclaim responsibility for any injury to people or property resulting from any ideas, methods, instructions or products referred to in the content.

Article

Diversity and Distribution of the Benthic Foraminifera on the Brunei Shelf (Northwest Borneo): Effect of Seawater Depth

Sulia Goeting [1,*], Huan Chiao Lee [2], László Kocsis [3], Claudia Baumgartner-Mora [1] and David J. Marshall [2]

[1] Faculté des géosciences et de l'environnement, Institut des sciences de la Terre, Université de Lausanne, Géopolis, CH 1015 Lausanne, Switzerland; claudia.baumgartner@unil.ch
[2] Faculty of Science, Environmental and Life Sciences, Universiti Brunei Darussalam, Jalan Tungku Link, Bandar Seri Begawan BE 1410, Brunei; 22m1413@ubd.edu.bn (H.C.L.); david.marshall@ubd.edu.bn (D.J.M.)
[3] Faculté des géosciences et de l'environnement, Institut des dynamiques de la surface terrestre, Université de Lausanne, Géopolis, CH 1015 Lausanne, Switzerland; laszlo.kocsis@unil.ch
* Correspondence: sulia.hajimohamedsalim@unil.ch

Abstract: The marine benthic diversity of the Palawan/North Borneo ecoregion is poorly known, despite its implied unique high species richness within the Coral Triangle. The present study investigated the diversity and distribution of benthic foraminifera on the Brunei shelf. The objectives were to determine the species composition of sediment samples collected from 11 sites, extending ~70 km from the Brunei coastline and along a depth gradient of 10–200 m. We retrieved a total of 99 species, belonging to 31 families and 56 genera, out of which 52 species represented new records for Brunei and probably the ecoregion. Using presence/absence data, analyses were also performed to compare species diversity patterns (species richness, occupancy, taxonomic distinctness) and species assemblage similarity across the sites. For further insight into the relationship between distribution and depth-associated environmental conditions, we undertook stable isotope analyses of selected species of Rotaliida, Miliolida, and Lagenida. Oxygen isotope values were positively correlated with depth and species distribution, confirming cooler temperatures at greater depth. The carbon isotope data revealed species differences relating to habitat and food source specificity and a biomineralization effect. Close to one-third of the species were recorded from single sites, and species richness and taxonomic distinctness increased with depth and were greatest at the second deepest site (144 m). Together, these findings suggest data underrepresentation of diversity, habitat disturbance in shallower water, and species specialization (adaptation) in deeper water. Importantly, assemblage similarity suggests the occurrence of at least three marine biotopes on the Brunei shelf (10–40 m, 40–150 m, and >150 m). This study contributes significantly to our understanding of the local and regional patterns of foraminiferal diversity and distribution.

Keywords: benthic foraminifera; Brunei shelf; stable isotopes; Indo-Pacific

1. Introduction

Foraminifera are single-celled protists that are common in marine sediments from intertidal to abyssal zones, continental marshes, and freshwater environments. Benthic foraminifera can be divided into two main groups: the larger benthic foraminifera (LBF), which have algal photosymbionts, and the smaller benthic foraminifera (SBF), which are suspension or detritus feeders. Both of these groups live in eutrophic to oligotrophic conditions in different environmental settings. Benthic foraminifera is a good indicator of environmental conditions and is used in applications such as biostratigraphy, palaeoceanographic and paleoclimatic reconstructions, environmental biomonitoring, and petroleum exploration e.g., [1–8]. The present study focused on documenting and understanding the distribution of the benthic foraminifera of Brunei. The most comprehensive biogeographical analysis suggests that Brunei belongs to the Palawan/North Borneo ecoregion of the Western Coral Triangle province of the Central Indo-Pacific realm [9]. Compared to other regions of the South China Sea (Singapore, Vietnam, Malaysia), the benthic fauna of this region is relatively poorly known [10]. Marine faunas recently studied for Northwest Borneo include reef corals, sponges, echinoderms, gastropods, and foraminifera [11–19].

Regionally, multiple studies have been conducted on modern benthic foraminifera of the South China Sea (and Southeast Asia). These include studies from the Challenger Expedition [20–23], from the International Indian Ocean Expedition [24], and more recently in Indonesia, Malaysia, the Philippines, and Brunei Darussalam [4,7,17–19,25–38]. Local benthic foraminiferal distributions are influenced by several parameters, such as seafloor substrate type and grain size, trophic conditions, temperature, availability of nutrients, and depth [2,4,5,7,33,36,39–42]. The first insight into the benthic foraminifera of Brunei was gained from the work of Ho (1971), who studied diversity and abundance variation along a transect spanning from the inner Brunei Bay to the middle Sungai Brunei estuarine system. Previous studies on the Brunei shelf benthic foraminiferal assemblages focused on coral reefs, wrecks, and local transects of the shallower waters (<60 m; [17–19]).

The Brunei shelf, from the coastline to the edge at 200 m depth, extends about 60 to 70 km and presents a significant gradient of environmental conditions with which benthic communities are associated. The seafloor of the shelf constitutes sandy patches of reefs and wrecks, scattered in extensive siliciclastic mud, deriving from river transportation into Brunei Bay [43]. The shelf waters are mesotrophic to eutrophic and have a relatively narrow photic zone [15,44–46]. However, the waters further offshore are not influenced to the same degree by muddy terrestrial input and are therefore much clearer. Reduced water turbidity covaries with seawater depth on the Brunei shelf and, thus, with greater distance from the coastline. Marine water temperature also typically covaries with depth [47,48]. Although benthic surveys of the Brunei shelf are routinely conducted by the oil and gas industry in line with environmental impact assessment requirements, these are mostly confined to drilling sites, and there is currently no published work documenting the effect of seawater depth and turbidity on the shelf benthic fauna, including the foraminifera.

The objectives of the study were: (1) to determine the benthic foraminiferal species composition on the Brunei shelf, (2) to compare diversity patterns (species richness, occupancy, taxonomic distinctness) across the sites varying in depth, and (3) to assess species assemblage similarity across the sites. Additionally, to gain insight into the relationship between foraminifera habitat and environmental parameters, stable carbon and oxygen isotope analyses were carried out on selected taxa.

2. Materials and Methods

2.1. Sample Collection and Species Identification

Seafloor sediment samples were collected across the Brunei shelf as part of a Marine Environmental Survey (MES) conducted under the auspices of an environmental consultancy company (Environmental Resources Management, ERM). The benthic foraminifera were, however, not resolved or reported in this survey. Of the samples collected from 30 stations in April 2013, only eleven sites were available for foraminiferal investigation (locations, coordinates, and depths are given in Figure 1 and Table 1). Figure 1 also includes study areas from previous studies by Ho [16] and Goeting et al. [17–19] which cover Brunei Bay and the transect samples M and T, respectively. The details of the transects, reefs, and wreck sites can be seen in Goeting et al. [17–19].

Sediment samples were collected using Van Veen Grab, with a sample surface area of 0.1 m^2. As a standard protocol for the company surveying the macrobenthic fauna, the sediments were preserved in formalin at sea and later washed and sieved (500 μm mesh) in the laboratory at Universiti Brunei Darussalam (UBD). Importantly, the sampling did not follow the typical approach used in foraminiferal studies and forewent determining sediment particle size distribution. Consequently, the foraminifera of the present study were resolved for a sediment size of only >500 μm. Furthermore, the processing of the material meant that foraminiferal abundance data were likely unreliable, and thus all analyses and interpretations are based on presence/absence data. The samples were investigated at UBD and the Université de Lausanne.

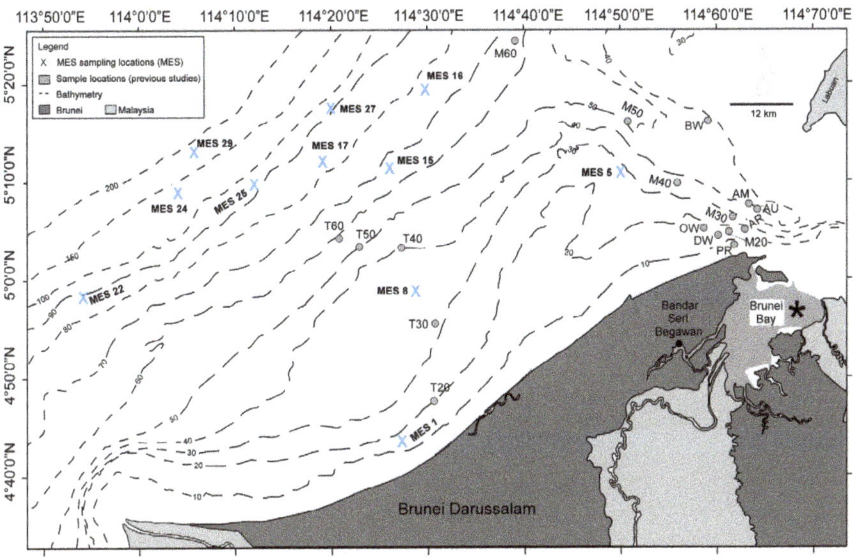

Figure 1. Study sites on the Brunei shelf for the current study of the MES sites (in bold) and previous studies by Ho [16] in Brunei Bay (*) and Goeting et al. [17–19]. Earlier studies focused on Brunei Bay and the reefs, wrecks, and transects mostly along the inner shelf and to depths of ~60 m, respectively. T20–60 (Tutong transect 20–60 mwd), M20–60 (Muara transect 20–60 mwd), PR (Pelong Rocks), AR (Abana Reef), OW (Oil Rig Wreck), DW (Dolphin Wreck), AM (American Wreck), AU (Australian Wreck), and BW (Bluewater Wreck).

Table 1. Details on the sampling sites and number of species present in each group from each site.

Sampling Station	Latitude	Longitude	Depth (m)	Tetulariida	Miliolida	Lagenida	Rotaliida
MES 1	4°44′30.52″ N	114°28′1.38″ E	14–15	4	11	0	7
MES 5	5°12′7.60″ N	114°51′0.73″ E	31	5	21	1	7
MES 8	4°59′56.80″ N	114°29′27.75″ E	34	3	18	1	8
MES 15	5°12′34.77″ N	114°26′44.96″ E	68	1	5	2	6
MES 16	5°20′42.72″ N	114°30′29.36″ E	77	2	3	2	5
MES 17	5°13′17.59″ N	114°19′42.50″ E	84	0	5	5	5
MES 22	4°59′18.18″ N	113°54′20.25″ E	93	5	3	8	8
MES 24	5°10′6.28″ N	114°4′22.84″ E	144	15	15	17	14
MES 25	5°10′56.63″ N	114°12′30.08″ E	94	1	12	7	11
MES 27	5°18′47.11″ N	114°20′32.54″ E	92	1	10	15	12
MES 29	5°14′21.78″ N	114°6′2.62″ E	190	2	0	2	5

Foraminifera specimens were picked from the samples using a LEICA EZ4 microscope and stored in small sample holders. Taxa identification was made to genus or species level with the aid of taxonomic references for the region based on Billman et al. [49], Loeblich and Tappan [50], Loeblich and Tappan [24], Jones [23], Parker [51], Debenay [52], Martin et al. [34], Förderer and Langer [33,37], Goeting et al. [19], Hohenegger [53], Renema [4], and Lei and Li [32]. Photographs of the benthic foraminifera were taken using a Keyence Digital Microscope VHX-7000. Plates of benthic foraminifera taxa that had not previously been recorded for Brunei were prepared using Adobe Photoshop CS6.

2.2. Species Richness, Distributions, and Assemblage Comparisons

From an Excel matrix comprising presence/absence data for the species at each study site, we assessed species richness in relation to depth. This was determined for the benthic foraminifera as a whole and separately for the four major groups (i.e., Textulariida, Rotaliida, Miliolida, and Lagenida). In addition to plotting the species richness at each site, we assessed the representativeness of the sampling of the real situation by plotting an occupancy frequency distribution (the number of species versus the number of sites). Because there are normally more common species than rare species, this relationship is usually slightly right-skewed, and a variation from this suggests an under-representation of the sampling [54].

Furthermore, we assessed the taxonomic (or phylogenetic) relatedness of the species at each site (assemblage) using a metric termed "taxonomic distinctness (Δ)". This is estimated from the path length (taxonomic distance) between two species organised in a taxonomic hierarchy (genus, family, order, and class; according to WORMS) to give a value of between 0 (same species) and 100 (different species), or at a more refined level, 0 = same species, 20 = different species in the same genera, 40 = different genera but same family, etc. (see Warwick and Clarke [55]). The average of the taxonomic distances of all species in the assemblage divided by the Simpson diversity index (Δo), gives the so-called average taxonomic distinctness (Δ^+), which was determined for each assemblage. Δ^+ is more informative of diversity than species richness, which does not account for the relatedness of the species in an assemblage.

Moreover, we assessed how the species assemblages across the Brunei shelf were related to one another. For this, we performed a Bray-Curtis similarity analysis based on presence or absence data, followed by a non-metric Multidimensional Scaling (nMDS) procedure (PRIMER ver. 6, Plymouth, UK). The nMDS produces a two-dimensional plot for each species assemblage, such that the distance between the assemblage points indicates the degree of similarity. Similar assemblages are plotted closer to one another, and more distantly related assemblages are further apart. This analysis also allows for post-hoc clustering to discriminate assemblage groups based on the level of similarity.

2.3. Stable Isotope Analyses

Three commonly occurring species were selected from Miliolida, Lagenida, and Rotaliida for stable carbon and oxygen isotope analyses (Supplementary Table S2). These taxa also cover the depth range of the MES sampling sites. Between ~80 and 200 μg of foraminifera test material were weighed in glass vials. Using a Gas Bench device, the closed vials were flushed with He, and then the test material was dissolved in phosphoric acid. The released CO_2 was analysed for stable isotopic compositions with a Thermo Delta-V mass spectrometer, in the stable isotope laboratory of the University of Lausanne (IDYST, UNIL). The procedure is described in Spötl and Vennemann [56]. The data are expressed in Permil (‰) in δ-notation, such as $\delta^{18}O$ and $\delta^{13}C$ relative to VPDB (Vienna Pee Dee Belemnite international reference standard). The analytical precision was better than ±0.1‰ standard deviation, determined from multiple analyses based on the laboratory's in-house standards. In addition, four filtered seawater samples were analysed from scuba diving sites for the stable isotope composition of oxygen and dissolved inorganic carbon (DIC) (Supplementary Table S3). The $\delta^{13}C_{DIC}$ was analysed using the same settings as used for the carbonates following the method of Spötl [57], while the $\delta^{18}O$ values (and δD) were obtained using a Picarro device, as described in Halder et al. [58]. The $\delta^{18}O$ values of the water data are expressed relative to VSMOW (Vienna Standard Mean Ocean Water).

3. Results

3.1. Species Richness and Distributions

A total of 99 species of benthic foraminifera belonging to 56 genera (40 hyaline, 37 porcelaneous, and 22 agglutinated forms) were identified (Table 1, Supplementary Table S1). Newly reported species found on the Brunei shelf are shown in Appendix A.

The species distributions of the four main taxonomic groups differed with respect to distance from the coast and, accordingly, with depth. Miliolida showed the greatest species richness closer to the coast, while Textulariida, Lagenida, and Rotaliida species were mainly distributed away from the coastline (Table 1). Textulariida (Table 1) were mainly distributed towards the outer shelf in MES 24 at 144 m water depth (mwd), where 15 species have been found before the numbers decrease at MES 29 at 190 mwd. MES 24 has the highest number of agglutinated species, including deeper species within this group. Miliolida and Rotaliida (Table 1) have similar distributions throughout the shelf, but Rotaliida has fewer species in shallower depths compared to Miliolida. Lagenida species are mainly distributed in abundance towards the outer shelf, including MES 27 and MES 24 (Table 1), which is opposite to the distribution patterns of Textulariida, Miliolida, and Rotaliida. The number of species within Lagenida increases with depth and is mainly concentrated within the middle to outer shelf areas (Table 1).

Collectively, species richness varied among the sites to a depth of 100 metres, then dramatically increased to the second most deep site (MES 24; 61 species; Figure 2a). The lowest species richness was found at the deepest site (MES 29; 9 species; Figure 2a). The occupancy frequency distribution was strongly left-skewed with most species ($n = 35$) found at only one site, and one species found at all the sites (*Pseudorotalia indopacifica*; Figure 2b, Supplementary Table S1).

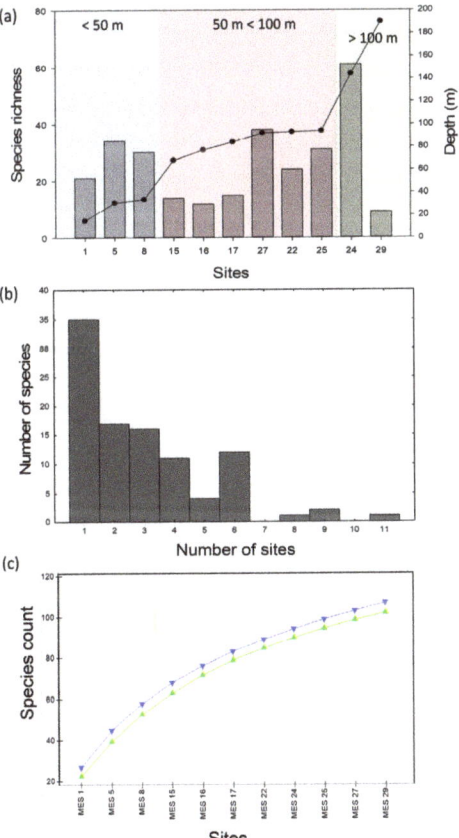

Figure 2. (a) Species richness for each MES site (bars). The secondary y-axis indicates site depth (filled circles) and colour coding defines broad arbitrary depth categories. (b) Occupancy frequency distribution. (c) Species accumulation (rarefaction) curve; upright triangles, MM, inverted triangles, UGE.

The phylogenetic relatedness of the species in the assemblages, indicated by average taxonomic distinctness (Δ^+), ranged between 76 and 89 (Figure 3). Greatest Δ^+ was observed at MES 24 (depth 144 m), and the least Δ^+ was observed in the shallowest and deepest water assemblages (MES 1, MES 5, MES 8, MES 29; Figure 3).

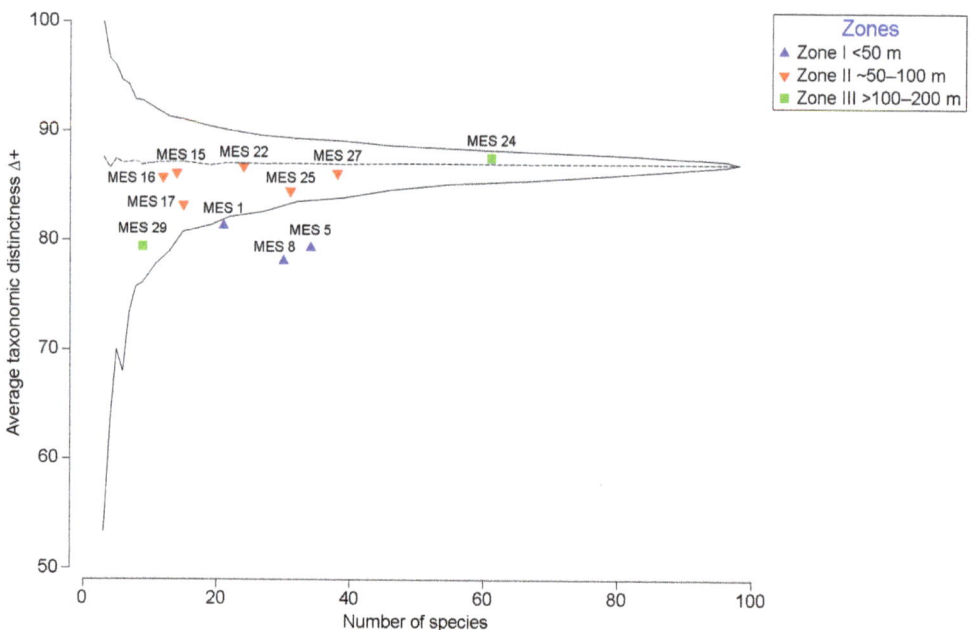

Figure 3. Average taxonomic distinctness (Δ^+) of species assemblage at each site plotted against a number of species. Colour codes refer to depth (see inset). The solid line refers to the limit within which 95% of simulated Δ^+ values lie, and the dashed line indicates mean Δ^+ (see Warwick and Clarke, 1995).

3.2. Species Assemblage Analysis

The nMDS analysis (Figure 4a) revealed clustering (no overlapping) of the species assemblages at 30 and 40% similarities. The 40% similarity level revealed five clusters, whereas the 30% similarity level was especially informative by showing three distinct clusters (groups I, II, and III; Figure 4a). The relationships between these clusters (i.e., assemblage group I being more closely related to group II and then to group III) coincide with their patterns of distribution in relation to seawater depth and distance from the coastline (Figure 4b). Group I consists of MES 1, 5, and 8, which share shallow water sites and are located on the inner shelf (0–50 m). Group II comprises assemblages MES 15, 16, 17, 22, 25, and 27 from the middle shelf zone (50–100 m), but also the deep-water site MES 24 (100–200 m; Figure 4a,b). Group III contains a single assemblage, MES 29, which is distantly related to the other assemblages and occurs in the deepest water sampled (190 m; Figure 4a,b).

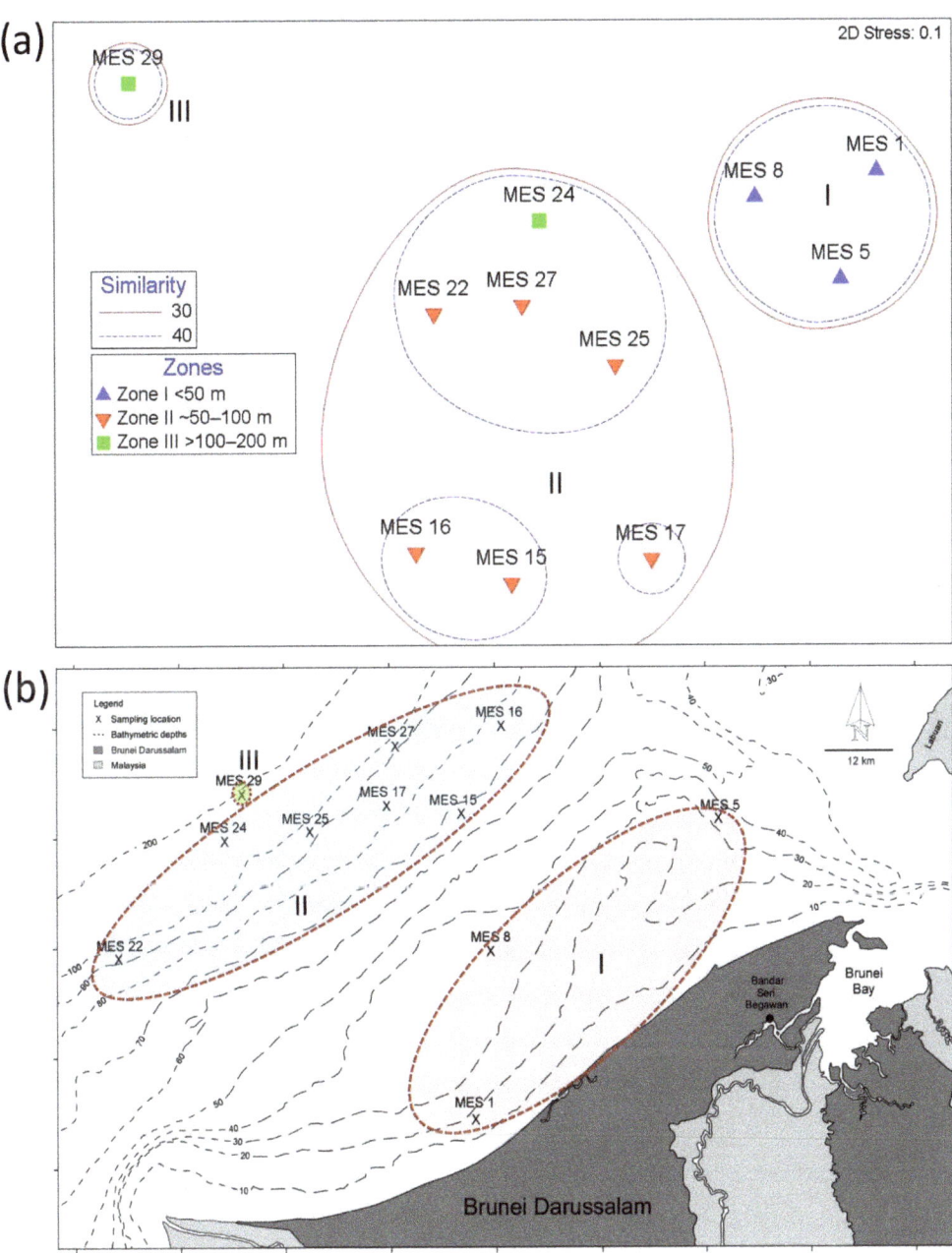

Figure 4. (a) nMDS showing clusters of the eleven foraminifera assemblages for two levels of similarity. Three clusters are formed at the 30% similarity level. Depth zones are indicated by different coloured symbols. (b) A bathymetric map showing that the nMDS pattern of assemblage clustering coincides with distribution relative to the shoreline and depth (i.e., assemblage cluster I is more closely related to II than to III).

3.3. Isotope Analyses

Altogether, 47 benthic foraminiferal samples were analysed. The $\delta^{18}O$ ranges from −3.62 to 0.70‰ (Figure 5), while the $\delta^{13}C$ varies between −1.18 and 1.57‰ (Figure 6). The full dataset and the taxon averages are listed in Supplementary Table S2. The seawater yielded an average $\delta^{18}O$ value of −0.64 ± 0.10‰, while the $\delta^{13}C_{DIC}$ has a mean value of −0.41 ± 0.19‰ in the depth range of 5–24.5 m (Supplementary Table S3).

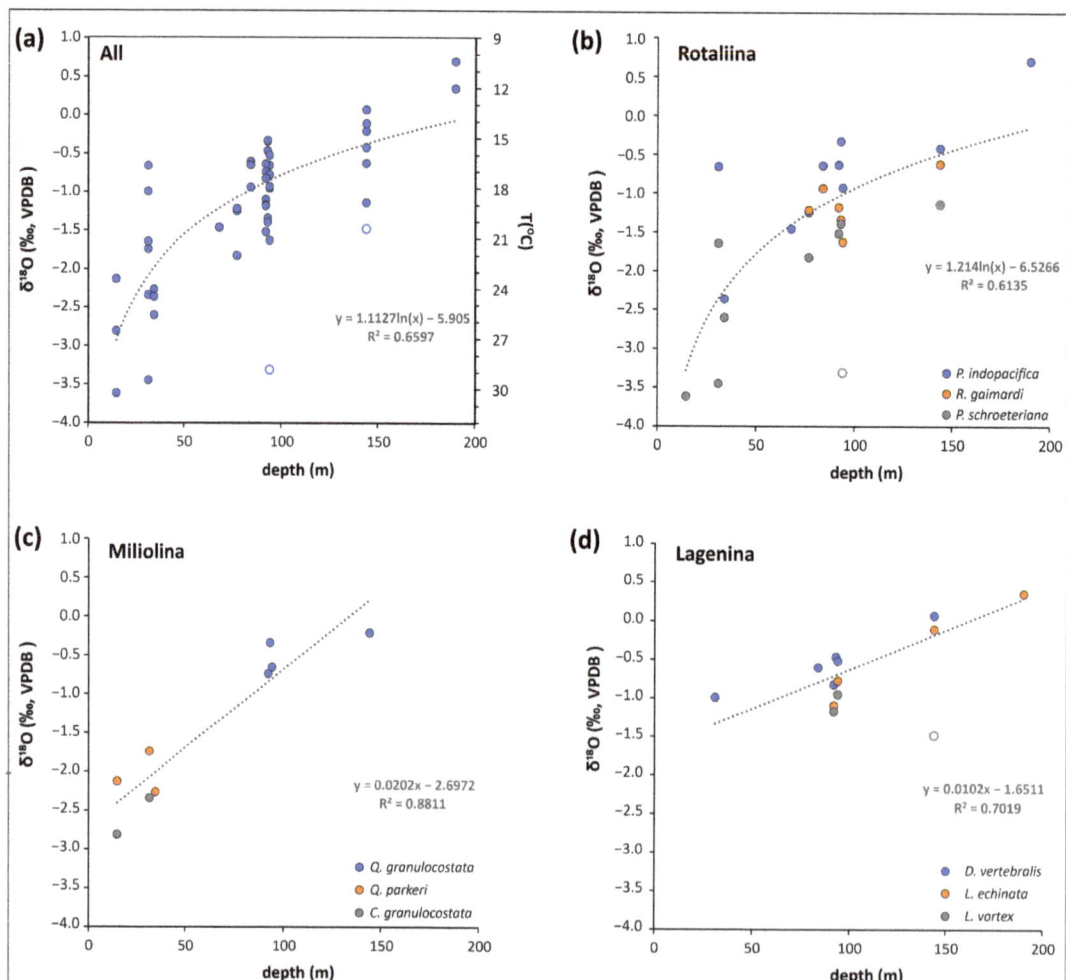

Figure 5. Oxygen isotope compositions of benthic foraminifera from Brunei as a function of their depth occurrence. (**a**) All the data and related regression lines. Note that the specimens with empty circles are not included in the best-fit function. These are possibly transported from shallower depths. The secondary y-axis indicates temperatures derived from Equation (1) at a $\delta^{18}O_{water}$ value of −0.64‰. (**b**–**d**) The $\delta^{18}O$ data according to different suborders and their given species. Note that for Miliolina and Lagenina the best fit is represented by a linear function, while for Rotaliina by logarithmic one.

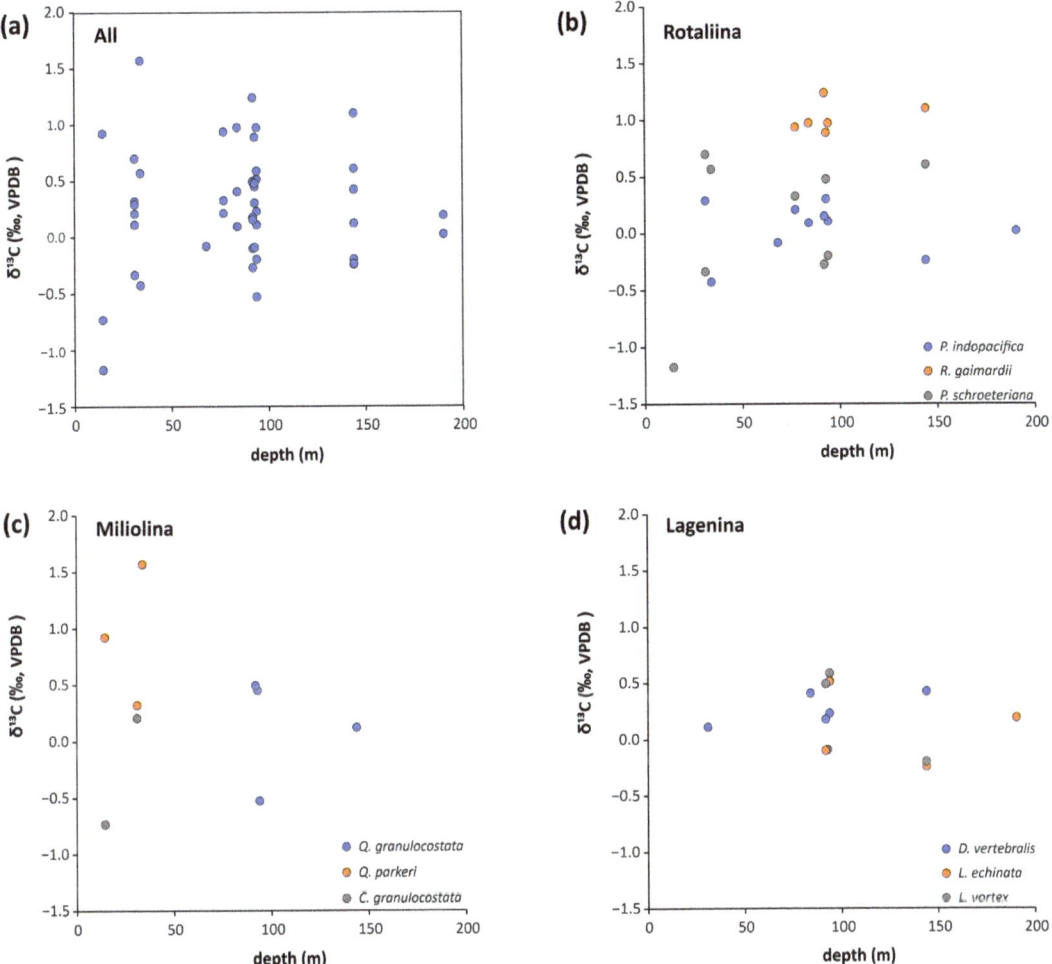

Figure 6. Carbon isotope compositions of benthic foraminifera from Brunei as a function of their depth occurrence. (**a**) All the data. Note the somewhat wider scatter at shallower depth (<50 m). (**b–d**) The $\delta^{13}C$ results according to the different suborders and their given species. Note that the rotaliid species of *R. gaimardii* and the miliolid *Q. parkeri* tend to have higher $\delta^{13}C$ values compared to other species within their respective group.

4. Discussion

4.1. Species Richness and Distributions

From Textulariida (Table 1), only one species is present (*Bigenerina nodosaria*) at MES 25 (Supplementary Table S1). This species is common at shelf settings from 60 mwd, which are also found within the region [27]. Then, starting at MES 24 (144 mwd), 15 species are present, together with the appearance of *Cyclammina subtrullissata, C. trullissata, Cribrostomoides subglobosus, Rhizammina algaeformis,* and *Tritaxilina caperata,* which are known to be usually found in deeper depths [24,27,52]. *Rhizammina algaeformis* is a deep-agglutinated species with a tubular form, and it is mainly found below 100 mwd [27]. Furthermore, the presence of *C. subtrullissata* and *C. trullissata* together with *T. caperata* is a significant marker of the outer shelf environments ([27]). Although *C. subglobosus* is a deeper species, it is also found within the inner shelf (see Goeting et al., [18]). This is most likely related to the sediment grain size as the inner shelf has finer sediments, a characteristic of deeper settings, and the availability of organic matter, which is their preference [5] from inputs of rivers to the Brunei shelf [18]. In the nearshore environment, the agglutinated foraminifera are more abundant, especially where higher clay content was detected [18]. The sediment grain size along the inner shelf varies from the northeast side (Muara region), which has a finer-grained seafloor compared to the southwest side (Tutong region) of the shelf [18]. *Textularia* cf. *agglutinans* (Supplementary Table S1) is a commonly found taxa in these sites, but it is also found in deeper waters [59] and is associated with muddy sand [38,60]. Agglutinated foraminifera are known to occur in salt marshes, marine shelf, bathyal zones, and abyssal zones [5] which are also found within the region especially the marine shelf, bathyal zones, and abyssal zones [24,27].

From Miliolida (Table 1), on the inner shelf, there is the presence of shallower, larger benthic miliolid species such as *Dendritina zhengae*, *Parasorites orbitolitoides*, *Peneroplis planatus*, and *Alveolinella quoyi* (Supplementary Table S1) that are mainly present in shallow waters in tropical seas [7,33,53,61]. The species *P. planatus* prefers shallow waters, clinging to small macroalgae [53]. The genus *Dendritina* is usually found avoiding shallow energetic waters and is hence mainly found in 10–40 mwd between sand grains. Furthermore, *A. quoyi* is found until 50 mwd and *P. orbitolitoides* until 60 mwd [53], but can occur much deeper at almost 80 mwd in Sesoko Island, Okinawa, Japan [62]. However, throughout the samples, the species *D. zhengae* is more common compared to the other larger benthic species, which are present but most likely concentrated in reef environments. Interestingly, the presence of *P. orbitolitoides* is usually found in oligotrophic settings [53], but it seems to be living in different trophic conditions [7,33], especially here in the shallower depths of the Brunei shelf which is known in previous studies [17,18]. Furthermore, towards the outer shelf zone, *Pyrgo vespertilio* and *Triloculina vespertilio* appear (Supplementary Table S1), which are also common species found within the Indo-Pacific region [24,33,52]. However, they are found at a much shallower depth from these studies which could be related to the transport of shallower miliolid species towards deeper depths. The smaller miliolids are mainly distributed towards the transect samples rather than at the reef and wreck sites [18], therefore being mainly distributed towards soft sediment substrate rather than at sites with harder substrate. The larger miliolid species are mainly concentrated in the reef sites in Brunei [17]. Miliolids are common in shallow environments [52] although they can still be found in greater depths [27], but they are mainly concentrated in shelf settings.

From Lagenida (Figure 2a), there are more lagenid species found compared to the previous study by Goeting et al. [18], where there is a higher diversity of this taxon, especially towards the outer shelf zone (>60 mwd). This is also the case in the Sunda shelf [27] and the Sahul shelf [24], where species such as *Marginulinopsis philippinensis*, *Saracenaria angularis*, and *Spincterules compressus* are also found and mainly distributed from the inner shelf to upper bathyal from 140 to 270 mwd. These species are reported for the first time in the Brunei shelf starting at MES 27 (92 mwd) until MES 29 (190 mwd), which is shallower than the usual distribution in the region, but since the samples are sieved through 500 micrometres, there is a possibility of loss of lagenid species. Lagenids are infaunal and usually occur in deeper marine environments below the photic zone (50 m) [28,63], feeding on the available organic matter within the sediment [18]. They are also found in sites with higher clay content, which is seen in one of the nearby wreck sites, (Dolphin wreck) DW [18], but does not have a higher species richness compared to lageninds found in much deeper depth within the MES sites below 60 mwd. Hence, there could be more lagenid species in the deeper setting that have yet to be reported on the Brunei shelf.

From Rotaliida, there are newly reported species within this group, such as *Neoassilina discoidalis*, *Operculina complanata*, and *Amphistegina papillosa* from the LBF group (Figure A8), while from the SBF group are *Uvigerina schwageri* and *Paracibicides edomica* (Figure A8). *N. discoidalis* and *O. complanata* can occur until depths of 100 to 120 mwd, respectively [53], but in the West Pacific [61], *O. complanata* has a deeper distribution compared to *N. ammonoides*; hence, these species can be seen still present in deeper sites until around

140 mwd (MES 24). Compared to the study by Goeting et al. [18], the only operculinid species found throughout the inner shelf is *N. ammonoides*, which is also found in the MES sites. This species is also commonly found in the shelf environment on the Sunda shelf [27], New Caledonia [52], and Sahul shelf up to 140 mwd [24]. However, the presence of more operculinid species suggests that the environment is suitable for their living conditions, as they prefer light for their symbionts and less energetic environments [5,61,64]. Therefore, they occur in deeper environments when terrestrial input is reduced, resulting in enhanced water clarity and hence deeper light penetration through the water column.

Uvigerina schwageri and *P. edomica* are present in the mid-to-outer shelf zone from MES 22 to MES 29 (92 to 190 mwd) (Supplementary Table S1). *P. edomica* is mainly distributed in deeper depths in the Sahul shelf at around 80 mwd [24] and from inner to upper bathyal in the Sunda shelf [27]. However, *U. schwageri* is found slightly deeper in the Sahul shelf at around 100–150 mwd [24] and is mainly distributed on the outer shelf in the Sunda shelf [27]. Both of these smaller benthic species are marker species towards deeper marine shelf environments.

In Goeting et al. [18], the rotaliids are mainly distributed around sunken wrecks and nearby reefal localities, and some have increased numbers along the Tutong transect. The reef sites are concentrated with LBF species [17], which are not commonly found in the MES sites except for *A. papillosa* at MES 1 and MES 5 (14–30 mwd) and the operculinids (Supplementary Table S1). Furthermore, *A. pulchella* is also found in the MES sites in addition to previous studies [18] until 144 mwd (MES 24) (Supplementary Table S1) but was also reported from the Sunda shelf within 100 mwd [27]. Two other species recovered in the MES sites, *Pseudorotalia indopacifica* and *P. schroeteriana*, are common within the inner shelf zone [18]. *P. indopacifica* is distributed deeper compared to *P. schroeteriana*, which is similar to what Gallagher et al. [65] observed in the Sunda shelf [27], such as *P. indopacifica* here, which is found up to 226 mwd, while *P. schroeteriana* is found up to 166 mwd.

A crucial finding of this study was the general pattern of increasing species richness with increasing depth along the Brunei shelf such that the greatest richness was observed at a depth of 144 m (MES 24; Figure 2a). There are several possible reasons for this pattern, including (1) greater habitat disturbance in the shallower waters; (2) greater time (ecological and evolutionary) for community establishment in the deeper water; (3) the introduction of novel deep water specialist taxa; and (4) a sampling artefact. Despite the less-than-perfect sampling protocol, the latter seems least likely, especially when considering the tendency towards flattening the species-accumulation curve (Figure 2c). Such a pattern suggests that the sampling is representative of the overall foraminiferal species richness (for the 500 μm size fraction) for the Brunei shelf. However, the occupancy analysis suggests that site-specific species richness might not be very well reflected by the sampling protocol (Figure 2b). Normally, occupancy frequency distributions tend towards being right-skewed unimodal curves, which indicate relatively fewer rare species, in contrast to the left-skewing curve observed for our data. This pattern might be explained by an inappropriate grain size (sampling unit or number of samples for each site) necessary to better capture site-specific species richness [54]. Nonetheless, the data are still highly interpretable when considering the variation in taxonomic distinctness across the sites and site assemblage groupings relative to the broad depth categories. A reduced taxonomic distinctness is commonly associated with habitat disturbance [55]. This was observed for the shallower water sites (MES 1, 5, and 8; Figure 3) and may be a function of natural instability of the seafloor (ground swells) or anthropogenic disturbance (fishing, oil drilling and oil rig deconstruction) in the shallower waters of the shelf. Greater taxonomic distinctness usually implies less disturbance and/or greater niche specialization (adaptation). In the present study, this may refer to greater colonization of deeper water by foraminiferal species, especially at MES 24. The details of the species contributing to this taxonomic distinctiveness across sites are given in the above discussion, and in summary, they suggest an introduction into the deeper water assemblages of agglutinated and lagenid foraminifera species (Table 1).

4.2. Species Assemblage Analysis

The most useful information derived from the nMDS was the formation of three assemblage clusters sharing 30% of their species (Groups I, II, and III; Figure 4a). Furthermore, the relationships of the assemblages in each group and the relationships between the groups conform with site depth. For example, Group I contained assemblages on the inner shelf (0–50 m), Group II contained assemblages between 50 and 100 m but also a deepwater assemblage (144 m, MES 24), and Group III contained the single deepest assemblage (190 m, MES 29). Additionally, the assemblages in Group I were more similar to those of Group II (slightly deeper) than to those of Group III (deepest).

Although we refer to depth, the distributions of species within assemblages must respond to a range of environmental factors that covary with distance from the shore and water depth, including greater water clarity (reduced suspended sediment effect), cooler temperature, and more stable oceanic salinities and pH in the deeper water [48,66]. Additionally, the seafloor and sediment properties must influence species distributions along the depth gradient of the Brunei shelf. The observed pattern of foraminiferal assemblage distributions in relation to inferred seawater parameters and distance from the shore bears similarity to that seen along the coastline of China [67]. Because foraminifera are good ecological indicators, our findings suggest that similar patterns may occur in other marine benthic taxa (e.g., polychaetes, amphipods). Importantly, these findings represent the first indication of three distinct possible marine biotopes on the Brunei shelf, which has far-reaching implications for future marine environmental conservation and resource management.

4.3. Stable Isotope Analyses

The $\delta^{18}O$ values of the benthic foraminiferal tests show a distinct decreasing trend from deeper to shallower depth. There are few clear outliers that can be explained by the downward transport of the foraminifera test, while the larger scatter of the dataset could be related to the bulk sediment sampling resulting in some temporal mixing. As the oxygen isotopic composition of the biogenic calcite is a function of the ambient water temperature and water isotopic composition (e.g., [68]), the decreasing trend is best explained by the anticipated colder temperature at greater depth than that of shallower setting. However, in the coastal region, freshwater runoff may also induce variation in the water isotopic composition ($\delta^{18}O_{water}$). The classic paleotemperature equation [68,69] is as follows:

$$T(°C) = 15.75 - 4.3 \times (\delta^{18}O_{carb\text{-}VPDB} - \delta^{18}O_{water\text{-}VSMOW}) - 0.14 \times (\delta^{18}O_{carb\text{-}VPDB} - \delta^{18}O_{water\text{-}VSMOW})^2 \qquad (1)$$

This equation can be used to obtain quantitative temperature data. Based on our foraminifera oxygen isotope results and using our average $\delta^{18}O_{water}$ value of −0.64‰, a temperature range of 10.2 to 29.8 °C can be calculated (Figure 5a). The array fits well with our observation in a shallow water environment where the top 20–30 metres of the seawater is 29–31 °C during July–September (southwest monsoon), meanwhile the temperature drops to as low as 24–25 °C during December–March (northeast monsoon). The deep-water temperature data are slightly lower than reported for the South China Sea [70]. However, our deepest water oxygen isotope data comes from only 24.5 mwd, and the deeper water body could have higher $\delta^{18}O$ composition as it is less affected by the continental runoffs of the tropical coastline. Using the global average seawater value of 0‰ would result in ~13 °C at a depth of 190 metres (MES29).

As a third factor, species-related isotope offset from the equilibrium precipitation of the tests might as well be expected (i.e., vital effect, see [71]). When the $\delta^{18}O$ data are separated based on the different taxa, the obtained isotope data versus depth relationships are strong and seem independent of the species (Figure 5b–d). For Rotaliina, there is a larger scatter, especially in the shallower environment (<50 m), that could be a result of greater variation in $\delta^{18}O_{water}$. Nevertheless, the obtained relationships may as well be used to estimate paleo-depth from analyses of fossil foraminifera of the same size range derived from shallow water deposits in northern Borneo.

The carbon isotope composition of marine carbonates is generally related to that of the dissolved inorganic carbon (DIC) in seawater (e.g., [71,72]), which is dominated by the bicarbonate ion. The average $\delta^{13}C_{DIC}$ for the ocean is 0–1‰, but on the water surface, higher values can be expected due to photosynthesis, while bottom water $\delta^{13}C_{DIC}$ is often lower due to organic matter degradation, especially in shelf areas. In shallow coastal settings, continental input may also induce some variation in the DIC composition. The fractionation between marine carbonate and DIC is small and relatively insensitive to temperature, but marine carbonates in equilibrium with DIC generally have 1 to 2 permil higher values than seawater DIC (e.g., [73]). The $\delta^{13}C$ composition of foraminiferal tests also depends on the habitat of the given taxa (epi- vs. infaunal, symbiont-bearing taxa, etc.), and some are also known about their vital effect on the isotopic composition when growing their test (e.g., [71]). Generally, the $\delta^{13}C$ results from the benthic foraminiferal tests and seawater fit well with the described carbon dynamics and habitat conditions in the shelf areas of Brunei. The overall data scatters around an average of 0.27‰, which variation is larger in shallower depth (<50 m) (Supplementary Table S2, Figure 6a). This latter may be explained by higher amounts and various nutrient sources from the land. When the data are grouped based on the different taxa, some further details can be detected. Lagenids have the most restricted variation in $\delta^{13}C$ with no differences among the investigated three species (0.19 ± 0.28‰, n = 13) (Figure 6d). Among the three rotaliids, the two *Pseudorotalia* species vary more than that of the lagenids, but on average around 0‰. However, the species *Rotalinoides gaimardii* yielded significantly higher $\delta^{13}C$ values of 1.02 ± 0.13‰ (n = 6) (post hoc Tukey test $p < 0.05$) (Figure 6b). The miliolids are the ones that scatter the most in their $\delta^{13}C$ (Supplementary Table S2, Figure 6c), which may indicate access to and alternate use of different carbon sources. Interestingly, the shallow-water species *Q. parkeri* tend to have higher, more positive $\delta^{13}C$ composition when compared to the co-occurring *C. granulocostata*, which yielded negative values too (Figure 6c). Though only few specimens have been analysed, the observed differences within the rotaliid and miliolid species could indicate different habitats and/or use of different food sources [6], but alternatively could also reflect a different way of biomineralization (i.e., vital effect). A wider scale of study together with in vitro growth of foraminifera could help further studies in these scenarios.

5. Conclusions

This study shows an increase in benthic foraminiferal (>500 microns) species richness in deeper waters of the Brunei shelf. While the study on the MES sites revealed unreported species from inner to outer shelf zones, the additional data from previous studies of the Brunei shelf have given a complete view of the changes in foraminiferal assemblages throughout the shelf, which are related to depth, distance from the shore, substrate type, and light intensity.

Species richness increases from 14 to 144 m of water depth (mwd) but drastically decreases at 190 mwd, suggesting a shift towards deeper assemblages. $\delta^{18}O$ results from selected species support the expected temperature variation with depth, but coastal sites may be influenced by continental runoffs. $\delta^{13}C$ data of certain rotaliid and miliolid species show variations reflecting specialized habitats, but this separation is not observed in the investigated lagenid species.

Supplementary Materials: The following supporting information can be downloaded at: https://www.mdpi.com/article/10.3390/d15080937/s1.

Author Contributions: Methodology, S.G., H.C.L., L.K. and D.J.M.; Software, H.C.L., L.K. and D.J.M.; Formal analysis, S.G., L.K. and D.J.M.; Investigation, S.G., H.C.L., L.K. and D.J.M.; Resources, L.K., C.B.-M. and D.J.M.; Writing—original draft, S.G.; Writing—review & editing, S.G., L.K., C.B.-M. and D.J.M.; Supervision, D.J.M. All authors have read and agreed to the published version of the manuscript.

Funding: This research was funded by a Universiti Brunei Darussalam research grant [UBD/RSCH/1.4/FICBF(b)/2021/033] and the Swiss Government Excellence Scholarship. In addition, the APC was funded by Université de Lausanne.

Institutional Review Board Statement: Not applicable.

Data Availability Statement: Data is contained within the article or supplementary material. The data presented in this study are available in [supplementary Table S1].

Acknowledgments: We thank David J. Lane as he helped in the initial organising and sorting of samples provided by Environmental Resources Management (ERM), as part of a marine benthic consultancy. Furthermore, we are grateful to the Swiss Government and University of Lausanne in providing all the materials and equipment required for this research.

Conflicts of Interest: The authors declare no conflict of interest.

Appendix A

Figures A1–A8 and explanation of plates. The scale bar is at 100 µm unless specified.

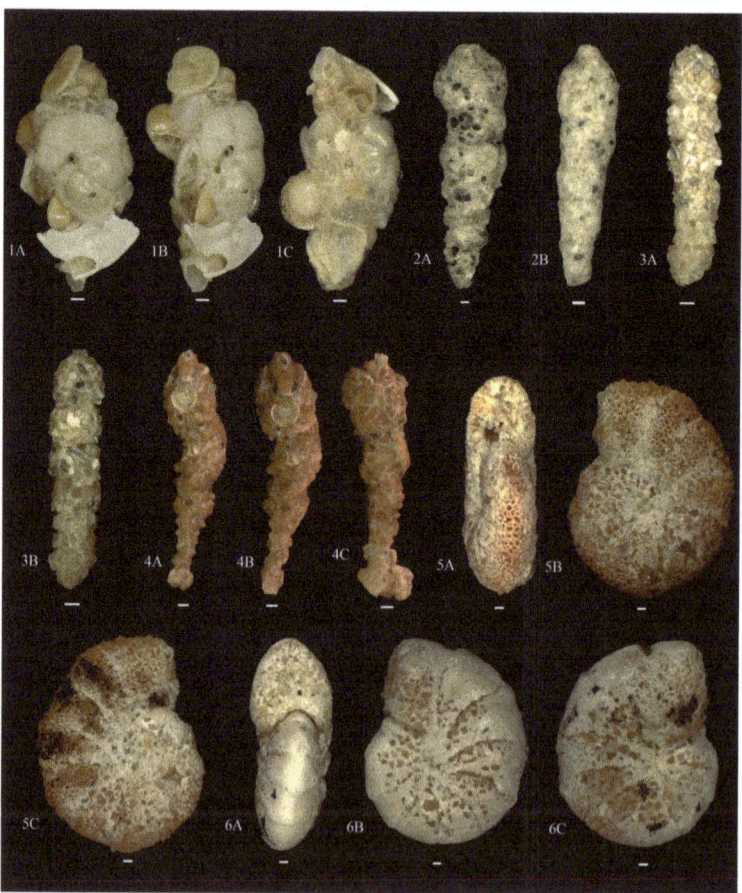

Figure A1. 1A–C. *Rhizammina algaeformis*; 2A,B. *Reophax* cf. *agglutinans*; 3A,B. *Reophax communis*; 4A–C. *Reophax scorpiurus*; 5A–C. *Cyclammina subtrullisata*; 6A–C. *Cyclammina trullisata*.

Figure A2. 1A–C. *Tritaxilina caperata*; 2A,B. *Cylindroclavulina* sp.; 3A,B. *Bigenerina agglutinans*; 4. *Bigenerina* cf. *agglutinans*?; 5A,B. *Bigenerina* sp.; 6. *Textularia goessi*; 7A,B. *Textularia porrecta*; 8A,B. *Textularia pseudogramen*; 9A,B. *Textularia stricta*.

Figure A3. 1A,B. *Planispirinella exigua*/sp.; 2A–C. *Spiroloculina subimpressa*; 3A,B. *Spiroloculina* sp.1; 4A–C. *Lachlanella* sp.; 5A–C. *Quinqueloculina* cf. *Bicarinata*; 6A–C. *Quinqueloculina granulocostata*.

Figure A4. 1A–C. *Quinqueloculina* sp. 2; 2A–C. *Flintina (Triloculina) bradyana*; 3A–C. *Pyrgo vespertilio*; 4A–C. *Triloculina vespertilio*.

Figure A5. 1A–H. *Dendritina zhengae*; 2A,B. *Parasorites orbitolitoides*.

Figure A6. 1. *Dentalina catelunata*; 2A,B. *Dentalina vertebralis*; 3A,B. *Dentalina* sp. 1; 4. *Dentalina* sp. 2; 5. *Nodosaria pyrula*; 6A,B. *Laevidentalina californica*; 7A,B. *Pyramidulina catesbyi*; 8A,B. *Pyramidulina obliquatus*; 9A,B. *Pyramidulina* sp. 1; 10A,B. *Lenticulina echinata*; 11A,B. *Lenticulina limbosa*; 12A,B. *Lenticulina vortex*.

Figure A7. 1A–F. *Marginulinopsis philippinensis*; 2A,B. *Saracenaria angularis*; 3. *Spincterules compressus*; 4. *Amphicoryna sublineata*; 5A,B. *Planularia gemmata*; 6A,B. *Lagena annulatacollare*?; 7. *Lagena spicata*.

Figure A8. 1A,B. *Uvigerina schwageri*; 2A–C. *Paracibicides edomica*; 3A–C. *Amphistegina papillosa*; 4A–C. *Operculina complanata*; 5A,B. *Neoassilina discoidalis*.

References

1. Phleger, F. *Ecology and Distribution of Recent Foraminifera*; Johns Hopkins Press: Baltimore, MA, USA, 1960.
2. Murray, J.W. *Ecology and Palaeoecology of Benthic Foraminifera*; John Wiley and Sons Inc.: New York, NY, USA, 1991; pp. 1–341.

3. Hayward, B.; Hollis, C. Brackish Foraminifera in New Zealand: A Taxonomic and Ecologic Review. *Micropaleontology* **1994**, *40*, 185–222. [CrossRef]
4. Renema, W. *Larger Foraminifera as Marine Environmental Indicators*; Nationaal Natuurhistorisch Museum: Leiden, The Netherlands, 2002; p. 263.
5. Murray, J.W. *Ecology and Applications of Benthic Foraminifera*; Cambridge University Press: Cambridge, UK, 2006; pp. 1–426.
6. Jones, R.W. *Foraminifera and Their Applications*; Cambridge University Press: Cambridge, UK, 2014; p. 391.
7. Renema, W. Terrestrial influence as a key driver of spatial variability in large benthic foraminiferal assemblage composition in the Central Indo-Pacific. *Earth-Sci. Rev.* **2018**, *177*, 514–544. [CrossRef]
8. Suriadi, R.; Shaari, H.; Culver, S.; Husain, M.; Vijayan, V.; Parham, P.; Sapon, N. Inner Shelf Benthic Foraminifera of the South China Sea, East Coast Peninsular Malaysia. *J. Foraminifer. Res.* **2019**, *49*, 11–28. [CrossRef]
9. Spalding, M.D.; Fox, H.E.; Allen, G.R.; Davidson, N.; Ferdaña, Z.A.; Finlayson, M.; Halpern, B.S.; Jorge, M.A.; Lombana, A.; Lourie, S.A.; et al. Marine Ecoregions of the World: A Bioregionalization of Coastal and Shelf Areas. *BioScience* **2007**, *57*, 573–583. [CrossRef]
10. Li, J.; Asner, G.P. Global analysis of benthic complexity in shallow coral reefs. *Environ. Res. Lett.* **2023**, *18*, 1–9. [CrossRef]
11. Lane, D.J.W.; Marsh, L.M.; Vanden Spiegel, D.; Rowe, F.W.E. Echinoderm fauna of the South China Sea: An inventory and analysis of distribution patterns. *Raffles Bull. Zool.* **2000**, *8*, 459–493.
12. Hoeksema, B.W.; Lane, D.J.W. The mushroom coral fauna (Scleractinia: Fungiidae) of Brunei Darussalam (South China Sea) and its relation to the Coral Triangle. *Raffles Bull. Zool.* **2014**, *62*, 566–580.
13. Setiawan, E.; Relex, D.; Marshall, D.J. Shallow-water sponges from a high-sedimentation estuarine bay (Brunei, northwest Borneo, Southeast Asia). *J. Trop. Biodivers. Biotechnol.* **2021**, *6*, 66435. [CrossRef]
14. Mustapha, N.; Baharuddin, N.; Tan, S.K.; Marshall, D.J. The neritid snails of Brunei Darussalam: Their geographical, ecological and conservation significance. *Ecol. Montenegrina* **2021**, *42*, 45–61. [CrossRef]
15. Lane, D.J.W.; Lim, G.P.C. Reef corals in a high sedimentation environment on the 'mainland' coast of Brunei, Northwest Borneo. *Galaxea. J. Coral Reef Stud.* **2013**, *15*, 166–171. [CrossRef]
16. Ho, K.F. Distribution of Recent Benthonic foraminifera in the "inner" Brunei Bay. *Brunei Mus. J.* **1971**, *2*, 124–137.
17. Goeting, S.; Briguglio, A.; Eder, W.; Hohenegger, J.; Roslim, A.; Kocsis, L. Depth distribution of modern larger benthic foraminifera offshore Brunei Darussalam. *Micropaleontology* **2018**, *64*, 299–316. [CrossRef]
18. Goeting, S.; Ćosović, V.; Benedetti, A.; Fiorini, F.; Kocsis, L.; Roslim, A.; Briguglio, A. Diversity and depth distribution of modern benthic foraminifera offshore Brunei Darussalam. *J. Foraminifer. Res.* **2022**, *52*, 160–178. [CrossRef]
19. Goeting, S.; Fiorini, F.; Benedetti, A.; Kocsis, L.; Roslim, A.; Zaini, N.; Briguglio, A. Catalogue of modern smaller benthic foraminifera from offshore Brunei Darussalam. *Palaeontogr. Abt. A* **2021**, *318*, 129–223. [CrossRef]
20. Brady, H.B. Notes on some of the reticularian Rhizopoda of the Challenger Expedition, Part I. On new or little-known arenaceous types, part II. Additions to the Knowledge of porcellaneous and hyaline types. *Q. J. Microsc. Sci. New Ser.* **1879**, *19*, 261–299.
21. Brady, H.B. Notes on some of the reticularian Rhizopoda of the Challenger Expedition, Part III. 1. Classification. 2. Further notes on new species. 3. Note on Biloculina mud. *Q. J. Microsc. Sci. New Ser.* **1881**, *21*, 31–71.
22. Brady, H.B. *Report of the Foraminifera Dredged by H. M. S. Challenger during the Years 1873–1876*. Reports of the Scientific Results of the Voyage of H. M. S. Challenger during the Years 1873–1876; Zoology: Jena, Germany, 1884; Volume 9, pp. 1–814.
23. Jones, R.W. *The Challenger Foraminifera*; Oxford University Press: Oxford, UK, 1994; pp. 1–149.
24. Loeblich, A.R.; Tappan, H. *Foraminifera of the Sahul Shelf*; Cushman Foundation for Foraminiferal Research: Cambridge, MA, USA, 1994; Volume 31, pp. 1–661.
25. Lesslar, P. Computer-assisted interpretation of depositional palaeoenvironments based on foraminifera. *Bull. Geol. Soc. Malays.* **1987**, *21*, 103–119. [CrossRef]
26. Renema, W. Large benthic foraminifera from the deep photic zone of a mixed siliciclastic-carbonate shelf off East Kalimantan, Indonesia. *Mar. Micropaleontol.* **2006**, *58*, 73–82. [CrossRef]
27. Szarek, R. Biodiversity and Biogeography of Recent Benthic Foraminiferal Assemblages in the South-Western South China Sea (Sunda Shelf). (Unpublished Ph.D. Thesis). Faculty of Mathematics and Natural Sciences, Kiel, Germany, 2001.
28. Szarek, R.; Kuhnt, W.; Kawamura, H.; Kitazato, H. Distribution of recent benthic foraminifera on the Sunda Shelf (South China Sea). *Mar. Micropaleontol.* **2006**, *61*, 171–195. [CrossRef]
29. Hoeksema, B.W. Delineation of the Indo-Malayan Centre of Maximum Marine Biodiversity: The Coral Triangle. In *Biogeography, Time and Place: Distributions, Barriers and Islands*; Renema, W., Ed.; Springer: Dordrecht, The Netherlands, 2007; pp. 117–178.
30. Veron, J.E.N.; De Vantier, L.M.; Turak, E.; Green, A.L.; Kininmonth, S.; Stafford-Smith, M.G.; Peterson, N. Delineating the Coral Triangle: Galaxea. *J. Coral Reef Stud.* **2009**, *11*, 91–100. [CrossRef]
31. Veron, J.E.N.; DeVantier, L.M.; Turak, E.; Green, A.L.; Kininmonth, S.; Stafford-Smith, M.G.; Peterson, N. The Coral Triangle. In *Coral Reefs: An Ecosystem in Transition*; Dubinsky, Z., Stambler, N., Eds.; Springer: Dordrecht, The Netherlands, 2011; pp. 47–55.
32. Lei, Y.; Li, T. *Atlas of Benthic Foraminifera from China Seas*; Science Press Ltd.: Beijing, China, 2016; p. 399.
33. Förderer, M.; Langer, M.R. Atlas of benthic foraminifera from coral reefs of the Raja Ampat Archipelago (Irian Jaya, Indonesia). *Micropaleontology* **2018**, *64*, 1–170. [CrossRef]

34. Martin, S.Q.; Culver, S.J.; Leorri, E.; Mallinson, D.J.; Buzas, M.A.; Hayek, L.A.; Shazili, N.A.M. *Distribution and Taxonomy of Modern Benthic Foraminifera of the Western Sunda Shelf (South China Sea) off Peninsular Malaysia*; Cushman Foundation, Special Publication: Kansas, KS, USA, 2018; Volume 47, p. 108.
35. Minhat, F.; Husain, M.; Sulaiman, A. Species composition and distribution data of benthic foraminifera from the straits of Malacca during the early Holocene. *Data Brief* **2019**, *25*, 104214. [CrossRef] [PubMed]
36. Novak, V.; Renema, W. Ecological tolerances of Miocene larger benthic foraminifera from Indonesia. *J. Asian Earth Sci.* **2018**, *151*, 301–323. [CrossRef]
37. Förderer, M.; Langer, M.R. Exceptionally species-rich assemblages of modern larger benthic foraminifera from nearshore reefs in northern Palawan (Philippines). *Rev. Micropaléontol.* **2019**, *65*, 100387. [CrossRef]
38. Azmi, N.; Minhat, F.I.; Hasan, S.S.; Rahman Abdul Manaf, O.A.; Abdul A'ziz, A.N.; Wan Saelan, W.N.; Suratman, S. Distribution of benthic foraminifera off Kelantan, Peninsular Malaysia, South China Sea. *J. Foraminifer. Res.* **2020**, *50*, 89–96. [CrossRef]
39. Hallock, P.; Lidz, B.H.; Cockey-Burkhard, E.M.; Donnelly, K.B. Foraminifera as bioindicators in coral reef assessment and monitoring: The FORAM Index. *Environ. Monit. Assess.* **2003**, *81*, 221–238. [CrossRef]
40. Armstrong, H.; Brasier, M. Foraminifera. In *Microfossils*, 2nd ed.; Armstrong, H., Brasier, M., Eds.; Blackwell Publishing: Malden, UK, 2005; pp. 142–187.
41. Pignatti, J.; Frezza, V.; Benedetti, A.; Carbone, F.; Accordi, G.; Matteucci, R. Recent foraminiferal assemblages and mixed carbonate-siliciclastic sediments along the coast of southern Somalia and northern Kenya. *Ital. J. Geosci.* **2012**, *131*, 66–75.
42. Benedetti, A.; Frezza, V. Benthic foraminiferal assemblages from shallow-water environments of northeastern Sardinia (Italy, Mediterranean Sea). *Facies* **2016**, *62*, 14. [CrossRef]
43. Zaini, N.; Briguglio, A.; Goeting, S.; Roslim, A.; Kocsis, L. Sedimentological characterization of sea bottom samples collected offshore Muara and Tutong, Brunei Darussalam. *Bull. Geol. Soc. Malays.* **2020**, *70*, 139–151. [CrossRef]
44. Hossain, M.B. *Macrobenthic Community Structure from a Tropical Estuary*; LAP Lambert Academic Publishing GmbH & Co.: Saarbruecken, Germany, 2011.
45. Hossain, M.B. Trophic functioning of microbenthic fauna in a tropical acidified Bornean estuary (Southeast Asia). *Int. J. Sediment Res.* **2019**, *34*, 48–57. [CrossRef]
46. Hossain, M.B.; Marshall, D.J.; Venkatramanan, S. Sediment granulometry and organic matter content in the intertidal zone of the Sungai Brunei estuarine system, northwest coast of Borneo. *Carpathian J. Earth Environ. Sci.* **2014**, *9*, 231–239.
47. Talley, L.D. *Descriptive Physical Oceanography: An Introduction*; Academic Press: Cambridge, MA, USA, 2011. [CrossRef]
48. Hee, Y.Y.; Weston, K.; Suratman, S.; Akhir, M.F.; Latif, M.; Valliyodan, S. Biogeochemical and physical drivers of hypoxia in a tropical embayment (Brunei Bay). *Environ. Sci. Pollut. Res.* **2023**, *30*, 65351–65363. [CrossRef]
49. Billman, H.; Hottinger, L.; Oesterle, H. Neogene to Recent Rotaliid Foraminifera from the Indo-Pacific Ocean, their canal system, their classification and their stratigraphic use. *Schweiz. Palaeontol. Abh.* **1980**, *101*, 7–113.
50. Loeblich, A.R.; Tappan, H. *Foraminiferal Genera and Their Classification*; Van Nostrand Reinhold Co.: New York, NY, USA, 1987; p. 869.
51. Parker, J.H. *Taxonomy of Foraminifera from Ningaloo Reef, Western Australia*; Memoirs of the Association of Australasian Palaeontologists: Canberra, Australia, 2009; Volume 36, pp. 1–810.
52. Debenay, J.P. *A Guide to 1,000 Foraminifera from the Southwestern Pacific New Caledonia*; IRD Editions: Paris, France, 2012; pp. 1–383.
53. Hohenegger, J. *Large Foraminifera: Greenhouse Constructions and Gardeners in the Oceanic Microcosm*; Kagoshima University Museum: Kagoshima, Japan, 2011; p. 81.
54. McGeogh, M.; Gaston, K.J. Occupancy frequency distributions: Patterns, artefacts and mechanisms. *Biol. Rev.* **2002**, *77*, 311–331. [CrossRef]
55. Warwick, R.M.; Clarke, K.R. New 'biodiversity' measures reveal a decrease in taxonomic distinctness with increasing stress. *Mar. Ecol. Prog. Ser.* **1995**, *129*, 301–305. [CrossRef]
56. Spötl, C.; Vennemann, W.T. Continuous-flow IRMS analysis of carbonate minerals. *Rapid Commun. Mass Spectrom* **2003**, *17*, 1004–1006. [CrossRef] [PubMed]
57. Spötl, C. A robust and fast method of sampling and analysis of $\delta^{13}C$ of dissolved inorganic carbon in ground waters. *Isot. Environ. Health Stud.* **2005**, *41*, 217–221. [CrossRef] [PubMed]
58. Halder, J.; Decrouy, L.; Vennemann, W.T. Mixing of Rhône River water in Lake Geneva (Switzerland–France) inferred from stable hydrogen and oxygen isotope profiles. *J. Hydrol.* **2013**, *477*, 152–164. [CrossRef]
59. Bidgood, M.D.; Simmons, M.D.; Thomas, C.D. Agglutinated Foraminifera from Miocene sediments of northwest Borneo. In Proceedings of the Fifth International Workshop on Agglutinated Foraminifera, Plymouth, UK, 6–16 September 1997; Hart, M.B., Kaminski, M.A., Smart, C.W., Eds.; Grzybowski Foundation Special Publication: Krakow, Poland, 2000; Volume 7, pp. 41–58.
60. Haunold, T.G.; Baal, C.; Piller, W.E. Benthic foraminiferal associations in the Northern Bay of Safaga, Red Sea, Egypt. *Mar. Micropaleontol.* **1997**, *29*, 185–210. [CrossRef]
61. Hohenegger, J. Coenoclines of Larger Foraminifera. *Micropaleontology* **2000**, *46*, 127–151.
62. Hohenegger, J. Depth coenoclines and environmental considerations of Western Pacific larger foraminifera. *J. Foraminifer. Res.* **2004**, *34*, 9–33. [CrossRef]

63. Hayward, B.W.; Grenfell, H.R.; Reid, C.M.; Hayward, K.A. *Recent New Zealand Shallow-Water Benthic Foraminifera: Taxonomy, Ecologic Distribution, Biogeography, and Use in Paleoenvironmental Assessments*; Institute of Geological and Nuclear Sciences Limited: Lower Hutt, New Zealand, 1999; p. 264.
64. Seddighi, M.; Briguglio, A.; Hohenegger, J.; Papazzoni, C.A. New results on the hydrodynamic behaviour of fossil Nummulites tests from two nummulite banks from the Bartonian and Priabonian of northern Italy. *Boll. Della Soc. Paleontol. Ital.* **2015**, *54*, 103–116.
65. Gallagher, S.; Waccace, M.W.; Li, C.L.; Kinna, B.; Bye, J.T.; Akimoto, K.; Torii, M. Neogene history of the West Pacific Warm Pool, Kuroshio and Leeuwin currents. *Paleoceanography* **2009**, *24*, 1–27. [CrossRef]
66. Belanger, C.L.; Jablonski, D.; Roy, K.; Berke, S.K.; Krug, A.Z.; Valentine, J.W. Global environmental predictors of benthic marine biogeographic structure. *Proc. Natl. Acad. Sci. USA* **2012**, *109*, 14046–14051. [CrossRef] [PubMed]
67. Jiang, F.; Fan, D.; Zhao, Q.; Wu, Y.; Ren, F.; Liu, Y.; Li, A. Comparison of alive and dead benthic foraminiferal fauna off the Changjiang Estuary: Understanding water- mass properties and taphonomic processes. *Front. Mar. Sci.* **2023**, *10*, 1114337. [CrossRef]
68. Epstein, S.; Buchsbaum, R.; Lowenstam, H.A.; Urey, H.C. Revised carbonate water isotopic temperature scale. *Geol. Soc. Am. Bull.* **1953**, *64*, 1315–1326. [CrossRef]
69. Sharp, Z. *Principles of Stable Isotope Geochemistry*, 2nd ed.; University of New Mexico: Albuquerque, NM, USA, 2017; p. 416. [CrossRef]
70. Zeng, L.; Wang, D.; Chen, J.; Wang, W.; Chen, R. SCSPOD14, a South China Sea physical oceanographic dataset derived from in situ measurements during 1919–2014. *Sci. Data* **2016**, *3*, 160029. [CrossRef]
71. Wefer, G.; Berger, W.H. Isotope palaeontology: Growth and composition of extant calcareous species. *Mar. Geol.* **1991**, *100*, 207–248. [CrossRef]
72. Shackleton, N.J.; Kennett, J.P. Paleotemperature history of the Cenozoic and initiation of Antarctic glaciation: Oxygen and carbon isotope analyses in DSDP sites 277, 279 and 281. *Initial Rep. DSDP* **1975**, *29*, 743–755.
73. Emrich, K.; Ehhalt, D.H.; Vogel, J.C. Carbon isotope fractionation during the precipitation of calcium carbonate. *Earth Planet. Sci. Lett.* **1970**, *8*, 363–371. [CrossRef]

Disclaimer/Publisher's Note: The statements, opinions and data contained in all publications are solely those of the individual author(s) and contributor(s) and not of MDPI and/or the editor(s). MDPI and/or the editor(s) disclaim responsibility for any injury to people or property resulting from any ideas, methods, instructions or products referred to in the content.

Article

Seasonal Variations in Invertebrates Sheltered among *Corallina officinalis* (Plantae, Rodophyta) Turfs along the Southern Istrian Coast (Croatia, Adriatic Sea)

Moira Buršić [1,*], Andrej Jaklin [2], Milvana Arko Pijevac [3], Branka Bruvo Mađarić [4], Lucija Neal [5], Emina Pustijanac [1], Petra Burić [1], Neven Iveša [1], Paolo Paliaga [1] and Ljiljana Iveša [2]

[1] Faculty of Natural Sciences, Juraj Dobrila University of Pula, Zagrebačka 30, 52100 Pula, Croatia; emina.pustijanac@unipu.hr (E.P.); petra.buric@unipu.hr (P.B.); neven.ivesa@unipu.hr (N.I.); paolo.paliaga@unipu.hr (P.P.)
[2] Center for Marine Research, Ruđer Bošković Institute, G. Paliage 5, 52210 Rovinj, Croatia; jaklin@cim.irb.hr (A.J.); ivesa@cim.irb.hr (L.I.)
[3] Natural History Museum Rijeka, Lorenzov Prolaz 1, 51000 Rijeka, Croatia; milvana@prirodoslovni.com
[4] Molecular Biology Division, Ruđer Bošković Institute, Bijenička 54, 10000 Zagreb, Croatia; branka.bruvo.madjaric@irb.hr
[5] Kaplan International College, Moulsecoomb Campus, University of Brighton, Watts Building, Lewes Rd., Brighton BN2 4GJ, UK; lucija.neal@gmail.com
* Correspondence: moira.bursic@unipu.hr

Abstract: Available research on invertebrates in *Corallina officinalis* settlements shows a high level of biodiversity due to a complex habitat structure. Our aim was to examine seasonal changes in the invertebrate population, considering the algae's growth patterns. Nine locations with over 90% algal coverage were selected in southern Istria, where quantitative sampling was performed using six replicates of 5 × 5 cm quadrats in each location. Results showed that 29,711 invertebrates were found during winter (maximum algae growth) and 22,292 during summer (minimum algae growth), with an extrapolated average density of 220,000 and 165,200 individuals per square meter, respectively. The total number of individuals showed a linear increase as the algae biomass increased. The highest density, 586,000 individuals, was recorded in the Premantura area during winter. Dominant groups such as amphipods, polychaetes, bivalves and gastropods made up over 80% of the invertebrates. Our study confirms high invertebrate richness in the *C. officinalis* settlements, with the maximum density being the highest when compared to previously published data.

Keywords: invertebrates; *Corallina officinalis*; seasonal variation; coastal area; north Adriatic

Citation: Buršić, M.; Jaklin, A.; Arko Pijevac, M.; Bruvo Mađarić, B.; Neal, L.; Pustijanac, E.; Burić, P.; Iveša, N.; Paliaga, P.; Iveša, L. Seasonal Variations in Invertebrates Sheltered among *Corallina officinalis* (Plantae, Rodophyta) Turfs along the Southern Istrian Coast (Croatia, Adriatic Sea). *Diversity* **2023**, *15*, 1099. https://doi.org/10.3390/d15101099

Academic Editor: Renato Mamede

Received: 7 September 2023
Revised: 16 October 2023
Accepted: 20 October 2023
Published: 22 October 2023

Copyright: © 2023 by the authors. Licensee MDPI, Basel, Switzerland. This article is an open access article distributed under the terms and conditions of the Creative Commons Attribution (CC BY) license (https://creativecommons.org/licenses/by/4.0/).

1. Introduction

The algae *Corallina officinalis* Linnaeus is the most commonly found species of its genus in temperate regions worldwide, forming dense settlements in the intertidal zone [1]. Previous mapping of macroalgae communities in the eastern Adriatic area using the CARLIT method showed that communities dominated by species of the genus *Corallina* are located in 13% of the investigated coast [2]. The unique structure of this alga modifies its surroundings, serving as an example of how algae can shape the composition and relationships of associated organisms. Its structure provides shelter for both vagile and sessile organisms during adverse weather [3–8]. Many benthic invertebrates found in algal forests have a larval stage that floats or swims in the plankton, making the macroalgae significant as larval collectors [9]. Macroalgae are crucial in supporting marine invertebrate communities, attracting diverse organisms [10–14]. These inhabitants form intricate associations, relying on macroalgae for various purposes, creating behavioral patterns that keep them within the algae [8]. Algal epibionts adapt to abiotic factors and exhibit varying degrees of dependence on algae resources [15–17].

According to previous research, several main groups of invertebrates have been recognized as the most abundant within *C. officinalis* settlements. These are mollusks (classes Gastropoda and Bivalvia), polychaetes and higher crustaceans (orders Amphipoda, Isopoda and Tanaidacea) [18–26]. Gastropods that usually live associated with macroalgae, whether they are species from the genus *Corallina* or some other macroalgae (such as fucalean species), are mainly herbivores and feed on the algae itself or on diatoms and epiphytic algae. Other species are detritivores, which feed on sediment trapped inside the algal thallus and use organic matter and microorganisms from that sediment [27,28]. The bivalves that live within *C. officinalis* settlements are diverse and can survive at various depths, temperatures and salinity [29]. In addition, there are also species that can survive in conditions in which the quality of sea water is lower [30]. Polychaetes are mainly marine invertebrates that inhabit a wide range of habitats, from the intertidal zone down to deep-sea sediments. They are one of the most diverse groups of invertebrates that constitute a significant percentage of the total number of organisms living associated with *C. officinalis* [24]. Peracarids are the most dominant crustaceans found within coralline algae [31]. Their feeding habits have been studied, and it has been found that, while detritus is one of their primary food sources, carnivorous species often feed on other types of crustaceans, and herbivores feed on macroalgae tissue [32]. This explains why they are a frequent group found in *C. officinalis* habitats. Other organisms that can be found in higher abundances in this type of habitat include pycnogonids, mites and nematodes [24–26,33–35].

As *C. officinalis* is a turf-forming algae, the effect of its biomass on the richness of invertebrate species has proven to be an important factor. Previous research has shown a positive correlation between the number of invertebrate species and algae biomass [36]. In the same research, another positive effect was recorded, namely the fact that turf-forming alga retain a larger amount of sediment that serves as a secondary habitat for many macrofauna, meiofauna and microfauna species [37]. Based on this observation, our research aimed to investigate how seasonal variations in *C. officinalis* algae, specifically its minimum and maximum growth phases, influence the dynamics of invertebrate structure and composition within its habitat. We hypothesized that changes in algal biomass during its maximum and minimum growth phases would result in differences in the number of individuals within the algae settlements.

2. Materials and Methods

2.1. Research Area and Sampling Methods

The research was conducted in the coastal area of southern Istria and the Brijuni National Park (Croatia, Adriatic Sea). Based on preliminary research by Buršić et al., 2019 [24] and essential coastal features (percentage of *Corallina officinalis* cover, slope of the coast and wind exposure), four areas were selected for seasonal sampling. In three of these areas (Pula, Banjole and Premantura), two locations were selected, while in NP Brijuni there were three locations (Figure 1). Within each location, two sites were chosen (minimum distance of 100 m between them) from which three subsamples were collected. Sampling was done within a 5 × 5 cm quadrat, as this was determined, after our preliminary research [24], to be the most appropriate size to provide a more accurate description of the research area, and also considering the number of invertebrates collected and further laboratory analyses to be conducted.

A total of 108 quadrats were sampled in the four areas during two seasons, 54 quadrats during the algae's maximum growth (from November 2017 to April 2018, the "winter" samples) and 54 quadrats during the algae's minimum growth (from June to August 2018, the "summer" samples). According to Croatian Meteorological and Hydrological Service data (https://meteo.hr/index_en.php, accessed on 12 October 2023), winters in Croatia's northern Adriatic have average temperatures of 5 °C to 10 °C, occasionally dropping below freezing, with frequent rainfall ranging from 20 mm to 130 mm. Snowfall is rare and usually light, not leading to lasting snow cover. In contrast, summers in Istria are warm to hot, with temperatures averaging from 22 °C to 25 °C and the possibility of heatwaves. Summers are typically dry, with low monthly rainfall between 20 mm and 60 mm, occasionally interrupted by thunderstorms. Quadrats were sampled in areas where algae coverage was evenly distributed and exceeded 90% along the entire coastline. The coverage was evaluated in a coastal belt with a width of approximately 50 cm. Samples were taken during low tide when *C. officinalis* was completely out of the water to minimize the loss of associated organisms. Collection was performed with a hammer and chisel in order to gather the entire algal thallus. Immediately after collection, samples were stored in plastic bottles and fixed with alcohol. The next day, samples were washed in the laboratory through a sieve with a mesh size of 0.5 mm and stored in alcohol for further processing.

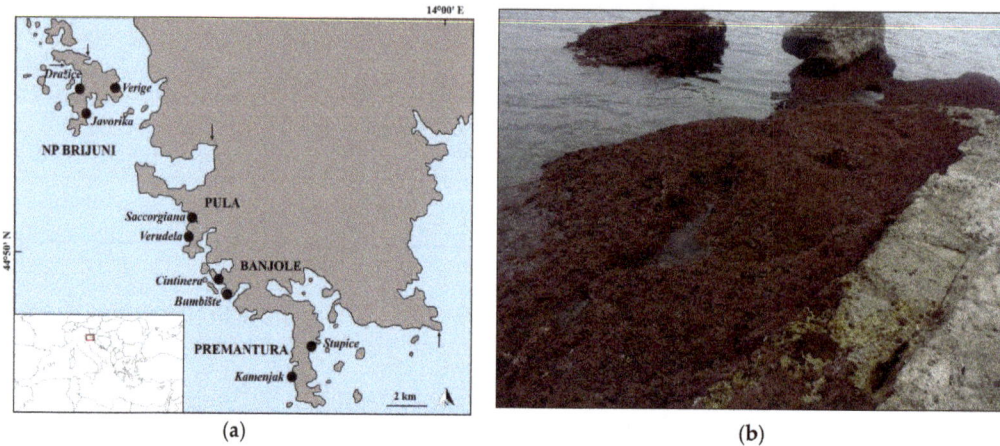

Figure 1. *Corallina officinalis* sampling locations (black dots) within four investigated areas in south Istria, namely Pula, Banjole, Premantura and NP Brijuni—Brijuni National Park; arrows indicate initial and final mapping points (**a**). *C. officinalis* habitat in Javorika Bay in NP Brijuni area (**b**).

Invertebrates were separated manually from algae samples under a stereomicroscope with 10× to 60× magnification. Each algal thallus was examined in detail to isolate all the invertebrates present. After the separation of all invertebrates, the algae were dried at 80 °C for 24 h and weighed. By using dry weight, the water component is removed, leading to a more consistent and comparable measurement. Organisms were determined to the lowest accessible taxonomic group according to various resources available in the literature [38–46].

2.2. Analysis of the Invertebrate Structure and Composition

The total number of individuals (N) per investigated area was recorded and expressed per unit area of 150 cm^2 (two sets of three replicate quadrats 5 × 5 cm in size). In addition, relative abundance and frequency of occurrence were calculated. These values were calculated separately for all taxonomic groups depending on sampling season, i.e., for 54 sampled quadrats during the algae's maximum growth ("winter") and for 54 sampled quadrates during the algae's minimum growth ("summer"). Based on the relative abundance and frequency of occurrence, the taxonomic groups found can be classified into several different categories. To assess the dominance structure, the relative abundance was employed using the formula $A = (n_i/N) \times 100$, where n_i represents the count of individual representatives for each taxonomic group and N indicates the total count of all individuals. The classification system introduced by Tischler, 1949 [47] for invertebrates and used by Travizi, 2010 [48] for marine invertebrates was adopted, comprising five categories as follows: eudominant (>10% of all individuals); dominant (5 to 10%); subdominant (2 to 5%); recedent (1 to 2%) and subrecedent (<1%). Frequency of occurrence was calculated using the formula $Fa = N_a/N \times 100$, where N_a represents the number of samples in which taxonomic group *a* occurs and N represents the total number of samples. Taxonomic groups were classified into 4 categories: very frequent—present in 75–100% of samples, frequent—present in 50–75% of samples, widespread—present in 25–50% of samples and rare—present in 0–25% of samples [48,49].

The linear dependence of the total number of individuals of higher taxonomic groups against the dry weight of *C. officinalis* was verified by linear regression analysis in Microsoft Office Excel 2016. The linear regression analysis was used to analyze the relationship between two variables: the number of individuals and the dry weight (biomass) of *C. officinalis*. The seasonal variation in invertebrates within the *Corallina officinalis* turf was analyzed using non-metric multidimensional scaling (nMDS) based on square root transformation and Bray–Curtis similarity.

3. Results

The southern Istrian coast was mapped from Pula in the north-west to Ližnjan in the south-east. A total of 29,711 individuals of invertebrates were isolated in the winter samples of *Corallina officinalis*, and 22,292 individuals in the summer samples. In the non-metric multidimensional scaling (nMDS) analysis, clear seasonal variations were observed between periods of minimum and maximum algae growth within the invertebrate taxonomic groups associated with *C. officinalis* settlements (Figure 2). The average number of individuals per sampled quadrat was 550 for winter samples and 413 for summer samples, which, calculated per square meter, amounts to 220,000 individuals (ind m^{-2}) for "winter" and 165,200 ind m^{-2} for "summer". As in the preliminary sampling [24], the density of individuals associated with *C. officinalis* settlements varied between different sampling locations. Thus, locations Kamenjak and Stupice in Premantura had the highest recorded abundance of invertebrates. The maximum recorded number of individuals within a 5 × 5 cm quadrat was 1465 (amounting to 586,000 ind m^{-2}) and was sampled during the period of maximum algae growth at the Stupice location (Figure 3).

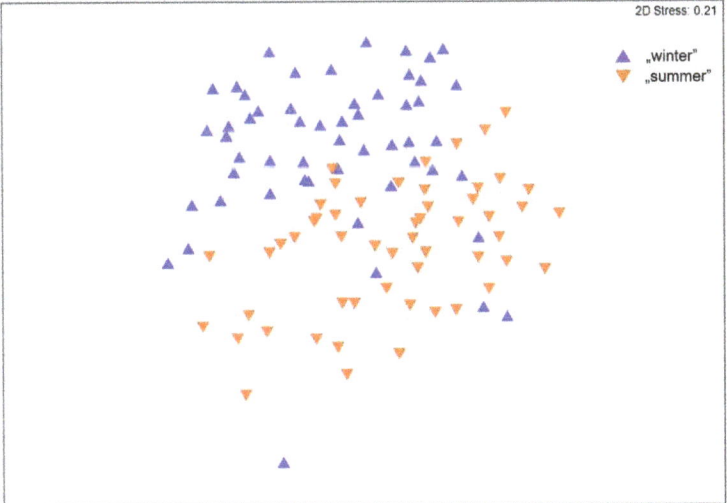

Figure 2. Non-metric MDS ordination based on square root transformation and Bray-Curtis similarity of the total number of individuals during both sampling seasons in nine locations ("winter"—samples taken during the algae's maximum growth; "summer"—samples taken during the algae's minimum growth).

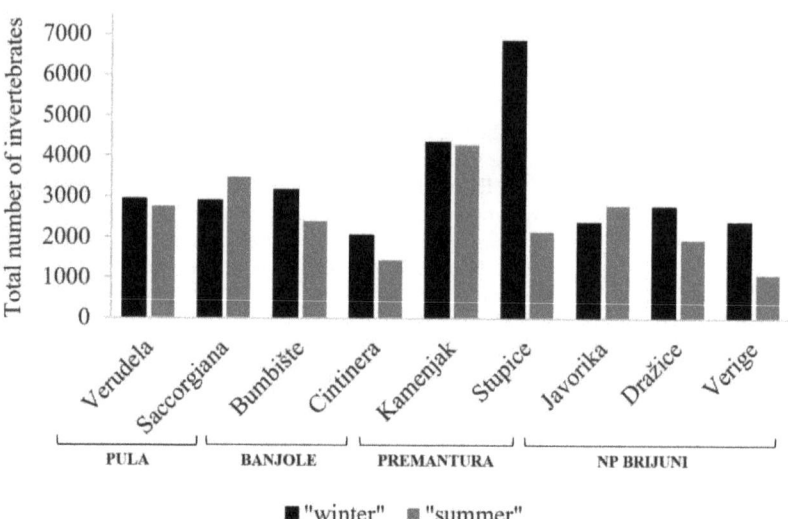

Figure 3. Seasonal variations in the total number of invertebrates, expressed per sampling location described by six replicates, i.e., area of 150 cm² ("winter" and "summer" having the same definitions as in Figure 2).

The recorded dry weight of algae from 108 samples was compared for both seasons (Figure 4). During the period of maximum growth, the average dry weight per sampling quadrat was 9.18 g, while during the minimum growth, it was 7.35 g.

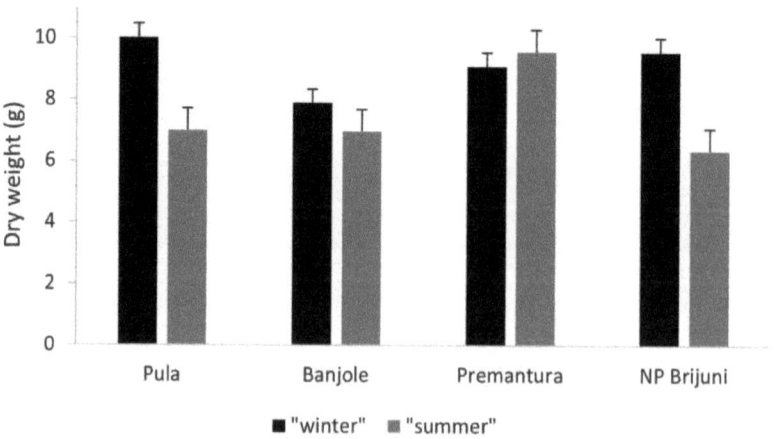

Figure 4. Comparison of *C. officinalis* average dry weight (g/25 cm²) during both sampling seasons ("winter" and "summer" having the same definitions as in Figure 2).

Isolated invertebrates were classified within 20 different taxonomic groups, whose relative abundance and frequency of occurrence varied depending on the sampling season. During the algae's maximum growth, relative abundance ranged from 0.01% to 31.72%, and frequency of occurrence from 7.41% to 100%, while during algae's minimum growth, relative abundance ranged from 0.01% to 55.24%, and frequency of occurrence from 1.85% to 100% (Table 1).

Table 1. Relative abundance (A) and frequency of occurrence (F) of the taxonomic groups recorded in this investigation ("winter"—samples taken during the algae's maximum growth; "summer"—samples taken during the algae's minimum growth).

Taxonomic Group	"Winter" A%	F%	"Summer" A%	F%
Platyhelminthes	0.01	7.41	0.01	1.85
Nemertea	0.12	31.48	0.13	18.52
Nematoda	3.08	98.15	2.17	81.48
Sipuncula	0.24	38.89	0.20	37.04
Gastropoda	7.69	100.00	10.21	100.00
Bivalvia	31.72	100.00	55.24	100.00
Polyplacophora	0.31	62.96	0.75	77.78
Polychaeta	13.62	100.00	9.27	100.00
Decapoda	0.08	33.33	0.01	1.85
Cumacea	0.11	22.22	0.02	7.41
Tanaidacea	5.67	87.04	6.02	79.63
Isopoda	3.66	87.04	1.44	75.93
Amphipoda	26.95	100.00	9.71	100.00
Caprellidae	0.80	51.85	0.06	14.81
Copepoda	1.96	90.74	0.87	75.93
Pantopoda	0.78	79.63	0.15	33.33
Acari	2.02	94.44	1.92	85.19
Diptera	0.43	50.00	0.73	50.00
Echinoidea	0.02	7.41	0.22	51.85
Ophiuroidea	0.72	64.81	0.89	62.96

The calculation for the relative abundance of taxonomic groups during both sampling seasons showed that the category of subrecedent was the most represented, including 11 taxons out of 20 (Table 2). Regarding frequency of occurrence, the category of euconstant groups was the most represented, and 50% of the recorded taxonomic groups in both sampling seasons belonged to this category (Table 3).

Table 2. Relative abundance of taxonomic groups in investigated areas (%—percentage of groups belonging to a certain category of relative abundance).

Relative Abundance	"Winter" No. of Taxonomic Groups	%	"Summer" No. of Taxonomic Groups	%
eudominant (>10%)	3	15.00	2	10.00
dominant (5–10%)	2	10.00	3	15.00
subdominant (2–5%)	3	15.00	1	5.00
recedent (1–2%)	1	5.00	2	10.00
subrecedent (<1%)	11	55.00	12	60.00
TOTAL	20	100.00	20	100.00

Table 3. Frequency of occurrence of taxonomic groups in investigated areas (%—percentage of groups belonging to a certain category of frequency of occurrence).

Frequency of Occurrence	"Winter" No. of Taxonomic Groups	%	"Summer" No. of Taxonomic Groups	%
very frequent (75–100%)	10	50.00	10	50.00
frequent (50–75%)	4	20.00	3	15.00
widespread (25–50%)	3	15.00	2	10.00
rare (0–25%)	3	15.00	5	25.00
TOTAL	20	100.00	20	100.00

The taxonomic groups of invertebrates that dominated in samples were amphipod crustaceans, polychaetes, bivalves and gastropods, which constituted 80% of all isolated invertebrates in "winter" samples and 84% in "summer" samples (Figure 5). Certain taxonomic groups were represented only by a few individuals, and their relative abundance was less than 0.1%. These were flatworms, decapods, sea urchins in "winter" samples and cumaceans in "summer" samples.

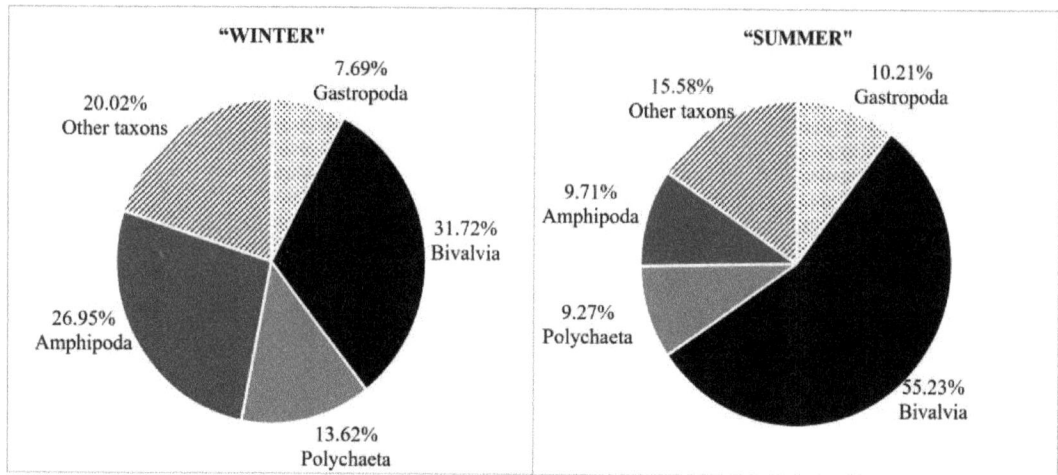

Figure 5. Dominant taxonomic groups within *C. officinalis* settlements in all sampling areas during both seasons ("winter" and "summer" having the same definition as in Figure 2).

The four previously mentioned dominant taxonomic groups had the highest value of 100% for both sampling seasons in terms of frequency of occurrence. The remaining euconstant taxonomic groups were almost always the same during both sampling seasons. The exceptions were the polyplacophorans and pantopods, whose frequency of occurrence was classified into two different categories depending on the sampling season. The Polyplacophora group had a frequency of occurrence of 62.96% in the winter samples, which placed it in the lower category, while the Pantopoda group had a frequency of occurrence of 33.33% during the summer (Tables 1 and 3).

The detailed qualitative and quantitative composition of invertebrates for each separate sampling area during both seasons is shown in Table 4. In addition, the frequency of occurrence and relative abundance of the taxonomic groups were calculated for each area to see the seasonal fluctuation in organisms with respect to different sampling areas. Detailed species lists for mollusks and pycnogonids are available in previously published papers [25,26,33]. In terms of relative abundance, the same four taxonomic groups were again the most dominant. Brijuni National Park is the only area where a higher relative abundance of amphipods over bivalves was recorded. Although in all other areas, regardless of the sampling season, bivalves always dominated, an increase in the relative abundance of this group can be observed if both sampling seasons are compared. During summer, no taxonomic group had a relative abundance higher than 50%, while in winter, bivalves were present in all locations with a relative abundance higher than 50%, i.e., in the range from 54% to 68%.

The relationship between the number of individuals and the dry weight of *C. officinalis* was studied for the taxonomic groups with the highest number of individuals (gastropods, bivalves, amphipods, tanaidaceans, isopods, mites, polychaetes and nematodes). No correlation was found between the number of bivalves and gastropods and the dry weight of *C. officinalis*, with low coefficients of determination ($R^2 = 0.011$ for Bivalvia and $R^2 = 0.005$ for Gastropoda) and high probabilities that the regression coefficients were not different from zero (P = 0.318 for Bivalvia and P = 0.511 for Gastropoda) (Figure 6). In contrast, other taxonomic groups showed a significant correlation with the dry weight of *C. officinalis*, with regression lines explaining up to 12.1% of the total variability in the number of individuals for amphipods (P = 0.001) and 7.9% for nematodes (P = 0.005) (Figure 7). The total number of individuals across all taxonomic groups increased linearly with the algae biomass (P = 0.002), explaining 9.3% of the variability in the number of fauna individuals (Figure 8).

Table 4. Number of individuals in each taxonomic group during "winter" (a) and "summer" (b) in the areas of Pula, Banjole, Premantura and Brijuni National Park. Locations are indicated as follows: A—Verudela, B—Saccorgiana, C—Bumbište, D—Cintinera, E—Kamenjak, F—Stupice, G—Javorika, H—Dražice, I—Verige. Each location is described with three replicate quadrats 5 × 5 cm in size.

(a)

Taxonomic Group	Pula		Banjole		Premantura		Brijuni National Park			Total
	A	B	C	D	E	F	G	H	I	
Platyhelminthes	0	0	1	1	1	1	0	0	0	4
Nemertea	2	2	5	8	12	0	1	7	0	37
Nematoda	89	125	151	148	78	134	57	71	61	914
Sipuncula	3	1	8	4	29	18	1	4	2	70
Gastropoda	294	194	408	129	789	187	86	117	82	2286
Bivalvia	1158	1367	468	862	1298	3269	422	450	131	9425
Polyplacophora	14	20	12	2	19	17	1	4	3	92
Polychaeta	332	289	612	390	306	1044	343	378	353	4047
Decapoda	1	0	7	0	7	2	1	3	2	23
Cumacea	0	0	6	0	0	1	0	4	22	33
Tanaidacea	96	10	64	6	42	258	387	669	154	1686
Isopoda	4	135	293	39	207	32	29	55	293	1087
Amphipoda	587	585	821	216	1315	1638	939	871	1035	8007
Caprellidae	20	3	11	0	8	87	4	3	103	239
Copepoda	62	58	65	230	11	63	5	38	50	582
Pantopoda	47	24	84	2	10	21	30	3	12	233
Acari	185	59	104	15	94	44	30	46	22	599
Diptera	4	14	15	3	50	13	0	5	23	127
Echinoidea	0	0	2	0	0	1	0	1	1	5
Ophiuroidea	31	1	33	3	62	6	30	20	29	215
Total	2929	2887	3170	2058	4338	6836	2366	2749	2378	29,711

(b)

Taxonomic Group	Pula		Banjole		Premantura		Brijuni National Park			Total
	A	B	C	D	E	F	G	H	I	
Platyhelminthes	0	0	0	0	0	3	0	0	0	3
Nemertea	1	17	0	1	2	0	1	4	3	29
Nematoda	23	125	30	12	110	88	34	17	44	483
Sipuncula	0	4	2	3	10	9	13	3	1	45
Gastropoda	132	358	339	117	269	28	568	279	186	2276
Bivalvia	1773	2459	1430	649	2613	1103	1122	796	368	12,313
Polyplacophora	19	22	13	14	24	5	42	18	11	168
Polychaeta	84	196	226	229	261	318	344	144	264	2066
Decapoda	0	0	0	0	2	0	0	0	0	2
Cumacea	0	0	0	0	1	1	0	1	1	4
Tanaidacea	180	2	8	15	383	202	280	246	25	1341
Isopoda	23	19	57	23	125	20	5	31	19	322
Amphipoda	327	90	167	326	329	286	267	304	68	2164
Caprellidae	3	1	1	0	0	2	1	5	0	13
Copepoda	22	20	9	15	11	14	20	29	55	195
Pantopoda	1	2	3	0	4	9	4	7	3	33
Acari	94	115	37	17	32	12	55	38	27	427
Diptera	62	39	31	0	18	5	2	4	1	162
Echinoidea	0	3	9	13	9	6	7	1	0	48
Ophiuroidea	8	16	28	9	72	30	22	12	1	198
Total	2752	3488	2390	1443	4275	2141	2787	1939	1077	22,292

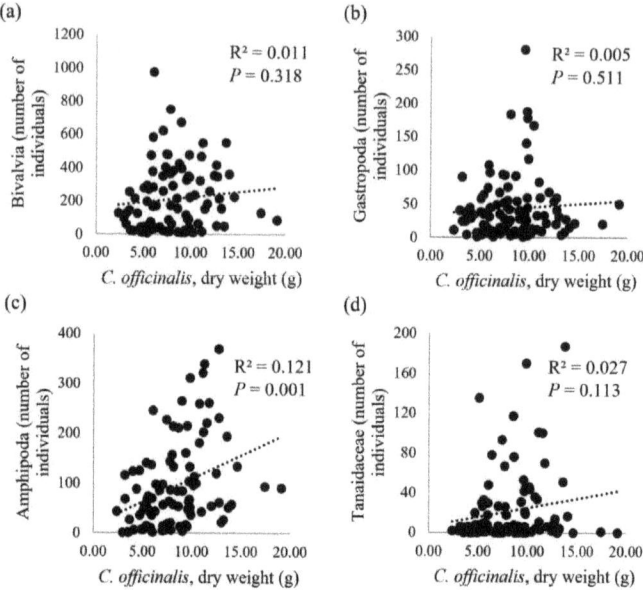

Figure 6. Scatter diagram of the total number of bivalves (**a**), gastropods (**b**), amphipods (**c**) and tanaidaceans (**d**) in relation to the dry weight of *C. officinalis* (g/25 cm^2). With the corresponding regression lines (n = 96), the values of the coefficient of determination (R^2) and the statistical significance of the slope coefficient (*P*) are displayed.

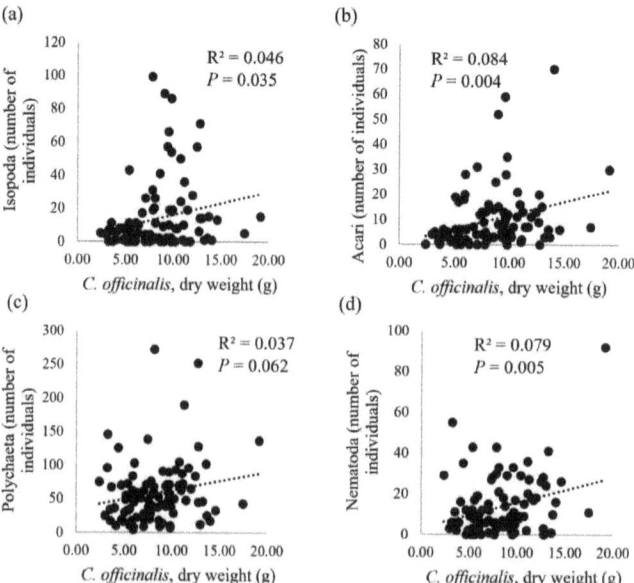

Figure 7. Scatter diagram of the total number of isopods (**a**), mites (**b**), polychaetes (**c**) and nematodes (**d**) in relation to the dry weight of *C. officinalis* (g/25 cm^2). With the corresponding regression lines (n = 96), the values of the coefficient of determination (R^2) and the statistical significance of the slope coefficient (*P*) are displayed.

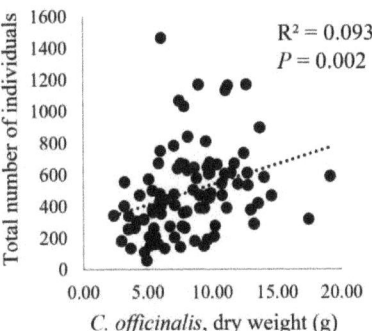

Figure 8. Scatter diagram of the total number of all recorded invertebrates in relation to the dry weight of *C. officinalis* (g/25 cm^2). With the corresponding regression lines (n = 96), the values of the coefficient of determination (R^2) and the statistical significance of the slope coefficient (P) are displayed.

4. Discussion

Previous research on the influence of turf-forming algal biomass on invertebrate species richness shows a positive correlation between the number of invertebrate species and the algal biomass [36]. The greatest diversity and abundance of fauna is always associated with thalli of higher density, which have the most complex structure [50]. Comparing the recorded dry weight of algae in this research with the total number of invertebrates, these numbers follow the same positive trend. Another positive effect of such a habitat has been recorded, which is the fact that algae with this specific structure retain a larger amount of sediment that serves as a secondary habitat for many macrofauna and meiofauna species [37]. In addition, the effect of habitat loss and changes in habitat structure on the number and distribution of species has previously been investigated. It was concluded that the reduction in habitat diversity has a negative effect on the diversity of taxonomic groups, considering that the habitat structure itself greatly affects the spatial distribution of species [8,51]. Research was conducted in an artificial habitat that simulates the appearance of algae to determine the extent to which these morphological changes affect the composition of the macrofauna [22,52,53]. As part of this current research, sampling was performed twice a year, i.e., during the minimum and maximum biomass of *C. officinalis*, and thus the results from this research confirm the previous findings. Regarding the total number of isolated individuals, 30% more individuals were recorded in samples taken during the algae's maximum biomass.

Seasonal variability in the composition and structure of the fauna within the algal community is influenced by the structure of the alga itself, its growth rate and its substrate complexity. It also depends on the availability of food, i.e., the amount of detritus and the abundance of epiphytic algae, as well as seasonal changes in population density due to migration of species [54]. The importance of detritus in benthic communities has often been mentioned in the literature, because much more energy and matter are exchanged in detritus than in the food web of organisms that feed on macroalgae. Many organisms can use detritus directly as food, because shortly after microorganisms colonize that detritus, its nutritional value increases significantly [55]. It was observed that differences in density, branching and overall compactness of the genus *Corallina* have different effects on the accumulation of detritus, and thus on the species living in the intertidal sediments [56]. Small changes in such morphological characteristics, as well as changes in water currents, also affect the abundance of fauna. With its structure and densely branched thallus, *Corallina* is a trap for sediment and food. Microgastropods that feed on microalgae in the sediment can obtain sufficient food from the sediment located within the algal turf. Sediment in exposed coastal areas is coarser than densely packed sediment in sheltered locations. This makes it susceptible to disturbance caused by seawater movement, which can limit the primary production of microalgae. Grain size can be a limiting factor for primary production, because microphytobenthos bind more strongly to coarser sediment. All this can prevent the survival of organisms that feed on microalgae on such sediment [57] and affect the overall abundance and composition of invertebrates.

Various factors, including diverse reproductive strategies, can also explain fluctuations in species abundance [58]. Species without pelagic larvae, like amphipods and isopods, experience significant abundance variations during recruitment. Additionally, species with larvae that prefer substrate settlement, such as *Spirorbis (Spirorbis) corallinae*, contribute to increased abundance during settlement periods [59]. The timing of reproduction also plays a role, especially when it is seasonal or intensified during certain times of the year. Many of these species exhibit seasonal reproductive timing, notably gastropods, like those within the *Littorina* genus [59]. Reproduction may take place throughout the year, with potential reductions in fecundity during the summer [60]. Variations in the timing of reproduction among these direct-developing species can significantly influence their abundance and occurrence, creating distinct seasonal patterns. Another example is the gastropod *Rissoella*, which develops directly from eggs in the spring and summer, when it can reach high abundance. Eggs are then laid at the end of the summer, which over-winter within the turf and hatch the following year [60,61].

The seasonal dynamics of invertebrates within this research show that the distribution of dominant groups is more uniform during the algae's maximum growth than during its minimum growth. To be more specific, during summer, bivalves become dominant and, considering the total number of individuals recorded at all research locations, constitute over 55% of the total. Given that the species *Mytilus galloprovincialis* was by far the most dominant bivalve, and that previous studies recorded the highest density of juvenile individuals of that species in May [16], it is understandable that in the summer period the competition for space within *C. officinalis* settlements will be on the side of bivalves. They can also eliminate larval/juvenile stages of other potential "settlers" through filtration. A study in southern Italy documented a high occurrence of juvenile *M. galloprovincialis* in algal settlements in coastal rocky areas, accounting for 96.6% of the total abundance within three fucalean species [28]. The research found adult forms of *M. galloprovincialis* located under or inside the algal branches, attached to the rocky bottom, where they compete for space with the algae [28].

Analysis of the linear dependence of the total number of individuals of higher taxonomic groups against the dry weight of *C. officinalis* indicates that while there are certain noteworthy correlations between the abundances of invertebrates and the algal biomass, the explained variance, as reflected by the R^2 values, remains relatively low. Consequently, approximately 90% of the variance remains unaccounted for by the regression models. This implies that factors other than the algal biomass are likely to contribute significantly to the observed patterns. One plausible explanation for this unexplained variance is the presence of patchiness in the distribution of algae, both within individual sampling locations and among different sampling locations. It is conceivable that variations in environmental conditions, habitat characteristics or other factors lead to heterogeneity in the distribution and abundance of both invertebrates and algae. The relative scatter of data points within and between sampling locations further supports this notion, offering indications of the potential influence of algal patchiness on the relationships with invertebrate abundances. These findings underscore the need for further investigation into the specific factors driving the unexplained variance and the role of algal patchiness in shaping the relationships between invertebrates and algal biomass.

The comprehensive analysis of invertebrates in this study not only provides valuable insights into the community structure and composition within *C. officinalis* turfs but also sheds light on the overall biodiversity of the coastal zone of southern Istria, where this alga occupies a significant area. This research recorded that the presence of the algae *C. officinalis* was almost three times higher than in a previously recorded study conducted in the eastern Adriatic Sea [2], and this indicates that the area of southern Istria is a suitable habitat for this species. The fact that the habitat is so widespread in the coastal area of southern Istria, and that the abundance of invertebrates is high therein, contributed to the need to conduct this research. Previous studies worldwide have documented the presence of numerous invertebrates within *C. officinalis* settlements, with reported values of up to 329,000 individuals per square meter [21]. Comparing those findings to the maximum (586,000 ind m^{-2}) and average (192,600 ind m^{-2}) number of individuals recorded in our research, it becomes evident that this habitat holds immense importance due to its high abundance and diversity of invertebrates.

Moreover, by considering clustering techniques to assess similarity in invertebrate assemblages among the research locations, further insights into the factors influencing the variation in invertebrate community composition can be gained. Exploring potential relationships between environmental variables, algal biomass, patchiness and other relevant habitat characteristics, as well as the observed clustering patterns, valuable information on invertebrate assemblage variations can be provided. These analyses may reveal the role of localized factors, spatial dynamics and ecological interactions in shaping the community structure within *C. officinalis* turfs and contribute to a deeper understanding of the ecological processes governing coastal ecosystems.

This study not only enhances our understanding of the community structure and composition within *C. officinalis* turfs but also highlights the significant biodiversity and the abundance of invertebrates in this habitat.

Author Contributions: Conceptualization, M.B. and L.I.; data curation, M.B., L.I., A.J. and M.A.P.; formal analysis, M.B. and L.I.; funding acquisition, E.P.; investigation, M.B., L.I., A.J., M.A.P., N.I. and P.P.; methodology, M.B. and L.I.; resources, M.B., L.I., A.J., M.A.P. and E.P.; writing—original draft preparation, M.B., L.I., A.J., M.A.P., B.B.M., L.N., E.P., P.B., N.I. and P.P.; writing—review and editing, M.B., L.I., A.J., B.B.M., P.B. and L.N. All authors have read and agreed to the published version of the manuscript.

Funding: This research received no external funding.

Institutional Review Board Statement: Not applicable.

Data Availability Statement: The authors can provide the data if needed.

Conflicts of Interest: The authors declare no conflict of interest.

References

1. Williamson, C.J.; Perkins, R.; Voller, M.; Yallop, M.L.; Brodie, J. The regulation of coralline algal physiology, an in situ study of *Corallina officinalis* (Corallinales, Rhodophyta). *Biogeosciences* **2017**, *14*, 4485–4498. [CrossRef]
2. Nikolić, V.; Žuljević, A.; Mangialajo, L.; Antolić, B.; Kušpilić, G.; Ballesteros, E. Cartography of littoral rocky-shore communities (CARLIT) as a tool for ecological quality assessment of coastal waters in the Eastern Adriatic Sea. *Ecol. Indic.* **2013**, *34*, 87–93. [CrossRef]
3. Bertness, M.D.; Crain, C.M.; Silliman, B.R.; Bazterrica, M.C.; Reyna, M.V.; Hildago, F.; Farina, J.K. The community structure of western Atlantic Patagonian rocky shores. *Ecol. Monogr.* **2006**, *76*, 439–460. [CrossRef]
4. Liuzzi, M.; Gappa, J.L. Macrofaunal assemblages associated with coralline turf: Species turnover and changes in structure at different spatial scales. *Mar. Ecol. Prog. Ser.* **2008**, *363*, 147–156. [CrossRef]
5. Nelson, W.A. Calcified macroalgae—Critical to coastal ecosystems and vulnerable to change: A review. *Mar. Freshw. Res.* **2009**, *60*, 787–801. [CrossRef]
6. Bracken, M.E.S.; Gonzalez-Dorantes, C.A.; Stachowicz, J.J. Whole-community mutualism: Associated invertebrates facilitate a dominant habitat-forming seaweed. *Ecology* **2007**, *88*, 2211–2219. [CrossRef] [PubMed]
7. Heck, K., Jr.; Crowder, L.B. habitat structure and predator—Prey interactions in vegetated aquatic systems. In *Habitat Structure: The Physical Arrangement of Objects in Space*; Springer: Dordrecht, The Netherlands, 1991; pp. 281–299.
8. Gallardo, D.; Oliva, F.; Ballesteros, M. Marine invertebrate epibionts on photophilic seaweeds: Importance of algal architecture. *Mar. Biodivers.* **2021**, *51*, 16. [CrossRef]
9. Orlando-Bonaca, M.; Trkov, D.; Klun, K.; Pitacco, V. Diversity of Molluscan Assemblage in Relation to Biotic and Abiotic Variables in Brown Algal Forests. *Plants* **2022**, *11*, 2131. [CrossRef]
10. Bégin, C.; Johnson, L.; Himmelman, J. Macroalgal canopies: Distribution and diversity of associated invertebrates and effects on the recruitment and growth of mussels. *Mar. Ecol. Prog. Ser.* **2004**, *271*, 121–132. [CrossRef]

11. Costa-Lotufo, L.V.; Colepicolo, P.; Pupo, M.T.; Palma, M.S. Bioprospecting macroalgae, marine and terrestrial invertebrates & their associated microbiota. *Biota Neotropica* **2022**, *22*, 1345. [CrossRef]
12. Choi, J.-Y.; Kim, S.-K.; Jeong, K.-S.; Joo, G.-J. Distribution pattern of epiphytic microcrustaceans in relation to different macrophyte microhabitats in a shallow wetland (Upo wetlands, South Korea). *Oceanol. Hydrobiol. Stud.* **2015**, *44*, 151–163. [CrossRef]
13. Guerra García, J.M.; Cabezas Rodríguez, M.D.P.; Baeza-Rojano Pageo, E.; Izquierdo, D.; Corzo, J.; Ros Clemente, M.; Sánchez, J.A.; Dugo Cota, Á.; Flores León, A.M.; Soler Hurtado, M.D.M. Abundance patterns of macrofauna associated to marine macroalgae along the Iberian Peninsula. *Zool. Baetica* **2011**, *22*, 3–17.
14. Schmidt, A.L.; Scheibling, R.E. Effects of native and invasive macroalgal canopies on composition and abundance of mobile benthic macrofauna and turf-forming algae. *J. Exp. Mar. Biol. Ecol.* **2007**, *341*, 110–130. [CrossRef]
15. Koehl, M.; Daniel, T.L. Hydrodynamic Interactions Between Macroalgae and Their Epibionts. *Front. Mar. Sci.* **2022**, *9*, 872960. [CrossRef]
16. Bulleri, F.; Pardi, G.; Tamburello, L.; Ravaglioli, C. Nutrient enrichment stimulates herbivory and alters epibiont assemblages at the edge but not inside subtidal macroalgal forests. *Mar. Biol.* **2020**, *167*, 1–15. [CrossRef]
17. Burnett, N.P.; Koehl, M.A.R. Ecological biomechanics of damage to macroalgae. *Front. Plant Sci.* **2022**, *13*, 981904. [CrossRef]
18. George, J.D. The polychaetes of Lewis and Harris with notes on other marine invertebrates. *Proc. R. Soc. Edinb. Sect. B Biol. Sci.* **1979**, *77*, 189–216. [CrossRef]
19. Johnson, S.B.; Attramadal, Y.G. Reproductive behaviour and larval development of *Tanais cavolinii* (Crustacea: Tanaidacea). *Mar. Biol.* **1982**, *71*, 11–16. [CrossRef]
20. López, C.A.; Stotz, W.B. Description of the fauna associated with *Corallina officinalis* L. in the intertidal of the rocky shore of Palo Colorado (Los Vilos, IV-region, Chile). *Oceanogr. Lit. Rev.* **1998**, *3*, 512.
21. Bussell, J.A.; Lucas, I.A.; Seed, R. Patterns in the invertebrate assemblage associated with *Corallina officinalis* in tide pools. *J. Mar. Biol. Assoc. United Kingd.* **2007**, *87*, 383–388. [CrossRef]
22. Kelaher, B.P.; Castilla, J.C.; Prado, L.; York, P.; Schwindt, E.; Bortolus, A. Spatial variation in molluscan assemblages from coralline turfs of Argentinean Patagonia. *J. Molluscan Stud.* **2007**, *73*, 139–146. [CrossRef]
23. Magill, C.L.; Maggs, C.A.; Johnson, M.P.; O'connor, N. Sustainable Harvesting of the Ecosystem Engineer *Corallina officinalis* for Biomaterials. *Front. Mar. Sci.* **2019**, *6*, 285. [CrossRef]
24. Buršić, M.; Iveša, L.; Jaklin, A.; Pijevac, M.A. A preliminary study on the diversity of invertebrates associated with *Corallina officinalis* Linnaeus in southern Istrian peninsula. *Acta Adriat.* **2019**, *60*, 127–136. [CrossRef]
25. Buršić, M.; Iveša, L.; Jaklin, A.; Pijevac, M.A.; Kučinić, M.; Štifanić, M.; Neal, L.; Mađarić, B.B. DNA Barcoding of Marine Mollusks Associated with *Corallina officinalis* Turfs in Southern Istria (Adriatic Sea). *Diversity* **2021**, *13*, 196. [CrossRef]
26. Buršić, M.; Iveša, L.; Jaklin, A.; Pijevac, M.A.; Mađarić, B.B.; Neal, L.; Pustijanac, E.; Burić, P.; Iveša, N.; Paliaga, P. Changes in Composition of Mollusks within *Corallina officinalis* Turfs in South Istria, Adriatic Sea, as a Response to Anthropogenic Impact. *Diversity* **2023**, *15*, 939. [CrossRef]
27. Terlizzi, A.; Scuderi, D.; Fraschetti, S.; Guidetti, P.; Boero, F. Molluscs on subtidal cliffs: Patterns of spatial distribution. *J. Mar. Biol. Assoc. United Kingd.* **2003**, *83*, 165–172. [CrossRef]
28. Chiarore, A.; Fioretti, S.; Meccariello, A.; Saccone, G.; Patti, F.P. Molluscs community associated with the brown algae of the genus *Cystoseira* in the Gulf of Naples (South Tyrrhenian Sea). *BioRxiv* **2017**, 160200. [CrossRef]
29. Laakkonen, H.M.; Strelkov, P.; Väinölä, R. Molecular lineage diversity and inter-oceanic biogeographical history in *Hiatella* (Mollusca, Bivalvia). *Zool. Scr.* **2015**, *44*, 383–402. [CrossRef]
30. Sánchez-Moyano, J.; Estacio, F.; García-Adiego, E.; García-Gómez, J. The molluscan epifauna of the alga *halopteris scoparia* in southern Spain as a bioindicator of coastal environmental conditions. *J. Molluscan Stud.* **2000**, *66*, 431–448. [CrossRef]
31. Izquierdo, D.; Guerra-García, J.M. Distribution patterns of the peracarid crustaceans associated with the alga Corallina elongata along the intertidal rocky shores of the Iberian Peninsula. *Helgol. Mar. Res.* **2011**, *65*, 233–243. [CrossRef]
32. Guerra-García, J.M.; de Figueroa, J.M.T.; Barranco, C.N.; Ros, M.; Sanchez-Moyano, J.E.; Moreira, J. Dietary analysis of the marine *Amphipoda* (Crustacea: Peracarida) from the Iberian Peninsula. *J. Sea Res.* **2014**, *85*, 508–517. [CrossRef]
33. Lehmann, T.; Spelda, J.; Melzer, R.; Buršić, M. Pycnogonida (Arthropoda) from Northern Adriatic *Corallina officinalis* Linnaeus, 1758 belts. *Mediterr. Mar. Sci.* **2020**, *22*, 102–107. [CrossRef]
34. Esquete, P.; Rubal, M.; Veiga, P.; Troncoso, J. New records of Sea Spiders (Arthropoda: Pycnogonida) for continental Portugal and notes on species distribution. *Mar. Biodivers. Rec.* **2016**, *9*, 1–5. [CrossRef]
35. Irwin, S.; Davenport, J. Oxygen microenvironment of coralline algal tufts and their associated epiphytic animals. In *Biology and Environment: Proceedings of the Royal Irish Academy*; Royal Irish Academy: Dublin, Ireland, 2010; pp. 185–193.
36. Matias, M.G.; Arenas, F.; Rubal, M.; Pinto, I.S. Macroalgal Composition Determines the Structure of Benthic Assemblages Colonizing Fragmented Habitats. *PLoS ONE* **2015**, *10*, e0142289. [CrossRef]
37. Airoldi, L.; Cinelli, F. Effects of sedimentation on subtidal macroalgal assemblages: An experimental study from a mediterranean rocky shore. *J. Exp. Mar. Biol. Ecol.* **1997**, *215*, 269–288. [CrossRef]
38. Nordsieck, F. *Die Europäischen Meeres-Gehäuseschnecken (Prosobranchia)*; Gustav Fischer Verlag: Stuttgart, Germany, 1968; pp. 1–273.
39. Nordsieck, F. *Die Europäischen Meeresmuscheln (Bivalvia)*; Gustav Fisher Verlag: Stuttgart, Germany, 1969; pp. 1–256.
40. Parenzan, P. Gasteropodi. In *Carta d'Identità delle Conchiglie del Mediterraneo*; Bios Taras: Taranto, Italy, 1970; Volume 1, pp. 1–283.

41. Parenzan, P. Bivalves. In *Carta d'Identità delle Conchiglie del Mediterraneo*; Bios Taras: Taranto, Italy, 1974; Volume 2, pp. 1–277.
42. Sabelli, B.; Gianuzzi-Savelli, R.; Bedulli, D. *Catalogo Annotato dei Molluschi Marini del Mediterraneo*; Libreria Naturalistica Bolognese: Bologna, Italy, 1990; Volume 1, pp. 1–348.
43. Poppe, G.T.; Goto, Y. Scaphopoda, Bivalvia, Cephalopoda. In *European Seashells*; Verlag Christa Hemmen: Wiesbaden, Germany, 1993; Volume 2, pp. 1–221.
44. Gianuzzi-Savelli, R.; Pusateri, F.; Palmeri, A.; Ebreo, C. *Atlante delle Conchiglie Marine del Mediterraneo*; La Conchiglia: Roma, Italy, 1996; pp. 1–258.
45. Gofas, S.; Moreno, D.; Salas, C. (Eds.) *Moluscos marinos de Andalucía*; Universidad de Málaga: Málaga, Spain, 2011; Volume 1, pp. 1–342.
46. Gofas, S.; Moreno, D.; Salas, C. (Eds.) *Moluscos marinos de Andalucía*; Universidad de Málaga: Málaga, Spain, 2011; Volume 2, pp. 343–798.
47. Tischler, W. *Grundzüge der Terrestrischen Tierökologie*; Friedrich Vieweg und Sohn: Braunschweig, Germany, 1949; 219p. [CrossRef]
48. Travizi, A. The nematode fauna of the northern Adriatic offshore sediments: Community structure and biodiversity. *Acta Adriat.* **2010**, *51*, 169–180.
49. Năstase, A.; Honț, S.; Iani, M.; Paraschiv, M.; Cernișencu, I.; Năvodaru, I. Ecological status of fish fauna from Razim Lake and the adjacent area, the Danube Delta Biosphere Reserve, Romania. *Acta Ichthyol. Piscat.* **2022**, *52*, 43–52. [CrossRef]
50. Kelaher, B.P. Changes in habitat complexity negatively affect diverse gastropod assemblages in coralline algal turf. *Oecologia* **2003**, *135*, 431–441. [CrossRef]
51. Matias, M.; Underwood, T.; Hochuli, D.; Coleman, R. Habitat identity influences species—Area relationships in heterogeneous habitats. *Mar. Ecol. Prog. Ser.* **2011**, *437*, 135–145. [CrossRef]
52. Kelaher, B. Influence of physical characteristics of coralline turf on associated macrofaunal assemblages. *Mar. Ecol. Prog. Ser.* **2002**, *232*, 141–148. [CrossRef]
53. Lavender, J.T.; Dafforn, K.A.; Bishop, M.J.; Johnston, E.L. Small-scale habitat complexity of artificial turf influences the development of associated invertebrate assemblages. *J. Exp. Mar. Biol. Ecol.* **2017**, *492*, 105–112. [CrossRef]
54. Urra, J.; Rueda, J.; Ramírez, M.; Marina, P.; Tirado, C.; Salas, C.; Gofas, S. Seasonal variation of molluscan assemblages in different strata of *photophilous algae* in the Alboran Sea (western Mediterranean). *J. Sea Res.* **2013**, *83*, 83–93. [CrossRef]
55. Mann, K.H. Production and use of detritus in various freshwater, estuarine, and coastal marine ecosystems. *Limnol. Oceanogr.* **1988**, *33*, 910–930. [CrossRef]
56. Olabarria, C.; Chapman, M. Comparison of patterns of spatial variation of microgastropods between 2 contrasting intertidal habitats. *Mar. Ecol. Prog. Ser.* **2001**, *220*, 201–211. [CrossRef]
57. Olabarria, C.; Chapman, M. Habitat-associated variability in survival and growth of three species of microgastropods. *J. Mar. Biol. Assoc. United Kingd.* **2001**, *81*, 961–966. [CrossRef]
58. Hicks, G.R.F. Meiofauna associated with rocky shore algae. In *The Ecology of Rocky Coasts*; Moore, P.G., Seed, R., Eds.; Hodder and Stoughton: London, UK, 1985; pp. 36–56.
59. Bussell, J.A. *Biodiversity of the Invertebrate Community Associated with the Turf-Forming Red Alga Corallina officinalis in Tide Pools*; Bangor University: London, UK, 2003; pp. 1–248.
60. Graham, A. *Molluscs: Prosobranch and Pyramidellid Gastropods: Keys and Notes for the Identification of the Species*; E.J. Brill/Dr. W. Backhuys: Leiden, Germany, 1988; pp. 1–662.
61. Hayward, P.J.; Ryland, J.S. *Handbook of the Marine Fauna of North-West Europe*; Oxford University Press: Oxford, UK, 1995; pp. 1–799.

Disclaimer/Publisher's Note: The statements, opinions and data contained in all publications are solely those of the individual author(s) and contributor(s) and not of MDPI and/or the editor(s). MDPI and/or the editor(s) disclaim responsibility for any injury to people or property resulting from any ideas, methods, instructions or products referred to in the content.

Brief Report

New Data on Exotic Muricid Species (Neogastropoda: Muricidae) from Spain Based on Integrative Taxonomy

Rafael Bañón [1,*], Juan Fariña [2] and Alejandro de Carlos [3,4]

1. Grupo de Estudos do Medio Mariño (GEMM), Edif. Club Naútico Bajo, 15960 Ribeira, Spain
2. Cofradía Santiago Apóstol de Barallobre, Paseo Portuario s/n, 15528 Fene, Spain; asistenciatecnica@cofradiabarallobre.org
3. Departamento de Bioquímica, Xenética e Inmunoloxía, Facultade de Bioloxía, Universidade de Vigo, Campus Universitario Lagoas-Marcosende s/n, 36310 Vigo, Spain; adcarlos@uvigo.es
4. Centro de Investigación Mariña, Edificio Filomena Dato, Campus Universitario Lagoas-Marcosende s/n, 36310 Vigo, Spain
* Correspondence: anoplogaster@yahoo.es

Abstract: The occurrence of *Ocinebrellus inornatus* and *Rapana venosa*, two exotic marine gastropods of the family Muricidae originating from the northwest Pacific, is reported in Spanish waters, specifically in the Galician waters (NW Spain) in 2023. Live specimens of *O. inornatus* were found on Illa de Arousa, in the Ría de Arousa, southern Galicia, where they are already established. Two new specimens of *R. venosa* are recorded in Galicia, one of them for the first time out of the Ría de Arousa, representing a range expansion for the species. The DNA barcoding analysis confirms the previous morphological identifications. It is suspected that both species may have been introduced through importation of clam spat and middle-sized oysters from countries such as France, the United Kingdom or Italy for subsequent culture, as has been the case with other exotic species that can currently be found in Galician waters. The continuous arrival of marine exotic species strongly supports the need to establish a monitoring program in Galician waters.

Keywords: mollusca; gastropods; *Ocinebrellus inornatus*; *Rapana venosa*; invasive species

Citation: Bañón, R.; Fariña, J.; de Carlos, A. New Data on Exotic Muricid Species (Neogastropoda: Muricidae) from Spain Based on Integrative Taxonomy. *Diversity* **2023**, *15*, 1185. https://doi.org/10.3390/d15121185

Academic Editor: Renato Mamede

Received: 10 November 2023
Revised: 24 November 2023
Accepted: 28 November 2023
Published: 29 November 2023

Copyright: © 2023 by the authors. Licensee MDPI, Basel, Switzerland. This article is an open access article distributed under the terms and conditions of the Creative Commons Attribution (CC BY) license (https:// creativecommons.org/licenses/by/ 4.0/).

1. Introduction

With over 1600 extant described species, the family Muricidae is one of the most species-rich and morphologically diverse families of molluscs living worldwide, from tropical to polar seas, and ranging from the intertidal zone down to more than 3000 m depth [1]. There are 52 muricid species reported in Spanish waters [2], of which 14 can be found in the waters off Galicia, including several exotic ones [3].

The introduction and spread of exotic (alien, non-native, non-indigenous) species is considered one of the main threats to biodiversity, as they displace native species, modify community structure and food webs, and alter fundamental processes such as nutrient cycling and sedimentation [4]. A total of 1369 marine alien species have been reported in European seas, including 110 cryptogenic and 139 questionable species [5]. Regarding Spanish marine waters, a total of 574 species have been identified with an alien, cryptogenic, crypto-expanding or debatable status [6]. The European Union (EU) recognises the need for strong action to control biological invasions and thus mitigate their impact on biodiversity, ecosystem services and human activities [7].

The Japanese oyster drill *Ocinebrellus inornatus* (Récluz, 1851) is a muricid gastropod mollusc native to the North Pacific, ranging from northern China, Korea and Japan to the Sakhalin and Kuril Islands in Russia [8,9]. In this native area, its distribution is largely overlapping with *Magallana gigas* (Thunberg, 1793), which is its main prey and on whose shells it tends to lay its eggs [9]. *Ocinebrellus inornatus* was first discovered outside its natural range on the west coast of the United States in 1924, linked to the importation of

M. gigas from Japan [10], spreading to all the states of the west coast of the United States and to British Columbia in Canada in a few years. In Europe, it was first reported from France in 1995 [11], probably due to the importation of *M. gigas* brood stock from British Columbia in the 1970s, as the resemblance of this species to the native *Ocenebra erinaceus* (Linnaeus, 1758) caused it to go unnoticed for years. It was subsequently reported from the coasts of the Netherlands [12], Portugal [13] and Denmark [9], usually associated with aquaculture activities of *M. gigas*.

The veined rapa whelk *Rapana venosa* Valenciennes 1846 is native to the Sea of Japan, the Yellow Sea, the Bohai Sea and the East China Sea to Taiwan [14]. This species was first discovered outside its native biogeographic range in the Black Sea in 1947 and later its range was extended to the Aegean and Adriatic seas, South America (Argentina and Uruguay), North America (eastern USA), France (Brittany coast) and the North Sea [15].

The aim of this paper is to report new data on the occurrence of two species of exotic muricid mollusc from Spanish Galician waters based on morphology and DNA barcoding.

2. Materials and Methods

2.1. Study Area

Galicia is an autonomous region of Spain located in the northwestern corner of the Iberian Peninsula, between the river Eo (43°32′ N, 7°01′ W) and the river Miño (41°50′ N, 9°40′ W) (Figure 1). The Galician coast has a length of 1498 km and is characterised by the presence of the Rías, which are tectonic estuaries penetrating the coast almost perpendicular to the coastline. The Ria of Arousa is the largest of the eighteen Galician Rías; it is located in the southern part of the Galician coast, belonging to the Rías Baixas group. It has an area of 230 km² and the maximum depth is 69 m at its mouth, which is open to the Atlantic Ocean along a northeast–southwest axis. Along the Ria, there are numerous commercial and fishing ports with important fishing and shellfishing activity.

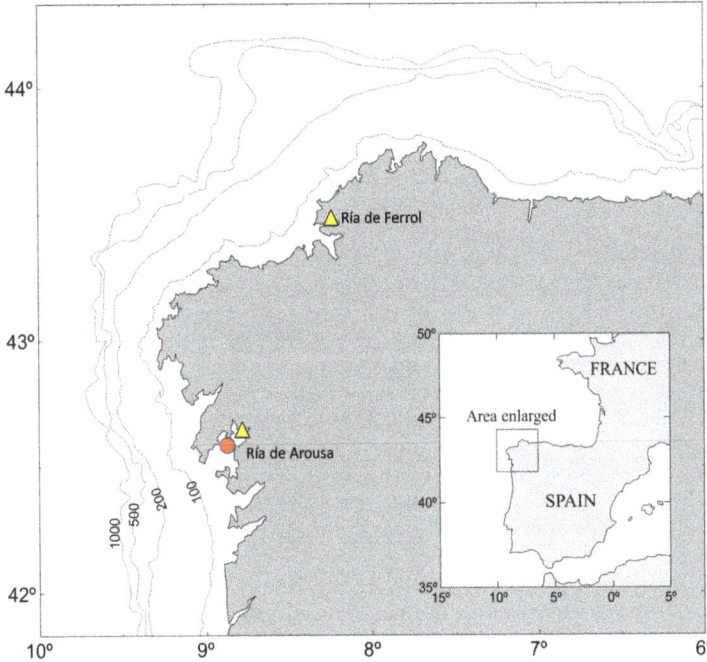

Figure 1. Map showing the location of the recorded specimens of *Ocinebrellus inornatus* (red circle) and *Rapana venosa* (yellow triangles).

2.2. Sampling Data

A total of 113 live specimens of *O. inornatus* were randomly collected from Xastelas beach, in the Illa de Arousa, in 2023 and transported to the laboratory to be measured. Two specimens of *R. venosa* were caught in 2023 by small-scale fishing vessels that regularly fish along the Galician coast. The identification was initially made by examination of morphological characters following molluscan literature. Total shell height and shell width were measured to 0.01 mm using a digital calliper (TesaCalIP65) and specimens were weighed with a digital scale (Mettler Toledo) to 0.1 g.

A subsample of 11 specimens of *O. inornatus* and 4 of *O. erinaceus*, used as comparative material, was reserved frozen for the purpose of molecular identification. Finally, 16 specimens (11 *O. inornatus*, 4 *O. erinaceus* and 1 *R. venosa*) were deposited at the Museo de Historia Natural da Universidade de Santiago de Compostela (MHN USC, Santiago de Compostela, Spain) under reference codes 25217-1 to 25217-11 for *O. inornatus*, MHN USC 25218-1 to 25218-4 for *O. erinaceus* and MHNUSC 25214 for *R. venosa*.

2.3. DNA Extraction, PCR, and Sequencing

DNA was extracted from a tissue sample from each individual using the E.Z.N.A Tissue DNA Kit. The 5′ region of the mitochondrial gene coding for subunit I of the cytochrome c oxidase enzyme (*COI*-5P) was amplified using the universal primers LCO1490 and HCO2198 [16]. The enzymatic reaction was carried out using the Horse Power Green Taq DNA Polymerase mix (Canvax, Valladolid, Spain) with a classical PCR thermal regime [17]. The resulting amplicons were sequenced with BigDye Terminator v3.1 Cycle Sequencing Kit, in both directions using the same primers as for amplification. The sequences were 658 nucleotides long, contained no insertions or deletions, and encoded a polypeptide of 219 amino acids. All information regarding the specimens as well as their barcodes, images, places of capture and other complementary data are available in the project "Marine Invertebrates" (code INVMA) at the Barcode of Life Data System (BOLD Systems, www.boldsystems.org, accessed on 30 August 2023). The sequences were also submitted to GenBank (https://www.ncbi.nlm.nih.gov/genbank/) having the accession numbers OR524177–OR524179, OR524182–OR524185 and OR524187–OR524190 for *O. inornatus*; OR524176, OR524180, OR524186 and OR524191 for *O. erinaceus* and OR524181 for *R. venosa*.

2.4. Sequences Alignment

Sequences were aligned with each other and with other sequences obtained from the BOLD database. The number of differences among sequences of *O. inornatus* and *O. erinaceus* samples was calculated using uncorrected p-distances [18] in order to obtain a phylogenetic tree to be used as a taxonomic dendrogram by the neighbor joining (NJ) method [19], including resampling values (1000) in bootstrap form [20]. Prior to phylogenetic analysis, the sequences were collapsed into haplotypes using FaBox 1.61 [21]. The analysis included 11 nucleotide sequences with 658 positions in the final dataset and was conducted in MEGA11 [22].

The *COI* sequence from *R. venosa* was used to: (a) double check the morphological identification of the voucher specimen via DNA barcoding and (b) to assess possible genetic structure across individuals from different sampling sites. For this, a median joining network [23] was constructed with PopART 1.7 [24], with $\varepsilon = 0$, including all sequences available in BOLD whose geographic origin was specified.

3. Results

3.1. Systematics

Class Gastropoda Cuvier, 1795
Order Neogastropoda
Family Muricidae Rafinesque, 1815
Genus *Ocinebrellus* Jousseaume, 1880
Ocinebrellus inornatus (Récluz, 1851) (Figure 2).

Genus *Rapana* H. C. F. Schumacher, 1817
Rapana venosa (Valenciennes, 1846)

Figure 2. Specimens of exotic *Ocinebrellus inornatus* (**A**) and native *Ocenebra erinaceus* (**B**), collected from the Illa de Arousa, Galicia (NW Spain).

3.2. Material Examined

A total of 113 live specimens of *O. inornatus* between 28.3 and 45.7 mm shell height (Table 1, Figure 2) were caught on 5 June 2023 in Xastelas beach, 42.531 N, −8.867 W, Ría de Arousa, Galicia, NW Spain, at a depth of 0–0.5 m. Two specimens of R. venosa were captured in the south and north of the Galician coast (Figure 3). The first specimen, of 140.7 mm shell height and 104.2 mm shell width, was caught on 6 June 2023 by local fishermen using fyke nets, near to Illa de Cortegada, Ria de Arousa, south of Galicia, 42.607 N, −8.778 W, at 5 m depth. The second specimen, of 160.4 mm shell height and 130.1 mm shell width, was caught on 14 June 2023 by local fishermen using a dredge for molluscs, in the Ría de Ferrol, north of Galicia, 43.465 N, −8.241 W, at 10 m depth.

Table 1. Morphometric data of *Ocinebrellus inornatus* collected from the northern coast of Spain. Size is expressed in millimetres and weight in grams.

Character	Number	Minimum	Maximum	Mean	SD
Shell height	113	28.3	45.7	37.1	3.8
Shell width	113	15.6	28.8	22	2.7
Weight	113	2	11.2	5.8	1.8

Figure 3. Specimens of *Rapana venosa* 140.7 mm shell height (**A**) and MHNUSC 25214 160.4 mm shell height (**B**), collected from Galicia (NW Spain).

3.3. Remarks

Ocinebrellus inornatus is very similar to the native species *O. erinaceus* from which it is very difficult to distinguish due to the morphological variability of both species [3]. The main differences are a light brown operculum in *O. inornatus* versus brown in *O. erinaceus* and the shoulders of whorls, which are more prominent and angulated giving a more pronounced turreted effect in *O. inornatus* [25].

3.4. Molecular Taxonomy

All sequences obtained from *O. inornatus* specimens collected in Illa de Arousa showed the same haplotype of the barcoding region of the *COI* gene. In contrast, the four sequences of *O. erinaceus* represent three different haplotypes. The NJ dendrogram shows three clades, with two of them grouping sequences named as *O. inornatus*, separated by a 16.41% distance between the closest sequences, the one representing the individual collected in Illa de Arousa OR524179, and a sequence from Korea HM180493. It should be noted that the latter contains a sequence named *Neptunea cumingi* Crosse, 1862 (Figure 4). The distance between the sequence of *O. inornatus* from the Illa de Arousa and its nearest neighbour in the *O. erinaceus* clade is 12.46%, to a sequence of a specimen collected at the same site, OR524176. Mean distances within each clade ranged from 0.33% to 0.43%.

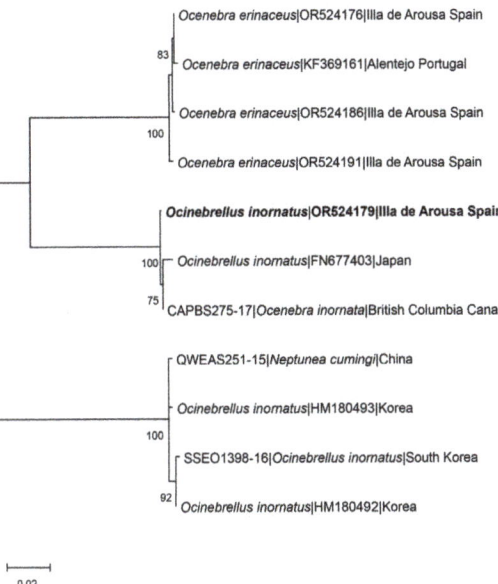

Figure 4. Uncorrected p-distance neighbor joining analysis of *COI*-5P barcode haplotypes of *Ocinebrellus inornatus* and *Ocenebra erinaceus*, including the one from specimens found in the Illa de Arousa, Spain (bolded).

Figure 5 shows a haplotype network of 43 sequences, 614 nucleotides long, of the mitochondrial *COI* marker of *R. venosa*. Individuals collected from China, Japan and Korea showed a wide variety of haplotypes, 12 out of a total of 13, whereas all non-Asian sequences clustered into a single haplotype H4.

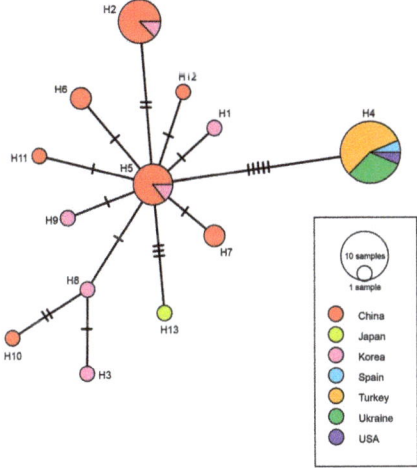

Figure 5. Haplotype network of *COI* sequences of *Rapana venosa* (n = 43). Median joining network (ε = 0) created in PopART v1.7. Each circle represents a haplotype; the size of circles corresponds to the number of individuals with the haplotype. Colours indicate sampling sites. Bars indicate the number of mutations between two haplotypes.

4. Discussion

The morphological features of the specimens of *O. inornatus* and *R. venosa* are in agreement with previous descriptions [14,15,26] and the taxonomic identification is confirmed by the similarity of the nucleotide sequences obtained, with others of the same nominal species present in the BOLD and GenBank databases.

The presence of *O. inornatus* in Galicia is not surprising given that it was introduced a long time ago into European waters, and it is widely distributed to the north and south of the reported area. Previously, this species had been recorded in Galician waters based on dead specimens [27]. *Ocinebrellus inornatus* has been transported around the world together with *M. gigas* transplants. The latter species was introduced into the southern Galician waters in the 1980s, and has been cultivated in the vicinity of the Illa de Arousa since the 1990s, from spat continuously imported from France [28], where *O. inornatus* is largely introduced [8,29].

In colonised areas in different parts of America and Europe, *O. inornatus* prefers the substrates of oyster beds naturally found between the intertidal zone and 5–6 m depth, of species such as *M. gigas*, *Ostrea lurida* P. P. Carpenter, 1864 and *O. edulis* [30]. However, there are no natural populations of *O. edulis* in the Ría de Arousa. Naturalised populations of *M. gigas* only exist on the north coast of Galicia, in the Rías altas [31]. Therefore, in the absence of oysters in the Galician occupied area, *O. inornatus* opts for a different diet, mainly other commercial and abundant bivalves present in Xastelas beach such as *Ruditapes philippinarum* (Adams and Reeve, 1850) and *Ruditapes decussatus* (Linnaeus, 1758), with the consequent economic loss for the shellfish sector. *Ocenebrellus inornatus* has 15 times higher reproductive effort and a better average growth rate than *O. erinaceus* which may explain the fast rapid invasive establishment of this species [32].

The settlement of *R. venosa* in Galicia has been comparatively slower than the other exotic muricid species. *Rapana venosa* was first introduced in the Ría de Arousa in 2003 [33]. Subsequently, only two specimens have been reported, one in 2005 [34] and one in 2007 [35], both also in the Ría de Arousa. *Rapana venosa* has a very showy and colourful shell, so empty shells are often kept as decorative elements, which have made it possible to backtrack the catches in time [33].

As for the two new reported specimens of *R. venosa*, although the first one also comes from the Ría de Arousa, the other was captured, for the first time, outside this ría, in the north of Galicia, extending the range of the invaded areas.

In the marine environment, the introduction of exotic species is mainly the result of aquaculture activities, especially shellfish farming, which represents a major cause of introduction, intentional or not, of exogenous species [36]. The most likely vector for the introduction of the exotic muricids, and other invasive molluscan species in Galician waters is also the unintentional co-transport of these species in commercial bivalve cultures, mainly clams and oysters, imported from Mediterranean or Atlantic areas, where they are already well established [35,37,38]. However, in the case of *R. venosa* the ship transport of egg cases or adults as biofouling is also plausible, especially in the case of the northern specimen, given the proximity to the site of capture of a major port of a liquefied natural gas company, with significant vessel movements.

Unlike the other exotic muricids already established in Galician waters [38], only six specimens of *R. venosa* have been reported in the last 20 years. Several factors may condition the low occurrence rate of this species, such as non-viable egg clutches or a low egg survival rate. The fact that the specimens are not accessible to different fishing gears, or that the pelagic larval stage causes high dispersion, may also play a role [33].

The DNA barcoding technique is a useful tool for marine species identification, including exotic molluscs [39,40]. Molecular taxonomy showed discordant results, with two distant groups containing sequences of *O. inornatus*. Although DNA barcoding is predominantly a tool for species identification, profound genetic divergences, when found, could be due to misidentification of specimens or involve cryptic or unrecognised speciation events [41].

The haplotype of *O. inornatus* from the Illa de Arousa grouped with others of the same species from Canada and China, showing a typical mean intraspecific distance value, and the distance to its nearest neighbour of *O. erinaceus* was within typical interspecific values, showing a classical barcoding gap. Surprisingly, the other clade also has haplotypes of *O. inornatus* from China and Korea, where this species also occurs. However, sequences of this clade contain one haplotype named as *N. cumingi*, a morphologically similar species, which would point to a misidentification of the *O. inornatus* specimens of this clade. Gaps and inconsistencies in reference DNA databases can make it difficult to accurately identify taxa to the species level, suggesting the need to strengthen the DNA barcoding reference datasets [42]. The comparison of newly generated barcodes with published data may help to detect misidentifications, taxonomic uncertainties or real cases of haplotype sharing among species [43].

In the case of *R. venosa*, collections from the native range, the far East, showed high levels of genetic variation, while collections from all introduced populations showed a total lack of genetic diversity; a single haplotype was common to all introduced individuals in the Mediterranean Sea and Atlantic Ocean. This finding has been observed before, and is consistent with the hypothesis of the introduction and establishment of a population in the Black Sea that would later serve as a source of secondary invasions mediated by vectors such as ballast water transport [44]. Although non-native *R. venosa* populations currently appear to be thriving in their new environments, the lack of genetic variability raises questions regarding their evolutionary persistence.

Galicia, but especially the Ría de Arousa, is a very important area of introduction for exotic marine organisms in Spain [35,38,39], and both *O. inornatus* and *R. venosa* are carnivorous gastropods whose main diet consists of a variety of native molluscs (shellfish and other bivalves), which could lead to important ecological and economic losses [32,33,35]. The aim of this report is to alert interested parties to the importance of establishing a monitoring program to assess new observations of alien species, in order to detect the increasing expansion of their range and to predict the impacts they may cause.

Author Contributions: Sampling: R.B. and J.F.; Conceptualization, R.B. and A.d.C.; methodology, R.B. and A.d.C.; formal analysis, R.B. and A.d.C.; writing—original draft preparation, R.B. and A.d.C.; writing—review and editing, R.B. and A.d.C.; funding acquisition, A.d.C. All authors have read and agreed to the published version of the manuscript.

Funding: A.d.C. was partially funded by the GRC programme of the Xunta de Galicia (ED431C 2019/28).

Institutional Review Board Statement: Not applicable.

Data Availability Statement: The datasets generated during the current study are deposited in public databases (NCBI GenBank, Barcode of Life Data System and Museo Luis Iglesias de Ciencias Naturais in Santiago de Compostela, Spain).

Acknowledgments: Thanks to Fernando Febrero (Xunta de Galicia) for his help with the measurements of *O. inornatus*.

Conflicts of Interest: The authors declare no conflict of interest.

References

1. Barco, A.; Claremont, M.; Reid, D.G.; Houart, R.; Bouchet, P.; Williams, S.T.; Cruaud, C.; Couloux, A.; Oliverio, M. A molecular phylogenetic framework for the Muricidae, a diverse family of carnivorous gastropods. *Mol. Phylogenet. Evol.* **2010**, *56*, 1025–1039. [CrossRef] [PubMed]
2. Gofas, S.; Luque, A.A.; Templado, J.; Salas, C. A national checklist of marine Mollusca in Spanish waters. *Sci. Mar.* **2017**, *81*, 241–254. [CrossRef]
3. Trigo, J.E.; Diaz Agras, G.J.; García Álvarez, O.L.; Guerra, A.; Moreira, J.; Pérez, J.; Rolán, E.; Troncoso, J.S.; Urgorri, V. *Guía de los Moluscos Marinos de Galicia*; Servicio de Publicacións da Universidade de Vigo: Vigo, Spain, 2018; pp. 1–836.
4. Molnar, J.L.; Gamboa, R.L.; Revenga, C.; Spalding, M.D. Assessing the global threat of invasive species to marine biodiversity. *Front. Ecol. Environ.* **2008**, *6*, 485–492. [CrossRef]
5. Katsanevakis, S.; Gatto, F.; Zenetos, A.; Cardoso, A.C. How many marine aliens in Europe? *Manag. Biol. Invasions* **2013**, *4*, 37–42. [CrossRef]

6. Png-Gonzalez, L.; Comas-González, R.; Calvo-Manazza, M.; Follana-Berná, G.; Ballesteros, E.; Díaz-Tapia, P.; Falcón, J.M.; García Raso, J.E.; Gofas, S.; González-Porto, M.; et al. Updating the National Baseline of Non-Indigenous Species in Spanish Marine Waters. *Diversity* **2023**, *15*, 630. [CrossRef]
7. Katsanevakis, S.; Genovesi, P.; Gaiji, S.; Nyegaard, H.; Roy, H.; Nunes, A.L.; Sánchez, F.; Bogucarskis, K.; Debusscher, B.; Deriu, I.; et al. Implementing the European policies for alien species—Networking, science, and partnership in a complex environment. *Manag. Biol. Invasions* **2013**, *4*, 3–6. [CrossRef]
8. Garcia-Meunier, P.; Martel, C.; Pigeot, J.; Chevalier, G.; Blanchard, G.; Goulletquer, P.; Robert, S.; Sauriau, P.G. Recent invasion of the Japanese oyster drill along the French Atlantic coast: Identification of specific molecular markers that differentiate Japanese, *Ocinebrellus inornatus*, and European, *Ocenebra erinacea*, oyster drills. *Aquat. Living Resour.* **2002**, *15*, 67–71. [CrossRef]
9. Lützen, J.; Faasse, M.; Gittenberger, A.; Glenner, H.; Hoffmann, E. The Japanese oyster drill *Ocinebrellus inornatus* (Récluz, 1851) (Mollusca, Gastropoda, Muricidae), introduced to the Limfjord, Denmark. *Aquat. Invasions* **2012**, *7*, 181–191. [CrossRef]
10. Galtsoff, P.S. Oyster industry of the Pacific coast of the United States. *Comm. Fish Rep.* **1929**, *8*, 367–400.
11. de Montaudouin, X.; Sauriau, P.G. Contribution to a synopsis of marine species richness in the Pertuis Charentais Sea with new insights in soft-bottom macrofauna of the Marennes Oléron Bay. *Cah. Biol. Mar.* **2000**, *41*, 181–222.
12. Faasse, M.; Ligthart, M. American (*Urosalpinx cinerea*) and Japanese oyster drill (*Ocinebrellus inornatus*) (Gastropoda: Muricidae) flourish near shellfish culture plots in The Netherlands. *Aquat. Invasions* **2009**, *4*, 321–326. [CrossRef]
13. Afonso, C.M. Non-indigenous Japanese oyster drill *Pteropurpura (Ocinebrellus) inornata* (Recluz, 1851) (Gastropoda: Muricidae) on the South-west coast of Portugal. *Aquat. Invasions* **2011**, *6*, S85–S88. [CrossRef]
14. Mann, R.; Occhipinti, A.; Harding, J.M. (Eds.) *Alien Species Alert: Rapana venosa (Veined Whelk)*; ICES Cooperative Research Reports; ICES: Copenhagen, Denmark, 2004; Volume 264, pp. 1–14.
15. Kerckhof, F.; Vink, R.J.; Nieweg, D.C.; Post, J.N.J. The veined whelk *Rapana venosa* has reached the North Sea. *Aquat. Inv.* **2006**, *1*, 35–37. [CrossRef]
16. Folmer, O.; Black, M.; Hoeh, W.; Lutz, R.; Vrijenhoek, R. DNA primers for amplification of mitochondrial cytochrome c oxidase subunit I from diverse metazoan invertebrates. *Mol. Mar. Biol. Biotechnol.* **1994**, *3*, 294–299. [PubMed]
17. Hebert, P.D.N.; Cywinska, A.; Ball, S.L.; de Waard, J.R. Biological identification through DNA barcodes. *Proc. Biol. Sci.* **2003**, *270*, 313–321. [CrossRef]
18. Nei, M.; Kumar, S. *Molecular Evolution and Phylogenetics*; Oxford University Press: New York, NY, USA, 2000; pp. 1–333.
19. Saitou, N.; Nei, M. The neighbour-joining method: A new method for reconstructing phylogenetic trees. *Mol. Biol. Evol.* **1987**, *4*, 406–425.
20. Felsenstein, J. Confidence limits on phylogenies: An approach using the bootstrap. *Evolution* **1985**, *39*, 783–791. [CrossRef]
21. Villesen, P. FaBox: An online toolbox for fasta sequences. *Mol. Ecol. Notes* **2007**, *7*, 965–968. [CrossRef]
22. Tamura, K.; Stecher, G.; Kumar, S. MEGA11: Molecular Evolutionary Genetics Analysis version 11. *Mol. Biol. Evol.* **2021**, *38*, 3022–3027. [CrossRef]
23. Bandelt, H.; Forster, P.; Röhl, A. Median-Joining networks for inferring intraspecific phylogenies. *Mol. Biol. Evol.* **1999**, *16*, 37–48. [CrossRef]
24. Leigh, J.W.; Bryant, D. PopART: Full-feature software for haplotype network construction. *Methods Ecol. Evol.* **2015**, *6*, 1110–1116. [CrossRef]
25. Smith, I.F. *Ocenebra erinaceus* (Linnaeus, 1758) Identification and Biology. Available online: https://flic.kr/s/aHBqjAx7Eg (accessed on 30 October 2023).
26. Amano, K.; Vermeij, G.J. Taxonomy and evolution of the genus *Ocinebrellus* (Gastropoda: Muricidae) in Japan. *Paleontol. Res.* **1998**, *2*, 199–212.
27. Trigo, J.E. *Ocinebrellus inornatus* (Recluz, 1851) (Gastropoda: Muricidae) en la ria de Arousa. Nueva especie exótica para aguas españolas. In Proceedings of the Foro Malacológico de la Sociedad Española de Malacología, Vigo, Spain, 19–21 September 2019.
28. Iglesias, D.; Rodríguez, L.; Gómez, L.; Azevedo, C.; Montes, J. Histological survey of Pacific oysters *Crassostrea gigas* (Thunberg) in Galicia (NW Spain). *J. Invertebr. Pathol.* **2012**, *111*, 244–251. [CrossRef] [PubMed]
29. Pigeot, J.; Miramand, P.; Garcia-Meunier, P.; Guyot, T.; Séguignes, M. Présence d'un nouveau prédateur de l'huitre creuse, *Ocinebrellus inornatus* (Récluz, 1851), dans le bassin conchylicole de Marennes-Oléron. *Comptes Rendus L'académie Sci.—Ser. III—Sci. Vie* **2000**, *323*, 697–703. [CrossRef] [PubMed]
30. Barbaran, D. *Japanese oyster Drills (Ocinebrellus inornatus): Exploring Prey Size and Species Preference in The Netherlands*; Final Report; HZ University op Applied Sciences: Flesinga, The Netherlands, 2017; pp. 1–39.
31. Des, M.; Gómez-Gesteira, J.L.; de Castro, M.; Iglesias, D.; Sousa, M.C.; ElSerafy, G.; Gómez-Gesteira, M. Historical and future naturalization of *Magallana gigas* in the Galician coast in a context of climate change. *Sci. Total Environ.* **2022**, *838*, 156437. [CrossRef]
32. Martel, C.; Guarini, J.M.; Blanchard, G.; Sauriau, P.G.; Trichet, C.; Robert, S.; Garcia-Meunier, P. Invasion by the marine gastropod *Ocinebrellus inornatus* in France. III. Comparison of biological traits with the resident species *Ocenebra erinacea*. *Mar. Biol.* **2004**, *146*, 93–102. [CrossRef]
33. Bañón, R.; Mascato, J. Nuevas citas de *Rapana venosa* (Valenciennes, 1846) (Gastropoda: Muricidae) en aguas de Galicia. *Noticiario SEM* **2014**, *62*, 39–41.

34. Trigo, J.E.; Vieites, N. Segunda cita de *Rapana venosa* (Valenciennes, 1846) (Gastropoda: Muricidae) para la Península Ibérica. *Noticiario SEM* **2013**, *59*, 77–78.
35. Rolán, E.; Bañón, R. Primer hallazgo de la especie invasora *Rapana venosa* y nueva información sobre *Hexaplex trunculus* (Gastropoda, Muricidae) en Galicia. *Noticiario SEM* **2007**, *47*, 57–59.
36. Pante, E.; Pascal, P.Y.; Becquet, V.; Viricel, A.; Simon-Bouhet, B.; Garcia, P. Evaluating the genetic effects of the invasive *Ocenebra inornata* on the native oyster drill *Ocenebra erinacea*. *Mar. Ecol.* **2015**, *36*, 1118–1128. [CrossRef]
37. Rolán, E.; Trigo, J.; Otero-Schmitt, J.; Rolán-Álvarez, E. Especies implantadas lejos de su área de distribución natural. *Thalassas* **1985**, *3*, 29–36.
38. Bañón, R.; Rolán, E.; García-Tasende, M. First record of the purple dye murex *Bolinus brandaris* (Gastropoda: Muricidae) and a revised list of non-native molluscs from Galician waters (Spain, NE Atlantic). *Aquat. Invasions* **2008**, *3*, 331–334. [CrossRef]
39. Bañón, R.; Fernández, J.; Trigo, J.E.; Pérez-Dieste, J.; Barros-García, D.; de Carlos, A. Range expansion, biometric features and molecular identification of the exotic ark shell *Anadara kagoshimensis* from Galician waters, NW Spain. *J. Mar. Biol. Assoc. UK* **2015**, *95*, 545–550. [CrossRef]
40. Pejovic, I.; Ardura, A.; Miralles, L.; Arias, A.; Borrell, Y.J.; Garcia-Vazquez, E. DNA barcoding for assessment of exotic molluscs associated with maritime ports in northern Iberia. *Mar. Biol. Res.* **2016**, *122*, 168–176. [CrossRef]
41. Bañón, R.; de Carlos, A.; Acosta-Morillas, V.; Baldó, F. Geographic range expansion and taxonomic notes of the shortfin neoscopelid *Neoscopelus cf. microchir* (Myctophiformes: Neoscopelidae) in the North-Eastern Atlantic. *J. Mar. Sci. Eng.* **2022**, *10*, 954. [CrossRef]
42. Ardura, A.; Morote, E.; Kochzius, M.; García-Vázquez, E. Diversity of planktonic fish larvae along a latitudinal gradient in the Eastern Atlantic Ocean estimated through DNA barcodes. *PeerJ* **2016**, *4*, e2438. [CrossRef]
43. Knebelsberger, T.; Landi, M.; Neumann, H.; Kloppmann, M.; Sell, A.F.; Campbell, P.D.; Laakmann, S.; Raupach, M.J.; Carvalho, G.R.; Costa, F.O. A reliable DNA barcode reference library for the identification of the North European shelf fish fauna. *Mol. Ecol. Resour.* **2014**, *14*, 1060–1071. [CrossRef]
44. Chandler, E.A.; McDowell, J.R.; Graves, J.E. Genetically monomorphic invasive populations of the rapa whelk, *Rapana venosa*. *Mol. Ecol.* **2008**, *17*, 4079–4091. [CrossRef]

Disclaimer/Publisher's Note: The statements, opinions and data contained in all publications are solely those of the individual author(s) and contributor(s) and not of MDPI and/or the editor(s). MDPI and/or the editor(s) disclaim responsibility for any injury to people or property resulting from any ideas, methods, instructions or products referred to in the content.

Communication

Benthic Biodiversity by Baited Camera Observations on the Cosmonaut Sea Shelf of East Antarctica

Jianfeng Mou [1,2], Xuebao He [1], Kun Liu [1], Yaqin Huang [1], Shuyi Zhang [1], Yongcan Zu [3], Yanan Liu [4], Shunan Cao [5], Musheng Lan [5], Xing Miao [1], Heshan Lin [1,*] and Wenhua Liu [2,*]

1. Laboratory of Marine Biodiversity, Third Institute of Oceanography, Ministry of Natural Resources, Xiamen 361005, China; moujianfeng@tio.org.cn (J.M.); hexuebao@tio.org.cn (X.H.); liukun@tio.org.cn (K.L.); huangyaqin@tio.org.cn (Y.H.); zhangshuyi@tio.org.cn (S.Z.); miaoxing@tio.org.cn (X.M.)
2. Guangdong Provincial Key Laboratory of Marine Disaster Prediction and Prevention, Institute of Marine Sciences, Shantou University, Shantou 515063, China
3. Center for Ocean and Climate Research, First Institute of Oceanography, Ministry of Natural Resources, Qingdao 266061, China; zuyongcan@fio.org.cn
4. Key Laboratory of Submarine Geosciences, Second Institute of Oceanography, Ministry of Natural Resources, Hangzhou 310012, China; ynliu@sio.org.cn
5. Key Laboratory for Polar Science, Polar Research Institute of China, Shanghai 200120, China; caoshunan@pric.org.cn (S.C.); lanmusheng@pric.org.cn (M.L.)
* Correspondence: linheshan@tio.org.cn (H.L.); whliu@stu.edu.cn (W.L.)

Abstract: A free-fall baited camera lander was launched for the first time on the Cosmonaut Sea shelf of East Antarctica at a depth of 694 m during the 38th Chinese National Antarctic Research Expedition (CHINARE) in 2022. We identified 31 unique taxa (23 were invertebrates and eight were fish) belonging to eight phyla from 2403 pictures and 40 videos. The Antarctic jonasfish (*Notolepis coatsi*) was the most frequently observed fish taxa. Ten species of vulnerable marine ecosystem (VME) taxa were observed, accounting for 32% of all species. The maximum number (MaxN) of *Natatolana meridionalis* individuals per image frame was ten, and they were attracted to the bait. The macrobenthic community type were sessile suspension feeders with associated fauna (SSFA), which was shaped by the muddy substrata with scattered rocks. Rocks served as the best habitats for sessile fauna. The study reveals the megafauna community and their habitat by image survey in the Cosmonaut Sea for the first time. It helped us obtain Antarctic biodiversity baselines and monitoring data for future ecosystem health assessment and better protection.

Keywords: lander; Cosmonaut Sea; megafauna; Antarctica; image survey

1. Introduction

The Southern Ocean is unique among the world's oceans in terms of its linkage with the other major ocean basins, its rich and unusual marine ecosystem, and its interaction with the physical climate system and the biogeochemistry of the region [1]. It comprises 15% of the world's oceans and is home to thousands of endemic species [2–4]. The Southern Ocean has become a hot area for global research on climate change and ecological evolution due to its harsh natural environment, fragile ecosystem, and sensitivity to environmental variations. There is a growing need for marine biodiversity baselines and monitoring data to assess ocean ecosystem health, especially around Antarctica, where data are rare [5–8].

The Cosmonaut Sea (30°~60° E) is located to the west of Enderby Land in east Antarctica and has been poorly explored [9]. Therefore, very few biological data have been recorded for the region [10–13], and existing ones mainly include the composition and distribution of phytoplankton, mesozooplankton, euphausiid larvae, krill, squid, and Antarctic jonasfish [14–16]. However, there are still no data about macrobenthos in the Cosmonaut Sea. Antarctic benthic ecosystems in the Cosmonaut Sea may be sentinels for monitoring the effects of climate change [17,18]. Macrobenthic communities in Antarctica

Citation: Mou, J.; He, X.; Liu, K.; Huang, Y.; Zhang, S.; Zu, Y.; Liu, Y.; Cao, S.; Lan, M.; Miao, X.; et al. Benthic Biodiversity by Baited Camera Observations on the Cosmonaut Sea Shelf of East Antarctica. *Diversity* **2024**, *16*, 277. https://doi.org/10.3390/d16050277

Academic Editor: Renato Mamede

Received: 21 March 2024
Revised: 29 April 2024
Accepted: 30 April 2024
Published: 6 May 2024

Copyright: © 2024 by the authors. Licensee MDPI, Basel, Switzerland. This article is an open access article distributed under the terms and conditions of the Creative Commons Attribution (CC BY) license (https://creativecommons.org/licenses/by/4.0/).

differ in biodiversity and ecosystem functioning and are shaped by a variety of physical and biological drivers [19]. Some of these communities are unique in their occurrence and proportions of species and life forms. Some are typical for the entire Antarctic shelf, but never occur with exactly the same proportions or compositions [18].

Trawls, sledges, and dredges were historically common facilities for sampling the shelf benthic marine communities of Antarctica [12,20,21]. These traditional methods can help in understanding the structure and function of the Antarctic benthic systems and are good for species identification of slow-moving macrobenthos. However, they cannot describe species behaviors and interactions and have difficulty in capturing more mobile species [4]. Video surveys are the emerging methods to estimate the relative abundances of scavenging fishes and invertebrates [8,22,23]. Such studies can help us develop a better understanding of ecosystem patterns and processes in Antarctica, such as the Antarctic Peninsula, Prydz Bay, and the Amundsen Sea, and the applicability has also been justified [4,24–27].

In this study, a free-fall baited camera lander was deployed for the first time on the Cosmonaut Sea shelf during the 38th Chinese National Antarctic Research Expedition (CHINARE) on 22 February 2022. The objective was to preliminarily survey fish and invertebrate communities and basically understand the structure of the benthic community and the interactions among species on the Cosmonaut Sea shelf.

2. Method and Equipment

2.1. Location

The lander was deployed on the shelf (48°50.980′ E, 66°22.226′ S) near the slope of the Cosmonaut Sea (Figure 1). The Cosmonaut Sea is the site of an important confluence in polar circulation, but poorly investigated, meriting more research efforts [9].

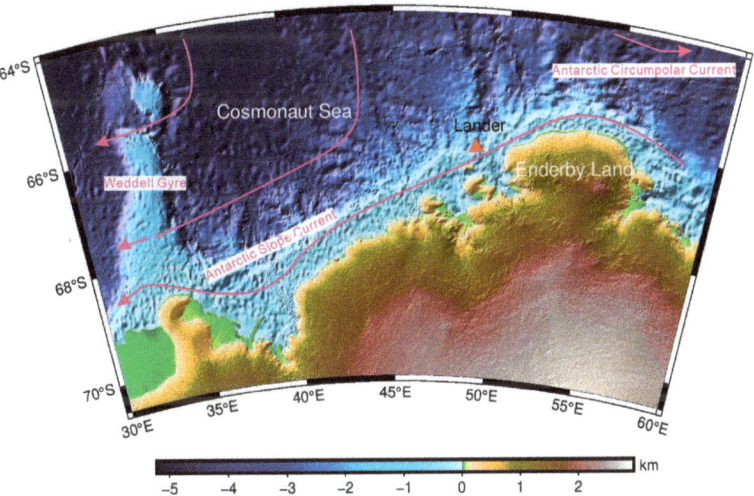

Figure 1. Sampling station of the lander.

2.2. Equipment

The lander system framework is made of titanium, consisting of batteries, an acoustic releaser, cameras, and traps (Figure 2). The batteries are rechargeable lithium batteries combined with capacity greater than 96 Ah. The releaser is manufactured by iXblue. The load and communication distance of the iXblue releaser are not less than 2500 kg and 10 km, respectively. iXblue is powered by a No. 1 alkaline battery or lithium battery. The Sea-Bird Scientific SBE 37 is equipped with high-accuracy temperature, conductivity, and pressure sensors with an RS-232 interface, internal batteries, data storage, and pump. The sampling interval was set to 120 s, and the maximum observation depth was 7000 m. Two cameras were configured: one (CO01-016E, 3648 × 2736) was used to take photos, and the other (CO02-016HE, HD1080P) was used to record videos. The lander used a high-efficiency LED lamp (CO04-50LI-6000AA), and the light was angled downward at a constant 45° from the horizontal plane. Two traps were used in the study: a large fish and invertebrate trap (82 cm × 47 cm × 47 cm) and an invertebrate trap (φ10 cm × 50 cm).

Figure 2. Structure of the free-fall baited camera lander.

2.3. Deployment Process

The lander was deployed and recovered during the 38th CHINARE on the R/V Xuelong2 icebreaker to a depth of 694 m. The equipment was deployed at 20:09 and landed on the sea floor at 20:26 on 22 February 2022 (UTC).

The lander deployment process had four stages: setup, deployment, recovery, and data exporting. Prior to deployment, the cameras were activated through underwater systems by a computer. The camera was preprogrammed to take 1 picture every 30 s, and the video camera was preprogrammed to take 10 min of video every 30 min. The survey time was set according to the battery charging time. Then, baits (chicken legs and *Silver sillago*) were loaded into the traps. The last step was connecting the acoustic releaser to the cement block. Deployment was completed, and the lander was dropped into the sea. The dropping rate was approximately 1 m/s. Location and bottom depths were based on triangulation of the acoustic releases after the deployment [28]. The lander was recovered in the daytime when the sea was calm. The ship traveled to the deployment area for lander recovery. To recover the lander, the deck unit was used to search for the signal from the acoustic releaser, and the distance between the equipment and the ship was measured. Then, the deck unit released the signal to recover the lander. The rising rate was approximately 1 m/s. The crew searched for the lander, which was attached to orange balls from the bridge. Finally, the equipment was found and brought up onto the deck.

The lander was washed using fresh water after being placed on the deck. Then, the samples were collected from the trap cage. Following retrieval, an external hard disk was used to store the images and videos from the cameras and the data from SBE 37 for further analysis.

2.4. Metrics and Biodiversity Analyses

Annotations were made from the video footage to identify species to the highest taxonomic resolution [8]. A relative abundance metric, the maximum number (MaxN), was used in the video survey. Counts of MaxN for individuals of each species in a video frame were performed rather than the total tally per deployment to avoid double-counting [29]. Other variables were also recorded, such as the time for a taxon's first arrival and the time when the maximum number of a taxon was observed upon the landing of the device on the seafloor [8,23,30].

3. Results

A total of 2403 photos and 40 videos were recorded underwater by the lander during approximately 20 h.

3.1. Substrate Type

The images were used to determine substrate characteristics. The area was dominated by muddy substrate containing some rocks. The rocks were scattered on the sea floor and served as substrata for different taxa, such as sponges, bryozoans, and corals. These taxa, in turn, served as living substrata for ophiuroids, asteroids, holothurians, and others (Figure 3).

Figure 3. Substrate type of the Cosmonaut Sea shelf.

3.2. Observation of the Hydrological Environment

The SBE 37 started to record data on salinity, temperature, and pressure at the bottom at 21:00. The linear trend of pressure is 0.0528 dbar per hour, which is significant at a 95% confidence level. The highest pressure was 701.545 dbar at 16:06, and the lowest one was 700.542 at 23:42. The average value of salinity and temperature were 34.6096 psu and 0.1091 °C, respectively. The highest salinity was 34.6446 psu at 5:42, and the highest value of temperature was 0.1584 °C at 8:20. Salinity and temperature varied strongly (Figure 4), with a sharp fall at 13:00. The salinity reached its lowest value (34.5511 psu) at 13:42. The temperature reached its lowest value (−0.0478 °C) at 13:32. The lower temperature and salinity lasted approximately one hour, and they quickly increased at 14:20. These data may imply that there was a low-temperature and low-salinity water mass passing over the bottom.

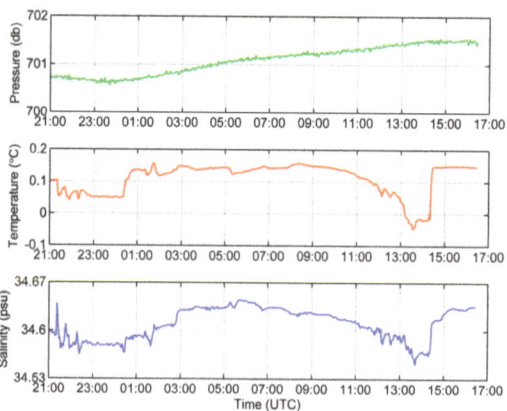

Figure 4. Data collected by the SBE 37 recorder.

3.3. Recorded Taxa

Thirty-one species were identified, representing eight phyla, twenty orders, and thirty families (Table 1). Invertebrates accounted for 23 species, and fishes accounted for 8 species (Figure 5). Bryozoa dominated in the images.

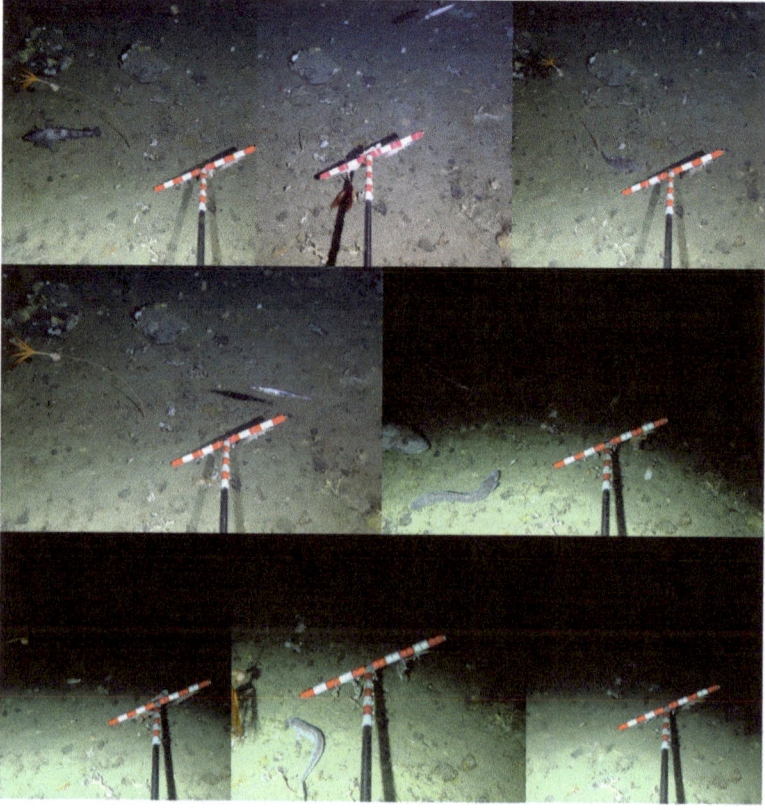

Figure 5. Fish images from the shelf of Cosmonaut Sea.

Three fish were the most frequently occurring fish taxa. *Notolepis coatsi* appeared in 37 videos with up to four individuals in one frame and in 29 of 2403 photos with up to three individuals in one photo. *Trematomus lepidorhinus* was found in 16 of 40 videos and 809 of 2403 photos. *Melanostigma gelatinosum* appeared in 117 of 2403 photos. *Dissostichus mawsoni* was recorded only once by the cameras at 3:35, 23 February 2022.

Ten species of VME (vulnerable marine ecosystem) taxa were observed, accounting for 32% of all species. VME taxa include sponges, sea mats, and cold-water corals, which can provide an important habitat for diversity of marine organisms [31]. Sea pens, sea anemones, and sea squirts were the VME taxa commonly observed (Table 1).

Table 1. Maximum number (MaxN) of individuals of each species.

Phylum	Order	Family	Genus Species	MaxN	VME
Porifera	Demospongiae		Demospongiae und.	1	✔
Cnidaria	Scleralcyonacea	Umbellulidae	*Umbellula carpenteri*	1	✔
		Mopseidae	*Primnoisis* sp.	1	✔
	Actiniaria	Actinostolidae	*Stomphia* sp.	2	✔
	Scleralcyonacea		Scleralcyonacea und.	3	✔
Mollusca	Neogastropoda	Cancellariidae	*Nothoadmete* sp.	4	
		Cochlespiridae	*Aforia* sp.	1	
	Elasipodida	Psychropotidae	*Psychroteuthis glacialis*	1	
Arthropoda	Euphausiacea	Euphausiidae	*Euphausia superba*	5	
	Decapoda	Thoridae	*Lebbeus* sp.	1	
		Nematocarcinidae	*Nematocarcinus lanceopes*	1	
	Isopoda	Cirolanidae	*Natatolana meridionalis*	10	
	Amphipoda	Uristidae	*Abyssorchomene* sp.	13 (trap)	
Echinodermata	Velatida	Pterasteridae	*Pteraster* sp.	1	
	Paxillosida	Astropectinidae	*Psilaster charcoti*	1	
	Ophiurida	Ophiopyrgidae	*Ophioplinthus* sp.	5	
		Ophiacanthidae	Ophiacanthidae und.	1	
	Crinoidea		Crinoidea und.	1	✔
Bryozoa	Cyclostomatida	Horneridae	*Horner* sp.	10	✔
	Cheilostomatida		*Austroflustra vulgaris*	6	✔
		Cellariidae	*Melicerita obliqua*	1	✔
Brachiopoda	Terebratulida	Terebratulinae	Terebratulinae und.	1	
Chordata	Aplousobranchia	Polyclinidae	Polyclinidae und.	2	✔
	Perciformes	Nototheniidae	*Dissostichus mawsoni*	1	
			Trematomus lepidorhinus	1	
		Zoarcidae	*Melanostigma gelatinosum*	2	
		Artedidraconidae	*Pogonophryne* sp.	1	
		Channichthyidae	*Chionobathyscus dewitti*	1	
			Perciformes und.	1	
	Myctophiformes	Myctophidae	*Gymnoscopelus* sp.	1	
	Aulopiformes	Paralepididae	*Notolepis coatsi*	4	

Nothoadmete sp. was first observed at 21:09, and the MaxN per image frame was four. *Natatolana meridionalis* was observed at 21:16, and the MaxN per image frame was ten. These two species were the main ones influencing the MaxN of individuals per image frame each hour.

The species caught in the traps were Isopoda (*N. meridionalis*), Amphipoda (*Abyssorchomene* sp.), and fish (*N. coatsi*), with nine, thirteen, and two individuals (Figure 6), respectively.

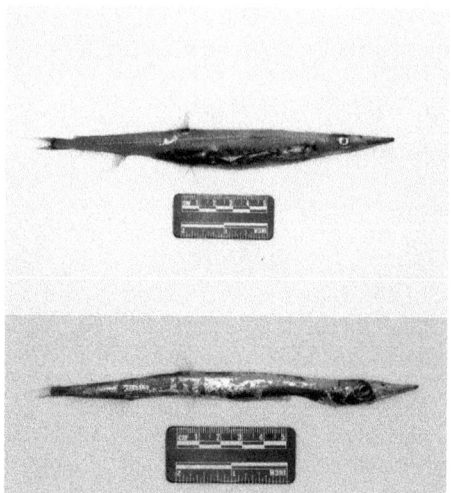

Figure 6. The pictures of *N. coatsi* caught.

4. Discussion

4.1. Substrate Type and Assemblages

Seabed imagery was used to determine substrate characteristics. The distribution of benthic biota on the Antarctic shelf have frequently been associated with substrate type and water depth [32,33]. Muddy substrates with scattered rocks were found in this study. Many rocks may be dropstones from iceberg scouring [33]. Rocks are important for building benthic communities and they provide a hard attachment site and an elevated position off the bottom, enhancing the food supply for sessile suspension feeders, such as sponges, bryozoans, pennatulaceun, and actiniae [32]. Deposit feeders are often associated with soft substrates, where they are able to feed on fine particles [34]. Amphipods were associated with muddy substrates from the images recorded, and the phytodetritus in muddy substrates may be the food source for them.

4.2. Hydrological Environment and Assemblages

Meijers et al. [9] found seasonal Antarctic Bottom Water (AABW) produced in the region (60° E), which moved down the slope and was deflected westward due to the Coriolis force. The mass water mixed with Antarctic Circumpolar Current (ACC) and Weddell Gyre waters above it as it moved westward across the regions 30° E, 40° E, and 50° E, eroding the strong characteristics observed in the region (60° E). Lowered acoustic Doppler current profiler (LADCP) data showed that the bottom water had large velocities near the deployment area [9]. The different postures of *U. carpenteri* showed high-strength and disordered current at the bottom. The current at the bottom brought a large number of organic debris, which provided the food source for sessile suspension feeders, such as sponges, colder-water corals, and sea pens. This was important for structuring the benthic community.

4.3. Comments on Fishes and Invertebrates in Images

Notolepis coatsi was the most frequently encountered fish taxon and the second-largest fish biomass in the Cosmonaut Sea of this study, like that in Prydz Bay [15,16]. It is a midsize bathypelagic fish widely distributed around the Antarctic continental shelf [16]. *N. coatsi* may feed only on krill [35] and account for 50% of the food consumed by Antarctic fur seals (*Arctocephalus gazella*) [36]. Therefore, *N. coatsi* plays an important role in the marine food web, which directly links krill and marine mammals in the Southern Ocean. *N. coatsi* is regarded as an ideal organism to evaluate the role of the Cosmonaut Sea because of its large population size, position in the food web, and specialized diet [16]. Additionally, we not only explored *N. coatsi* behavior but also trapped the individuals using the lander.

Dissostichus mawsoni was observed by photographic survey for the first time in the Cosmonaut Sea and is a large nototheniid species endemic to the Antarctic continental shelf. The Antarctic toothfish is an important fishery resource in the Southern Ocean and plays an important role in Antarctic ecosystems [11]. Sufficient data and stock information are needed to establish fisheries, but such information is rare in the Cosmonaut Sea. In the future, the lander can help us assess the relative abundance and population size structure of the Antarctic toothfish in the Cosmonaut Sea.

In total, 10 VME taxonomic categories were observed from the lander imagery, accounting for 32% of all species. The VME taxa were distributed sparsely, and most of them lived on rocks. The presence of benthic invertebrates contributes to the creation of complex three-dimensional structures and provides substrata for other organisms called sessile suspension feeders with associated fauna (SSFA) [32,33]. The main sessile suspension feeders recorded in this survey were Bryozoa. Associated fauna included Ophiuroidea, Asteroidea, Gastropoda, Isopoda, and others. Isopoda and Amphipoda were the main taxa in the trap, so the bait is possibly attractive to them. *Natatolana meridionalis* first arrived at the bait fifty-one minutes after the lander arrived at the seafloor. The bait was consumed for approximately eight hours, until it was fully exhausted. Data from the lander can help assess changes in benthic community diversity and their associated habitat structure.

4.4. Observation Advantages and Limitations of the Lander

A major advantage of the lander is that it can explore fauna behavior, including the interactions between species. It is a powerful tool for the detection of rare predatory fish species [37], which are difficult to find through physical sampling. The structure of the community should be visually determined to better understand vulnerable marine ecosystems. The lander, thus, allows us to clarify anthropogenic impacts, such as fishing and climate change. Also, the density of data for macrobenthos is low, especially in the Cosmonaut Sea. Although traditional trawling is advisable for the collection of all faunal types for higher resolution taxonomic investigations, the lander adds an existing monitoring program to extend biological observing capacity into the Antarctic Ocean [8]. Moreover, the lander is an unnamed vehicle that falls to the seafloor unattached to any cable, subsequently operating autonomously at the bottom, and it can carry diverse instruments for environmental parameter collections [38].

However, the lander cannot acquire images of small sessile fauna (meiofauna and infauna). Quantitative analysis like density per unit cannot be achieved, and only the relative measures of abundance (MaxN) can be applied at this moment. As the lander could help to better understand the current spatial variability on the Cosmonaut Sea shelf fauna, data here would serve as a supplement to the baseline for future comparisons [4].

Author Contributions: J.M.: investigation, data curation, writing—original draft, and validation. X.H.: conceptualization, software, and writing—review and editing. K.L.: data curation and software. Y.H.: data curation and writing—review & editing. S.Z.: data curation. Y.Z.: data curation and writing—review and editing. Y.L.: data curation and writing—review and editing. S.C.: investigation and supervision. M.L.: investigation and supervision. X.M.: investigation. H.L.: supervision and conceptualization. W.L.: supervision, conceptualization, and resources. All authors have read and agreed to the published version of the manuscript.

Funding: This work was supported by The National Natural Science Foundation of China under contract (No. 42076237, No. 42106227, No. 42106065) and Impact and Response of Antarctic Seas to Climate Change (IRASCC2020-2025-NO.01-02-02 & 02-02).

Institutional Review Board Statement: Although our research was about benthos in Antarctica, all the data were from the images of the lander. Therefore, we believe an ethical review process was not required for our study.

Informed Consent Statement: Informed consent was obtained from all individual participants included in the study.

Data Availability Statement: The raw data supporting the conclusions of this article will be made available by the authors on request.

Acknowledgments: We thank Ning Xu, Zhimin Xiao, Xiaodong Chen, Hao Xing, Tieyuan Li, Xian Su, Yaokui Fu, and all crew aboard the XUE LONG 2 Icebreaker for the security and logistical support during the 38th Chinese National Antarctic Research Expedition. We extend our heartfelt thanks to Pei-long Ju, Hai-bo Li, Ji-chang Zhang, Cai Zhang, and Li-wei Kou for their help in the sampling and data analysis processes.

Conflicts of Interest: The authors declare that they have no known competing financial interests or personal relationships that could have appeared to influence the work reported in this paper.

References

1. Busalacchi, A.J. The role of the Southern Ocean in global processes: An earth system science approach. *Antarct. Sci.* **2004**, *16*, 363–368. [CrossRef]
2. Convey, P.; Stevens, M.I. Antarctic Biodiversity. *Science* **2007**, *317*, 1877–1878. [CrossRef]
3. Griffiths, H.J. Antarctic marine biodiversity—What do we know about the distribution of life in the southern ocean? *PLoS ONE* **2010**, *5*, e11683. [CrossRef] [PubMed]
4. Friedlander, A.M.; Goodell, W.; Salinas-de-León, P.; Ballesteros, E.; Berkenpas, E.; Capurro, A.P.; Cárdenas, C.A.; Hüne, M.; Lagger, C.; Landaeta, M.F.; et al. Spatial patterns of continental shelf faunal community structure along the Western Antarctic Peninsula. *PLoS ONE* **2020**, *15*, e0239895. [CrossRef] [PubMed]
5. Costello, M.J.; Coll, M.; Danovaro, R.; Halpin, P.; Ojaveer, H.; Miloslavich, P. A census of marine biodiversity knowledge, resources, and future challenges. *PLoS ONE* **2010**, *5*, e12110. [CrossRef]
6. Levin, L.A.; Bett, B.J.; Gates, A.R.; Heimbach, P. Global observing needs in the deep ocean. *Front. Mar. Sci.* **2019**, *6*, 241. [CrossRef]
7. Rogers, A.; Aburto-Oropeza, O.; Appeltans, W.; Assis, J. *Critical Habitats and Biodiversity: Inventory, Thresholds and Governance*; World Resources Institute: Washington, DC, USA, 2020.
8. Giddens, J.; Turchik, A.; Goodell, W.; Rodriguez, M.; Delaney, D. The National Geographic Society Deep-Sea Camera System: A Low-Cost Remote Video Survey Instrument to Advance Biodiversity Observation in the Deep Ocean. *Front. Mar. Sci.* **2021**, *7*, 601411. [CrossRef]
9. Meijers, A.J.S.; Klocker, A.; Bindoff, N.L. The circulation and water masses of the Antarctic shelf and continental slope between 30 and 80°E. *Deep. Sea Res. Part II Top. Stud. Oceanogr.* **2010**, *57*, 723–737. [CrossRef]
10. Barnes, D.; Peck, L. Vulnerability of Antarctic shelf biodiversity to predicted regional warming. *Clim. Res.* **2008**, *37*, 149–163. [CrossRef]
11. Barnes, D.K.A.; Clarke, A. Antarctic marine biology. *Curr. Biol.* **2011**, *21*, 451–457. [CrossRef] [PubMed]
12. Linse, K.; Griffiths, H.J.; Barnes, D.K.A.; Brandt, A.; Davey, N.; David, B.; Grave, S.D.; D'Acoz, C.D.; Eléaume, M.; Glover, A.G.; et al. The macro- and megabenthic fauna on the continental shelf of the eastern Amundsen Sea, Antarctica. *Cont. Shelf Res.* **2013**, *68*, 80–90. [CrossRef]
13. Mou, J.F.; Liu, K.; Huang, Y.Q.; Lin, J.H.; He, X.B.; Zhang, S.Y.; Li, D.; Zu, Y.C.; Chen, Z.H.; Fu, S.J.; et al. Species Diversity and Community Structure of Macrobenthos in the Cosmonaut Sea, East Antarctica. *Diversity* **2023**, *15*, 1197. [CrossRef]
14. Hunt, B.; Pakhomov, E.; Trotsenko, B.J. The macrozooplankton of the Cosmonaut Sea, east Antarctica (30°E–60°E), 1987–1990. *Deep. Sea Res. Oceanogr. Res. Pap.* **2007**, *54*, 1042–1069. [CrossRef]
15. Van de Putte, A.P.; Jackson, G.D.; Pakhomov, E.; Flores, H.; Volckaert, F.A.M. Distribution of squid and fish in the pelagic zone of the Cosmonaut Sea and Prydz Bay region during the BROKE-West campaign. *Deep. Sea Res. Part II Top. Stud. Oceanogr.* **2010**, *57*, 956–967. [CrossRef]
16. Ran, Q.; Duan, M.G.; Wang, P.C.; Ye, Z.J.; Mou, J.F.; Wang, X.Q.; Tian, Y.J.; Zhang, C.; Qiao, H.J.; Zhang, J. Predicting the current habitat suitability and future habitat changes of Antarctic jonasfish *Notolepis coatsorum* in the Southern Ocean. *Deep. Sea Res. Part II Top. Stud. Oceanogr.* **2022**, *199*, 105077. [CrossRef]
17. Sahade, R.; Lagger, C.; Torre, L.; Momo, F.; Monien, P.; Schloss, I.; Barnes, D.K.A.; Servetto, N.; Tarantelli, S.; Tatián, M.; et al. Climate change and glacier retreat drive shifts in an Antarctic benthic ecosystem. *Sci. Adv.* **2015**, *1*, e1500050. [CrossRef]
18. Mou, J.F.; Liu, K.; Huang, Y.Q.; He, X.B.; Zhang, S.Y.; Wang, J.J.; Lin, J.H.; Lin, H.S.; Liu, W.H. The macro-and megabenthic fauna on the continental shelf of Prydz Bay, east Antarctica. *Deep. Sea Res. Part II Top. Stud. Oceanogr.* **2022**, *198*, 105052. [CrossRef]
19. Gutt, J. Some "driving forces" structuring communities of the sublittoral Antarctic macrobenthos. *Antarct. Sci.* **2000**, *12*, 297–313. [CrossRef]
20. Arnaud, P.M.; Lo'pez, C.M.; Olaso, I.; Ramil, F.; Ramos-Esplá, A.A.; Ramos, A. Semi-quantitative study of macrobenthic fauna in the region of the South Shetland Islands and the Antarctic Peninsula. *Polar Biol.* **1998**, *19*, 160–166. [CrossRef]
21. Malyutina, M. Russian deep-sea investigations of Antarctic fauna. *Deep. Res Part II Top. Stud. Oceanogr.* **2004**, *51*, 1551–1570. [CrossRef]

22. Linley, T.D.; Gerringer, M.E.; Yancey, P.H.; Drazen, J.C.; Weinstock, C.L.; Jamieson, A.J. Fishes of the hadal zone including new species, in situ observations and depth records of Liparidae. *Deep. Sea Res. Part I Oceanogr. Res. Pap.* **2016**, *114*, 99–110. [CrossRef]
23. Leitner, A.B.; Neuheimer, A.B.; Donlon, E.; Smith, C.R.; Drazen, J.C. Environmental and bathymetric influences on abyssal bait-attending communities of the Clarion Clipperton Zone. *Deep. Sea Res. Part I Oceanogr. Res. Pap.* **2017**, *125*, 65–80. [CrossRef]
24. Riddle, M.J.; Craven, M.; Goldsworthy, P.; Carsey, F. A diverse benthic assemblage 100 km from open water under the Amery Ice Shelf, Antarctica. *Paleoceanogr. Paleoclimatol.* **2007**, *22*, PA1204. [CrossRef]
25. Sumida, P.Y.G.; Bernardino, A.F.; Stedall, V.P.; Glover, A.G.; Smith, C.R. Temporal changes in benthic mega-faunal abundance and composition across the West Antarctic Peninsula shelf: Results from video surveys. *Deep. Res Part II Top. Stud. Oceanogr.* **2008**, *55*, 2465–2477. [CrossRef]
26. Eastman, J.T.; Amsler, M.O.; Aronson, R.B.; Thatje, S.; McClintock, J.B.; Vos, S.C.; Kaeli, J.W.; Singh, H.; Mesa, M.L. Photographic survey of benthos provides insights into the Antarctic fish fauna from the Marguerite Bay slope and the Amundsen Sea. *Antarct. Sci.* **2013**, *25*, 31–43. [CrossRef]
27. Ambroso, S.; Salazar, J.; Zapata-Guardiola, R.; Federwisch, L.; Richter, C.; Gili, J.M.; Teixidó, N. Pristine populations of habitat-forming gorgonian species on the Antarctic continental shelf. *Sci. Rep.* **2017**, *7*, 12251. [CrossRef] [PubMed]
28. Alberty, M.; Sprintall, J.; Mackinnon, J.; Germineaud, C.; Cravatte, S.; Ganachaud, A. Moored Observations of Transport in the Solomon Sea. *J. Geophys. Res. Oceans* **2019**, *124*, 8166–8192. [CrossRef]
29. Langlois, T.; Williams, J.; Monk, J.; Bouchet, P.; Currey, L.; Goetze, J.; Harasti, D.; Huveneers, C.; Lerodiaconou, D.; Malcolm, H.; et al. Marine sampling field manual for benthic stereo BRUVS (Baited Remote Underwater Videos). In *Field Manuals for Marine Sampling to Monitor Australian Waters*; Przeslawski, R., Foster, S., Eds.; National Environmental Science Programme (NESP): Canberra, Australia, 2018; pp. 82–104.
30. Linley, T.D.; Craig, J.; Jamieson, A.J.; Priede, I.G. Bathyal and abyssal demersal bait-attending fauna of the Eastern Mediterranean Sea. *Mar. Biol.* **2018**, *165*, 159. [CrossRef]
31. Brasier, M.J.; Grant, S.M.; Trathan, P.N.; Allcock, L.; Ashford, O.; Blagbrough, H.; Brandt, A.; Danis, B.; Downey, R.; Eléaume, M.P.; et al. Benthic biodiversity in the South Orkney Islands Southern Shelf Marine Protected Area. *Biodiversity* **2018**, *19*, 5–19. [CrossRef]
32. Smith, J.; O'Brien, P.E.; Stark, J.S.; Johnstone, G.J.; Riddle, M. Integrating multibeam sonar and underwater video data to map benthic habitats in an East Antarctic nearshore environment. *Estuar. Coast. Shelf Sci.* **2015**, *164*, 520–536. [CrossRef]
33. Post, A.L.; Lavoie, C.; Domack, E.W.; Leventer, A.; Shevenell, A.E.; Fraser, A.D. Environmental drivers of benthic communities and habitat heterogeneity on an East Antarctic shelf. *Antarct. Sci.* **2017**, *29*, 17–32. [CrossRef]
34. Jones, D.O.B.; Bett, B.J.; Tyler, P.A. Depth-related changes to 860 density, diversity and structure of benthic megafaunal assemblages 861 in the Fimbul ice shelf region, Weddell Sea, Antarctica. *Polar Biol.* **2007**, *862*, 1579–1592. [CrossRef]
35. Gon, O.; Heemstra, P.C. *Fishes of the Southern Ocean*; J.L.B. Smith Institute of Ichthyology Grahamstown: Grahamstown, South Africa, 1990.
36. Ciaputa, P.; Siciński, J. Seasonal and annual changes in Antarctic Fur seal (*Arctocephalus gazella*) diet in the area of Admiralty Bay, King George Island, South Shetland Islands. *Pol. Polar Res.* **2006**, *27*, 171–184.
37. Fujiwara, Y.; Tsuchida, S.; Kawato, M.; Masuda, K. Detection of the Largest Deep-Sea-Endemic Teleost Fish at Depths of Over 2000 m Through a Combination of eDNA Metabarcoding and Baited Camera Observations. *Front. Mar. Sci.* **2022**, *9*, 945758. [CrossRef]
38. Brandt, A.; Gutt, J.; Hildebrandt, M.; Pawlowski, J.; Schwendner, J.; Soltwedel, T.; Thomsen, L. Cutting the Umbilical: New Technological Perspectives in Benthic Deep-Sea Research. *J. Mar. Sci. Eng.* **2016**, *4*, 36. [CrossRef]

Disclaimer/Publisher's Note: The statements, opinions and data contained in all publications are solely those of the individual author(s) and contributor(s) and not of MDPI and/or the editor(s). MDPI and/or the editor(s) disclaim responsibility for any injury to people or property resulting from any ideas, methods, instructions or products referred to in the content.

 diversity

Article

Spatio-Temporal Structure of Two Seaweeds Communities in Campeche, Mexico

Cynthia Mariana Hernández-Casas, Ángela Catalina Mendoza-González, Deisy Yazmín García-López and Luz Elena Mateo-Cid *

Instituto Politécnico Nacional, Escuela Nacional de Ciencias Biológicas, Departamento de Botánica, Laboratorio de Ficología, Carpio y Plan de Ayala, Colonia Santo Tomás, Mexico City 11340, Mexico; mariana18biol@gmail.com (C.M.H.-C.); mcati8585@gmail.com (Á.C.M.-G.); dgarcial@ipn.mx (D.Y.G.-L.)
* Correspondence: luzmcyd@gmail.com

Abstract: Macroalgae populations are influenced by various factors that define their spatial and temporal distribution in different habitats and regions. In Mexico, studies addressing the abundance and diversity of macroalgae communities related to environmental factors are scarce. The objective is to determine the spatio-temporal variation of the structure of the community of seaweeds in Xpicob and Villamar, Campeche, during three climatic seasons. Sampling took place during each season using transects and quadrants; additionally, the type of substrate, water temperature, transparency, depth, salinity, and dissolved oxygen, were recorded. The total richness was 74 taxa, corresponding to three classes: Phaeophyceae (3), Florideophyceae (36), and Ulvophyceae (35). Filamentous algae dominate in species richness in the intertidal zone at low depths, while fleshy and calcareous algae predominate in number and biomass in the subtidal zone at higher depths (60–200 cm). Twenty-eight species were common to both sites; meanwhile, 46 taxa were exclusive of specific sites, including 13 found exclusively in Xpicob and 33 in Villamar. The most favorable climatic season for the macroalgae located in Xpicob was the winter rain. For the macroalgae community in Villamar, the most favorable climatic season was the dry. These differences are likely attributed to the predominant environmental and physicochemical characteristics of each site.

Keywords: marine macroalgae; benthos; diversity; seasonal variability; biomass

Citation: Hernández-Casas, C.M.; Mendoza-González, Á.C.; García-López, D.Y.; Mateo-Cid, L.E. Spatio-Temporal Structure of Two Seaweeds Communities in Campeche, Mexico. *Diversity* 2024, 16, 344. https://doi.org/10.3390/d16060344

Academic Editors: Bert W. Hoeksema and Renato Mamede

Received: 13 April 2024
Revised: 1 June 2024
Accepted: 7 June 2024
Published: 14 June 2024

Copyright: © 2024 by the authors. Licensee MDPI, Basel, Switzerland. This article is an open access article distributed under the terms and conditions of the Creative Commons Attribution (CC BY) license (https:// creativecommons.org/licenses/by/ 4.0/).

1. Introduction

On the littoral of Campeche, studies of macroalgae have mostly been floristic in approach, for example: Huerta-Múzquiz and Garza-Barrientos (1966) [1] published one of the first works carried out in the Campeche area, specifically, for the Terminus Lagoon; 20 years later, Huerta-Múzquiz et al. (1987) [2] carried out a floristic study in the Yucatan Peninsula, in which 102 species were recorded for the Campeche coast; on the other hand, Ortega (1995) [3] obtained 80 species collected between 1964 and 1966 in the Laguna de Términos. Callejas-Jiménez et al. (2005) [4] obtained 51 species from Santa Rosalía and Playa Preciosa, including 19 new records. Mateo-Cid et al. (2013) [5] listed 211 taxa of benthic algae from the Campeche coast, while Mendoza-González et al. (2013) [6] determined 85 species (74 macroalgae) associated with wrecks and other subtidal structures on the coast of Campeche. The information obtained in this specialized literature shows that the total number of marine algae for the Campeche coast is 271.

Regarding ecological studies of the Campeche coast, there is only the work carried out by Ortiz-Rosales (1988) [7], who, in his master's thesis, obtained a total of 47 species; in addition, he observed that the substrate and the depth are the factors that have the greatest influence on the distribution, composition, and richness of benthic algae. Biodiversity and marine ecosystems are intrinsically linked to a wide range of services that are essential for sustainable development; the ecological characterization of communities is a fundamental part of the use and exploitation of resources, including seaweed.

Among the biological factors that significantly influence the establishment of macroalgae at specific ecological niches are their physiology, morphology, life cycle, development, and the permanence of the algae in their habitat. The study of these phenological and demographic events allows us to partially understand the ecology of algae species [8–11]. To characterize algae communities, biogeographic affinity and epiphytism are also used, which allow us to understand an important part of the dynamics of species succession [12–15].

Benthic marine macroalgae communities are exposed to the influence of various physical factors both on a global and local scale; among these, there are physical factors, such as the type of substrate and its availability, temperature, exposure to waves, and depth (related to light availability); and also chemical factors such as salinity, nutrient availability, acidity, and dissolved oxygen [10,11,16–19].

The main factor influencing the geographical distribution of algae is the temperature. In addition, the salinity, tidal waves, and type of substrate are factors that have the highest influence on the local distribution of algae [20]. Therefore, these abiotic factors, together with the biotic ones, are thought to be responsible for the seasonal variation of algae in a specific region, affecting the composition of the species and, therefore, the structure of algal communities [21]. The effects of these biotic and abiotic factors have been observed in various populations of algae, affecting not only their presence or absence but also their morphology [22,23].

Ecological studies allow us to know the dynamics of the structure of a community over a period. Thus, this study will provide updated information about the attributes of the macroalgal communities of Campeche and their association with abiotic factors, as well as their behavior throughout the different seasons, allowing to strengthen conservation proposals, as well as the potential use of this resource in the study area.

2. Materials and Methods

The collection of biological material was performed in two locations on the coast of Campeche. Xpicob is located at the coordinates 19°43′13″ N and 90°40′09″ W. Due to its proximity to a turtle camp with the same name, Xpicob is a community dedicated mostly to fishing, with a smaller part dedicated to ecotourism. In Xpicob, the sandy substrate predominates with scattered rocky aggregates. Villamar is located 83 km northeast of Laguna de Terminos at the coordinates 19°16′13″ N and 90°47′54″ W; it is a sandy beach located very close to the Aak-bal hotel complex, with rocky platforms, rocky aggregates, and *Thalassia testudinum* meadows (Figure 1).

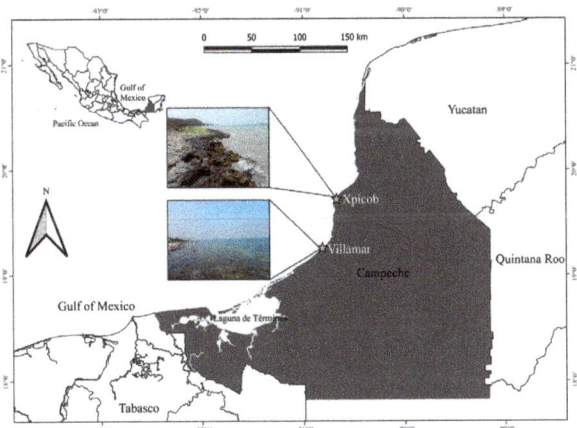

Figure 1. Location of the study area (Campeche) and sampling locations (Xpicob and Villamar) (Hernández-Casas C.M.).

According to the Köppen classification modified by García (2004) [24], the Campeche coastline has a variety of climates from very hot and warm semi-dry (BS1(h′)w(i′) and BS0(h′)w″(x′)) to the warm subhumid (Aw0(i′)gw″) and warm humid (Am), specifically, in the study locations, the warm subhumid occurs [24,25].

2.1. Fieldwork

At each location, three samplings were performed during October 2016 (winter rain season), April (dry season), and July 2017 (summer rain season).

Three 25 m transects were placed perpendicular to the coastline, separated by 25 m. Geographical coordinates of each transect were taken with an eXplorist 210 GPS Magellan® (Paris, France). A sampling unit consisted of 25 × 25 cm squares (0.0625 cm^2) placed every five meters on the transect [26–28].

For each square, the following data were recorded: type of substrate, water temperature (using a 100 °C mercury thermometer), transparency (using a Secchi disk), depth (using a depth meter), salinity (using a Hanna HI3835 Kit (Hanna Instruments®, Mexico City, Mexico), and dissolved oxygen (using a Hanna HI3810 kit).

The macroalgae located in each square were detached from the substrate and placed in transparent plastic bags, previously labeled with the name of the location, date of collection, transect, and quadrant number, as well as the name of the people who performed the collection. Algae were subsequently fixed with 5% formalin seawater and transferred to the Phycology laboratory of the Escuela Nacional de Ciencias Biológicas.

2.2. Laboratory Work

The identification of algae species was performed using the keys and descriptions of various relevant references [29–33].

The calculation of dry biomass of each species was estimated by placing each specimen for 15 days under a botanical press. Specimens were subsequently weighed on an analytical balance (Ohaus Scout® model H-7294, with 1 g precision, ULINE MX, Nuevo Leon, Mexico). The dry biomass value was extrapolated to m^2.

All epiphytic macroalgae were completely detached from their substrate before species determination and recording of biomass data.

Finally, each specimen was mounted on herbarium paper and added to the Phycological Collection of the Herbarium of the Escuela Nacional de Ciencias Biológicas (ENCB) of the IPN.

2.3. Species Composition

A floristic list of the marine macroalgae species located at both locations in the three sampling periods was prepared. The family, genera, and species were ordered according to the systematic criterion proposed by Wynne (2022) [34], while the nomenclatural update follows the scheme of Guiry and Guiry (2024) [35].

2.4. Biogeographic Affinity

Based on the floristic list observed, the biogeographic affinity was calculated using the Cheney index (1997) [36]: $(R + C)/P$, where R is the number of Rhodophyta species, C is the number of Chlorophyta species, and P is the number of Phaeophyceae species. Values > 6 indicate flora with tropical affinity, values < 3 indicate flora with temperate-cold affinity, and intermediate values indicate a mixed-type flora.

2.5. Distribution Profiles

With the data on the number of species per location and season, macroalgae distribution profiles were prepared along the beach, according to the variation of depth in the 25 m transect.

2.6. Temporal and Spatial Variation of Biomass

For both locations, the values of season–species biomass, total temporal biomass, temporal relative abundance, total species biomass, total annual biomass, and annual relative abundance were calculated using the formulas of Águila-Ramírez (1998) [37] shown below.

Specific temporality biomass (STB):

$$\sum_{i=1}^{15} Biomass\ of\ species\ n\ in\ each\ quadrant,\ for\ each\ transect\ for\ season\ k$$

Total temporality biomass (TTB):

$$\sum_{j=1}^{S} Specific\ temporary\ biomass\ for\ each\ season\ from\ all\ species$$

Temporal relative abundance (TRA): $TRA = \frac{STB}{TTB} \times 100$
Total species biomass (TSB): $TSB = \sum_{K=1}^{3} STB$
Total annual biomass (TAB): $TAB = \sum_{j=1}^{S} TTB$ of the three seasons of the year
Annual relative abundance (ARA): $ARA = \frac{TSB}{TAB} \times 100$

To determine if there were significant differences in biomass between locations or between seasons, a two-way PERMANOVA was applied using the Bray–Curtis similarity index, followed by a two-way ANOVA to identify which transects and/or squares were different according to their location or season.

In addition, to evaluate how similar the seasons and/or locations were in terms of biomass, the total biomass data of each species in each season (by location) were used to perform a grouping analysis (Cluster) using UPGMA (average linkage) and a Bray–Curtis matrix.

2.7. Index of Macroalgae Importance Value

Subsequently, the importance of the macroalgae species was determined according to the Importance Value Index (IVI), which correlates the biomass to the frequency of occurrence of a species [38] modified by [39], thus indicating the relative ecological importance of the species in a community.

$$IVI = RBi + RFi$$

where $RF = RB = \frac{Bi}{\sum Bi} \frac{Fi}{\sum Fi}$.

Bi is the biomass of each species of macroalgae, Fi is the proportion of each macroalgal species found in all quadrants.

2.8. Macroalgal Community Attributes

2.8.1. α (Alpha) Diversity

The spatial and temporal variation of community attributes (diversity, dominance, and evenness) was analyzed using the biomass data observed previously.

The α diversity for each season, at each location, was calculated using the Shannon–Wiener index. This index considers the fact that individuals are randomly sampled from an infinite population, and it is assumed that all species are represented in the sample [40].

$$H' = -\sum_{i=1}^{S} p_i log_2 p_i$$

where H' is the Shannon-Wiener index, p is ni/N, log2 is the base 2 logarithm, ni is the biomass in g/m^2 of each species, N is the total biomass of macroalgae, and S is the number of species.

2.8.2. β (Beta) Diversity

Spatial β-diversity (Xpicob and Villamar) was calculated in terms of macroalgal species turnover using the Whittaker index [41].

$$Bw = \frac{S}{\alpha - 1}$$

Similarly, to analyze the rate of change of algal communities over time, β diversity was analyzed on a temporal scale (winter rain, summer rain, and dry seasons). All statistical analyses were performed using the PAST 3.1 program [42].

2.8.3. Evenness Index

The distribution of the biomass of individuals between species was determined using the Pielou index (J') or Evenness index [43]. These indices assumes that all species in the community have been considered in the sample [40].

$$J' = \frac{H'}{H'max},$$

where J' is the Pielou Evenness index, H' is the Shannon-Wiener diversity index, H'max is the maximum value reached by H' (H'max = log2 S), and S is the total number of species recorded.

2.8.4. Dominance Index

The indices that are based on dominance are generally inverse to the uniformity or evenness of the community; they only consider the representativeness of the species with the highest importance value without evaluating the contribution of the rest of the species. The Simpson index is strongly influenced by the importance of the most dominant species [40,44]. In the present work, this index was used to determine the dominance of the species with the following formula:

$$\lambda = \sum p_i^2$$

where pi is the proportional abundance of species i; that is, the biomass of species i divided by the total biomass of the species in the sample.

2.8.5. Relationship between Environmental Variables and Species Abundance

First, to determine if there were differences in environmental variables, between seasons or between sampling locations, a two-way PERMANOVA analysis was applied using the Bray–Curtis similarity index. Likewise, a two-way ANOVA was performed for each of the specific environmental variables (type of substrate, water temperature, transparency, salinity, dissolved oxygen, and depth) recorded in the field.

To test if there is a relationship between the biomass (abundance) of the species and the environmental variables, a Canonical Correspondence Analysis (CCA) was used, using seven environmental variables: salinity, temperature, depth, transparency, dissolved oxygen, and substrate (sandy/rocky). A value of 1 was assigned to the substrate that dominated in the quadrant, and 0 to the one that was not dominant. Because the environmental variables have different units, they were standardized with (ln X + 1) before performing the CCA. The CCA was performed using the XLSTAT V2019.3.2 program.

3. Results

3.1. Species Composition of the Algae Communities

A total of 74 taxa were identified (Table 1). The class Florideophyceae was the best represented, in terms of richness, with 36 species, followed by Ulvophyceae with 35 species, and Phaeophyceae with three. Table 1 also shows the turnover of species in terms of presence/absence throughout the three seasons at each location. It should be noted that part of the material listed below has already been morphologically and molecularly characterized, as is the case with the *Udotea* specimens [45] and *Codiophyllum* [46].

Table 1. List of taxa located in Xpicob and Villamar, Campeche, during the three seasons (winter rain season, dry season, and summer rain season) during the sampling period (2016–2017). Also, its show the specific temporality biomass in a heat map where, deep red: lower values and deep blue: higher values.

	Species	Villamar			Xpicob		
		Winter Rains	Summer Rains	Dry	Winter Rains	Summer Rains	Dry
	Phylum Heterokontophyta						
	Class Phaeophyceae						
	Order Dictyotales						
	Family Dictyotaceae						
1	*Dictyota caribaea* Hörnig & Schnetter	0.78	1.55	0.07			0.01
2	*Padina gymnospora* (Kützing) Sonder			0.56			
3	*Padina sanctae-crucis* Børgesen			0.01			
	Phylum Rhodophyta						
	Class Florideophyceae						
	Order Corallinales						
	Family Corallinaceae						
4	*Jania capillacea* Harvey			0.66	49.03		
5	*Jania pedunculata* var. *adhaerens* (J. V. Lamouroux) A. S. Harvey, Woelkerling & Reviers	0.77	16.69		107.07	31.2	
	Family Lithophyllaceae						
6	*Amphiroa fragilissima* (Linnaeus) J. V. Lamouroux	0.01		2.81	3.99		
	Order Ceramiales						
	Family Ceramiaceae						
7	*Ceramium corniculatum* Montagne		54.25	2.5	0.46		1.56
	Family Rhodomelaceae						
8	*Acanthophora spicifera* (M. Vahl) Børgesen	4.61	3.27	1.69	8.97	0.26	0.97
9	*Alsidium seaforthii* (Turner) J. Agardh	2.22	1.74	7.98	10.86	2.38	
10	*Alsidium triquetrum* (S. G. Gmelin) Trevisan		0.33	0.2		0.62	
11	*Chondria curvilineata* Collins & Hervey						0.01
12	*Chondria floridana* (Collins) M. Howe		1.57	1.03			
13	*Chondria leptacremon* (Melvill ex G. Murray) De Toni			1.3			
14	*Digenea mexicana* G. H. Boo & D. Robledo	1.57	15.18	6.36	42.8	3	4.1
15	*Laurencia caraibica* P. C. Silva						0.16
16	*Laurencia filiformis* (C. Agardh) Montagne	0.17	0.82	0.9	2.28	0.01	
17	*Laurencia intricata* J. V. Lamouroux				2.36		0.02
18	*Laurencia obtusa* (Hudson) J. V. Lamouroux			3.21			
19	*Laurencia viridis* Gil-Rodríguez & Haroun			4.37			
20	*Lophosiphonia obscura* (C. Agardh) Falkenberg		0.01	0.01			
21	*Palisada corallopsis* (Montagne) Sentíes, Fujii & Díaz-Larrea	4.51	0.79	1.18	2.34		0.1
22	*Palisada perforata* (Bory) K. W. Nam	23.06		8.24	0.86		4.22
23	*Polysiphonia subtilissima* Montagne		4.44				0.08

Table 1. Cont.

		Villamar			Xpicob		
24	*Yuzurua poiteaui* (J. V. Lamouroux) Martin-Lescanne	8.8		4.43	14.1		
25	*Yuzurua poiteaui* var. *gemmifera* (Harvey) M. J. Wynne				0.84		
	Family Pterocladiaceae						
26	*Pterocladiella sanctarum* (Feldmann & Hamel) Santelices	0.01	0.64	1.14	0.8	0.52	
	Order Gigartinales						
	Family Cystocloniaceae						
27	*Hypnea cervicornis* J. Agardh	0.33	0.04	0.01			0.84
28	*Hypnea spinella* (C. Agardh) Kützing	0.28					
	Family Solieriaceae						
29	*Eucheumatopsis isiformis* (C. Agardh) Núñez-Reséndiz, Dreckmann & Sentíes			29.04			
30	*Meristotheca gelidium* (J. Agardh) E. J. Faye & M. Masuda	0.38					
31	*Wurdemannia miniata* (Sprengel) Feldmann & Hamel			0.68			
	Order Gracilariales						
	Family Gracilariaceae						
32	*Gracilaria caudata* J. Agardh		1.15				
33	*Gracilaria cervicornis* (Turner) J. Agardh	6.62	0.26		6.7		
34	*Gracilaria damicornis* J. Agardh			1.58			
35	*Gracilaria debilis* (Forsskål) Børgesen	83.51	5.35	26.05	0.86	2.88	
36	*Gracilaria flabelliformis* (P. Crouan & H. Crouan) Fredericq & Gurgel	0.5	0.1	0.08	0.83		0.06
37	*Gracilaria flabelliformis* subsp. *simplex* Gurgel, Fredericq & J. N. Norris					0.32	
	Order Halymeniales						
	Family Halymeniaceae						
38	*Codiophyllum mexicanum* Núñez-Resendiz, Dreckmann & Sentíes		2.04				
	Order Rhodymeniales						
	Family Champiaceae						
39	*Champia parvula* var. *prostrata* L. G. Williams			0.02			
	Phylum Chlorophyta						
	Class Ulvophyceae						
	Order Bryopsidales	4.07					
	Family Caulerpaceae					0.03	0.045
40	*Caulerpa ashmeadii* Harvey	0.09					
41	*Caulerpa fastigiata* Montagne						
42	*Caulerpa prolifera* (Forsskål) J. V. Lamouroux			0.32			
	Family Halimedaceae					0.82	
43	*Halimeda discoidea* Decaisne	43.69	20.72	33.29	6.85	0.14	
44	*Halimeda gracilis* Harvey ex J. Agardh				1.71		4.17
45	*Halimeda incrassata* (J. Ellis) J. V. Lamouroux		0.68		44.64		68.79
46	*Halimeda monile* (J. Ellis & Solander) J. V. Lamouroux	1.3	16.94	3.33			
47	*Halimeda opuntia* (Linnaeus) J. V. Lamouroux		2.4	0.51			
48	*Halimeda scabra* M. Howe			0.01	0.01		

Table 1. Cont.

		Villamar			Xpicob		
49	*Halimeda tuna* (J. Ellis & Solander) J. V. Lamouroux	4.51	0.79	1.18	2.34		0.1
50	*Penicillus capitatus* Lamarck		0.15	0.03			
51	*Udotea caribaea* D. S. Littler & Littler			7.46			
52	*Udotea conglutinata* (J. Ellis & Solander) J. V. Lamouroux			5.97			
53	*Udotea cyathiformis* Decaisne			1.92			
54	*Udotea cyathiformis* var. *flabellifolia* D. S. Littler & Littler			6.65			
55	*Udotea dixonii* D. S. Littler & Littler	14.28	9.11		3.95		
56	*Udotea dotyi* D. S. Littler & Littler			1.96			
57	*Udotea flabellum* (J. Ellis & Solander) M. Howe		2.63				
58	*Udotea looensis* D. S. Littler & Littler	0.6	0.2	0.04			0.16
59	*Udotea luna* D. S. Littler & Littler			0.3			
60	*Udotea spinulosa* M. Howe						0.25
61	*Udotea unistratea* D. S. Littler & Littler		0.72				
	Order Cladophorales						
	Family Boodleaceae						
62	*Cladophoropsis macromeres* W. R. Taylor						2.02
63	*Cladophoropsis membranacea* (Hofman Bang ex C. Agardh) Børgesen				2.04		10.61
	Family Cladophoraceae						
64	*Cladophora albida* (Nees) Kützing				1.14		
65	*Cladophora coelothrix* Kützing					0.01	
66	*Cladophora crispula* Vickers				0.94	1.77	11.33
67	*Cladophora flexuosa* (O. F. Müller) Kützing			5.75		8.05	1.81
68	*Cladophora laetevirens* (Dillwyn) Kützing	35.25		0.02	0.54	1.85	
69	*Cladophora sericea* (Hudson) Kützing	66.73			4.61	0.75	0.91
70	*Willeella brachyclados* (Montagne) M. J. Wynne			2.11	0.67	0.02	
	Order Dasycladales						
	Family Polyphysaceae						
71	*Acetabularia crenulata* J. V. Lamouroux	0.01	0.01	0.06			
72	*Acetabularia farlowii* Solms-Laubach			0.01			
	Order Ulvales						
	Family Ulvaceae						
73	*Ulva compressa* Linnaeus			0.02			
74	*Ulva flexuosa* Wulfen		0.13				

3.2. Biogeographic Affinity

The value of the Cheney index observed in this study was 23.6, which indicates that the flora of Campeche is tropical (Cheney > 6).

3.3. Spatial and Temporal Variation of Species Richness

The Florideophyceae class was the one with the highest species richness at both locations; however, Villamar had a higher species richness, with 61 taxa relative to Xpicob, where 41 species were identified (Figure 2).

The highest richness was observed in the dry season (58), followed by the summer rain (34) and winter rain seasons (34). During the winter rain and dry seasons, Florideophyceae was the class with the highest richness, with 20 and 29 species, respectively. In the summer rain season, the class with the highest richness was Ulvophyceae (19). Both Florideophyceae and Ulvophyceae had their highest richness during the dry season, with 29 and 26 species, respectively, and their lowest richness during the summer rain and winter rain seasons, with 17 and 13 species, respectively (Figure 3).

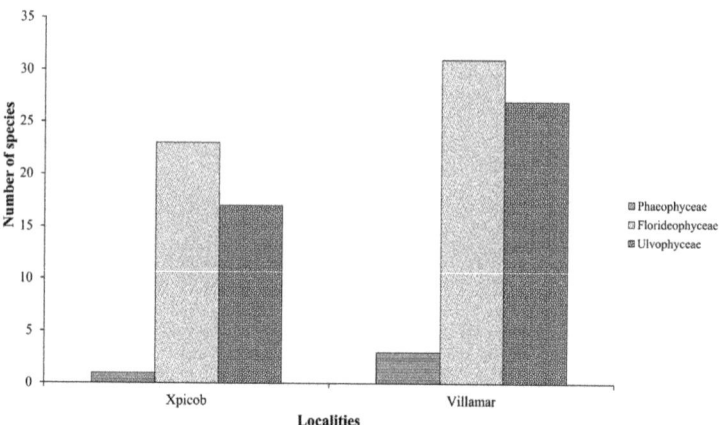

Figure 2. Total richness of macroalgae species in Xpicob and Villamar, Campeche, during the sampling periods performed in the winter rain (October 2016), dry (April 2017), and summer rain seasons (August 2017).

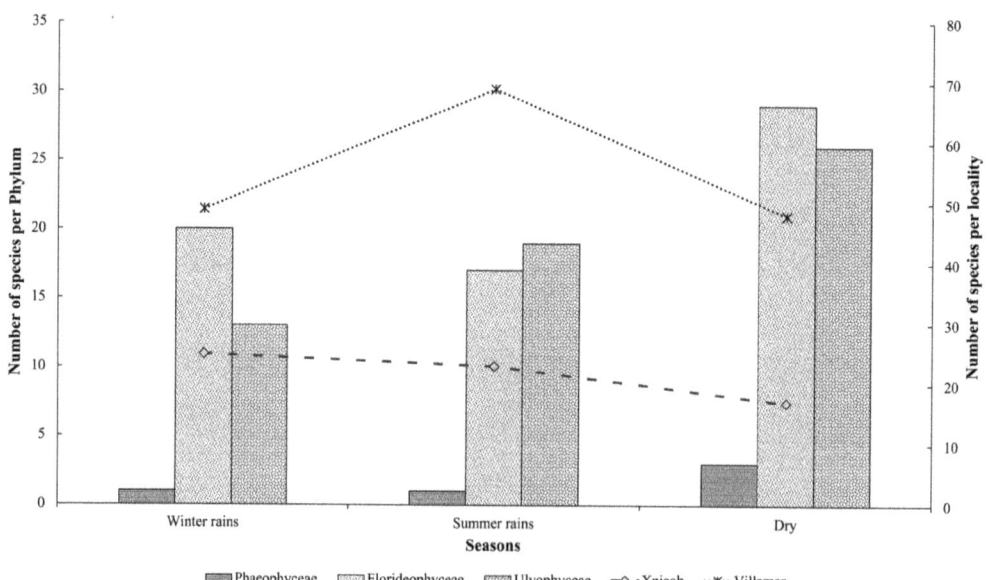

Figure 3. Seasonal species richness of each class of macroalgae in Villamar and Xpicob, Campeche, during the sampling periods performed in the winter rain (October 2016), dry (April 2017), and summer rain seasons (August 2017).

In the winter rain season, the highest richness was observed in Xpicob, with 25 species (only one more species than in Villamar); however, in the dry and summer rain seasons, Villamar recorded the highest richness, with 31 and 46 species, respectively (Figure 3).

Xpicob, in the winter rain season, presented two exclusive taxa, *Cladophora albida*, and *Yuzurua poiteaui* var. *gemmifera*. During the summer rain season, there were three exclusive taxa, *Gracilaria flabelliformis* subsp. *simplex*, *Halimeda gracilis*, and *Cl. coelothrix*. In the dry season, the exclusive species at this site were *Chondria curvilineata*, *Laurencia caraibica*, *Halimeda monile*, *Udotea spinulosa*, *Cladophoropsis macromeres* and *C. membranacea* (Table 1, Figure 4).

Figure 4. Seaweeds that are reported as exclusive for some climatic season in Xpicob and Villamar Campeche during sampling. (**A**) *Padina santae-crucis*, (**B**) *Acanthophora spicifera*, (**C**) *Meristotheca gelidium*, (**D**) *Caulerpa ashmeadii*, (**E**) *Halimeda scabra*.

In Villamar, four exclusive taxa were present in the winter rain season, *Hypnea spinella*, *Meristotheca gelidium*, *Caulerpa ashmeadii*, and *C. prolifera*. In the summer rain season, the exclusive species were *Gracilaria caudata* J. Agardh, *Udotea flabellum*, *U. unistratea*, and *Ulva compressa*. Finally, in the dry season, there were 19 exclusive species including *Padina gymnospora*, *P. sanctae-crucis*, *Wurdemannia miniata*, *Champia parvula* var. *postrata*, and *Eucheumatopsis isiformis* (Table 1, Figure 4).

In Xpicob, the taxa that were observed in all sampling periods are *Acanthophora spicifera*, *Digenea mexicana*, *Cladophora crispula*, and *C. sericea*. In Villamar, the taxa that were observed in all sampling periods are *Dictyota caribaea*, *Acanthophora spicifera*, *Alsidium seaforthii*, *D. mexicana*, *Laurencia intricata*, *Palisada corallopsis*, *Pterocladiella sanctarum*, *Hypnea cervicornis*, *Gracilaria debilis*, *G. flabelliformis*, *Halimeda scabra*, and *Acetabularia crenulata* (Table 1, Figure 4).

3.4. Vertical Distribution Profile of Macroalgae

Filamentous algae dominated both in richness and biomass in the intertidal zone at both locations. During the three seasons and in both locations, the fleshy, calcified algae were located at higher depths.

In Xpicob, the highest diversity was observed in the deepest areas (quadrants 4 and 5); this pattern was maintained in winter rains and summer rains and in the dry season; furthermore, quadrant 2 of transect one also presented high diversity (Table 2).

Table 2. Variation of Shannon–Wiener diversity for Xpicob and Villamar, in the three climatic seasons at the different depths sampled (quadrants).

	Xpicob			Villamar		
	Winter Rains	Summer Rains	Dry	Winter Rains	Summer Rains	Dry
T1C1	1.04				0.024	
T1C2	1.028	1.037	0.974	1.099	0.472	1.969
T1C3	0.936	0.946	0.753	0.666	0.538	1.503
T1C4	1.251		0.693	0.488	1.145	0.896
T1C5	1.166	1.228	1.041	0.365	1.807	
T2C1	1.063	0.951		0.816	0	0.494
T2C2	1.301	1.017		0.709		1.103

In contrast to Xpicob, Villamar had the highest diversity in the central quadrant (two, three, and four) in winter rains and dry seasons; in summer rain season, the highest diversity was that of quadrant 5 of transect one (Table 2).

In Villamar, we also found fleshy algae with economic importance (such as *Meristotheca* and *Eucheumatopsis*) that were not found in Xpicob. At both locations, we found *Gracilaria* species such as *G. cervicornis*, *G. debilis*, or *G. flabelliformis*; however, in Xpicob, this genus was less represented in terms of the number of species than in Villamar, where *G. caudata* and *G. damicornis* were also found.

3.5. Temporal and Spatial Variation of Biomass

In Xpicob, the highest TSB was observed in the winter rain season, and the lowest in the summer rain season (Figure 5, Table 1). In Villamar, the highest TSB was observed in the summer rain season, while the lowest was observed in the dry season (Figure 5, Table 1).

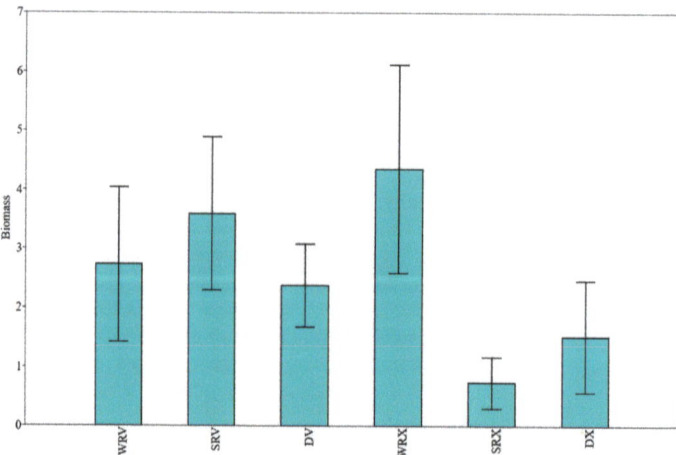

Figure 5. Average biomass (g/m^2) and standard error, of X: Xpicob and V: Villamar, Campeche, in each climatic season (N, Winter rain season; L, Summer rain season; and S, Dry season).

In the winter rain season, the highest biomass value was found in Xpicob; however, in the summer rain and dry seasons, the highest biomass value was found in Villamar. These results were corroborated by a two-way ANOVA, which showed that biomass varied significantly between seasons (F = 0.024^{29}, $p = 0.8805$) (Figure 5).

The grouping analysis (Cluster) revealed the existence of three groups. Group I was the most dissimilar, constituted only by the summer rain season from Xpicob (9% similarity). Group II was composed of the dry and winter rain seasons of Xpicob (38% similarity). Group III was formed by the three climatic seasons of Villamar (41% similarity) (Figure 6).

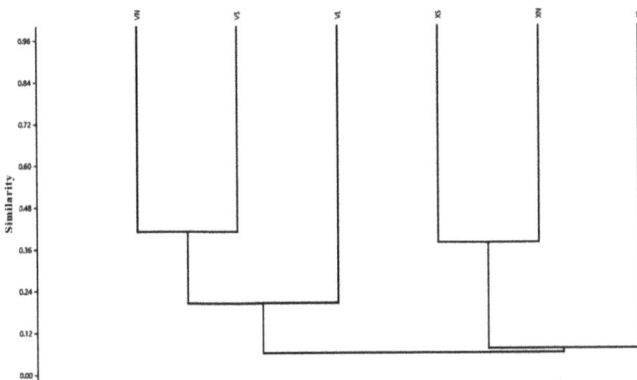

Figure 6. UPGMA clustering analysis using Bray–Curtis similarity. XS, Xpicob—dry season; XN, Xpicob—winter rain season; XL, Xpicob—summer rain season; VN, Villamar—winter rain season; VS, Villamar—dry; VL, Villamar—summer rain season.

Therefore, in terms of biomass, both locations are different from each other, since the cluster analysis showed a group of macroalgae from Villamar and two other separate groups for Xpicob (Figure 6).

3.6. Importance Value Index

3.6.1. Xpicob's Macroalgae Importance Value Index

During the winter rain season, the species with the highest importance values were: *Jania capillacea* (IVI = 0.4036), *Digenea mexicana* (IVI = 0.2328), *Halimeda opuntia* (IVI = 0.2099), *Jania pedunculata* var. *adhaerens* (IVI = 0.1664), *Laurencia intricata* (IVI = 0.1070), and *Alsidium seaforthii* (IVI = 0.1051). In the summer rain season, the taxa with the highest importance values were: *Cladophora flexuosa* (IVI = 0.3435), *D. mexicana* (IVI = 0.2280), *A. seaforthii* (IVI = 0.1730), *Gracilaria debilis* (IVI = 0.1372), *Cladophora crispula* (IVI = 0.1184), and *Laurencia intricata* (IVI = 0.1004). Finally, in the dry season, the taxa with the highest importance value were *H. opuntia* (IVI = 0.4935), *J. capillacea* (IVI = 0.2454), and *Cl. crispula* (IVI = 0.09294) (Figure 7).

3.6.2. Villamar's Macroalgae Importance Value Index

In the winter rain season, the species with the highest importance values were: *Gracilaria debilis* (IVI = 0.4947), *Halimeda incrassata* (IVI = 0.2773), *Udotea dixonii* (IVI = 0.1726), *Palisada perforata* (IVI = 0.1548), *Yuzurua poiteaui* (IVI = 0.1047), and *Gracilaria cervicornis* (IVI = 0.0939). In the summer rain season, the species with the highest importance values were: *Cladophora sericea* (IVI = 0.2860), *Ceramium corniculatum* (IVI = 0.2361), *Halimeda scabra* (IVI = 0.1638), *Cladophora laetevirens* (IVI = 0.1601), *Digenea mexicana* (IVI = 0.1568), and *Udotea dixonii* (IVI = 0.1325). In the dry season, the species with the highest importance value were *H. incrassata* (IVI = 0.2094), *Eucheumatopsis isiformis* (IVI = 0.1874), *Jania capillacea* (0.1476), *G. debilis* (0.1474), and *D. mexicana* (0.0940) (Figure 7).

Figure 7. Seaweeds that are reported as dominants for some climatic season in Xpicob and Villamar Campeche during sampling. (**A**) *Digenea mexicana*, (**B**) *Eucheumatopsis isiformis*, (**C**) *Gracilaria debilis*, (**D**) *Palisada perforata* and *Yuzurua poiteaui*, (**E**) *Halimeda incrassata*.

3.7. Macroalgal Community Attributes

3.7.1. Macroalgal Community Attributes

Xpicob

The highest diversity was observed in the winter rain season (H′ = 2.191), followed by the summer rain season (H′ = 2.084), and, finally, the dry season (H′ = 1.492). Because it is directly related to diversity, evenness followed the same pattern, with the lowest evenness value in the dry season (J′ = 0.51) and the highest in the summer rain season (J′ = 0.73). Conversely, dominance is indirectly related to diversity and evenness, so the highest dominance was observed in the dry season; in the two seasons with higher diversity, the dominance presented lower values than in the dry season (Figure 8).

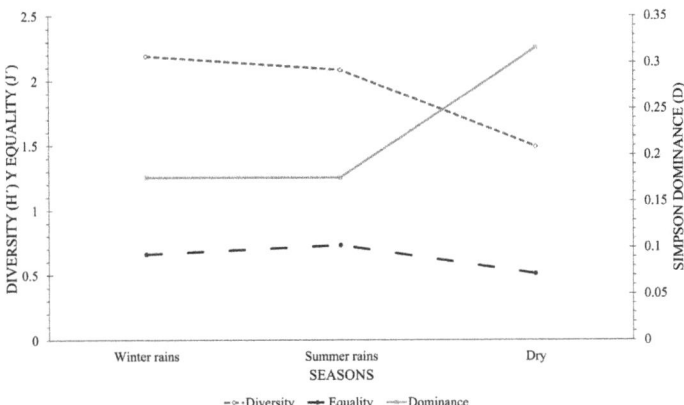

Figure 8. Estimation of alpha diversity (Shannon–Wiener), evenness, and dominance in Xpicob for the three climatic seasons.

Villamar

During the study period, the alpha diversity increased with time in Villamar; the lowest diversity was observed in the winter rain season ($H' = 1.866$) and the highest in the dry season ($H' = 2.823$). Evenness, being directly related to diversity, showed a similar pattern. The lowest evenness value was observed in the winter rain season ($J' = 0.587$) and the highest in the dry season ($J' = 0.737$). The lowest dominance was observed in the dry season ($D = 0.0903$) and the highest in the winter rain season ($D = 0.24$) (Figure 9).

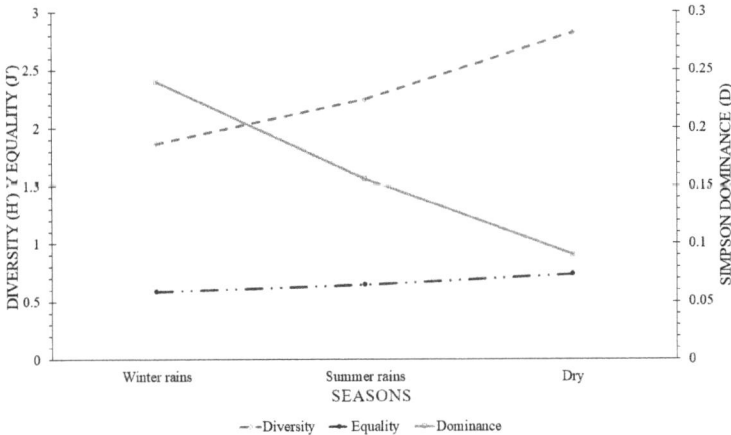

Figure 9. Estimation of alpha diversity (Shannon–Wiener), evenness, and dominance in Villamar for the three climatic seasons.

3.7.2. β (Beta) Diversity

Xpicob

The turnover of species between the three sampling periods was over 50%; however, the rate of turnover was very similar between the winter rain season and the summer rain season, and between the dry season and the winter rain season (0.5); however, the highest turnover corresponded to the change from the summer rain season to the dry season (0.65) (Table 3).

Table 3. Beta diversity (species turnover) of macroalgae species between climatic seasons for Xpicob and Villamar.

Localities	Seasons	β Diversity
Xpicob	Winter rains–Summer rains	0.5
	Summer rains–Dry	0.6571
	Dry–Winter rains	0.5111
Villamar	Winter rains–Summer rains	0.4285
	Summer rains–Dry	0.4615
	Dry–Winter rains	0.4857

Villamar

In this location, the species turnover was below 50% and was similar in the three climatic seasons (0.42, 0.46, and 0.48). The highest turnover rate was observed from the dry season to the winter rain season (Table 3).

3.8. Relationship between Environmental Factors and the Species Biomass

The PERMANOVA analysis showed that the environmental variables were significantly different both between locations and seasons (Table 4).

Table 4. Two-way PERMANOVA analysis for the environmental variables in each location.

	Sum of Squares	df	Mean Squares	F	p
Locality	0.8296	1	0.8296	19.416	01
Season	1.0302	2	0.51509	12.055	01
Interaction	0.050039	2	0.025019	0.58555	0.0653
Residual	2.6919	63	0.042728		
Total	4.6017	68			

The Analysis of Variance applied to each variable showed that the environmental variables that are significantly different between locations are: salinity ($F = 8.32^{30}$ $p = 0$), higher in Xpicob (35.5–39 ppm); depth ($F = 27.15$ $p = 2.21^{-06}$), higher in Villamar (240 cm); transparency ($F = 55.08$ $p = 4.36^{-10}$); dissolved oxygen ($F = 6.80^{31}$ $p = 0$), for which the highest value was observed in Villamar (8–13 mg/L); and the type of substrate, with the rocky substrate prevailing in Villamar ($F = 31.22$ $p = 5.69^{-7}$) and the sandy substrate in Xpicob ($F = 31.22$ $p = 5.69^{-7}$) (Figure 10).

Comparatively, when analyzing the environmental variables between seasons for each location, the following was observed:

In Xpicob, the environmental variables that were significantly different between seasons were as follows: salinity ($F = 1.53^{30}$ $p = 0$), with the lowest salinity value observed in the summer rain season (36 ppm), while in the winter rain and dry seasons, the values were very similar (39.9 and 40 ppm); temperature ($F = 51.96$ $p = 4.67^{-14}$), with the highest value observed in the summer rain season (36 °C) and the lowest in the dry season (28 °C); depth ($F = 15.22$ $p = 4.04^{-06}$), with the lowest value observed in the summer rain season and the highest in the dry season; transparency ($F = 17.04$ $p = 1.33^{-06}$), with the lowest value observed in the summer rain season and the highest in the dry season; and dissolved oxygen ($F = 2.18^{33}$ $p = 0$), with the lowest value observed in the summer rain season (5.3 mg/L) and the highest in the winter rain season (13 mg/L) (Figure 10).

In Villamar, the environmental variables that were significantly different between seasons were as follows: salinity ($F = 1.53^{30}$ $p = 0$), with the lowest value observed in the winter rain season (34.4 ppm) and the highest in the dry season (36 ppm); temperature ($F = 51.96$ $p = 4.67^{-14}$), with the highest value (36 °C) observed in the summer rain season and the lowest in the winter rain season (29 °C); depth ($F = 15.22$ $p = 4.04^{-06}$), with the lowest value observed in the summer rain season and the highest in the winter rain season;

transparency (F = 17.04 $p = 1.33^{-06}$), with the lowest value observed in the summer rain season and the highest in the winter rain season; and dissolved oxygen (F = 2.18^{33} $p = 0$), with the lowest value in the summer rain season (8 mg/L) and the highest in the winter rain season (13 mg/L) (Figure 10).

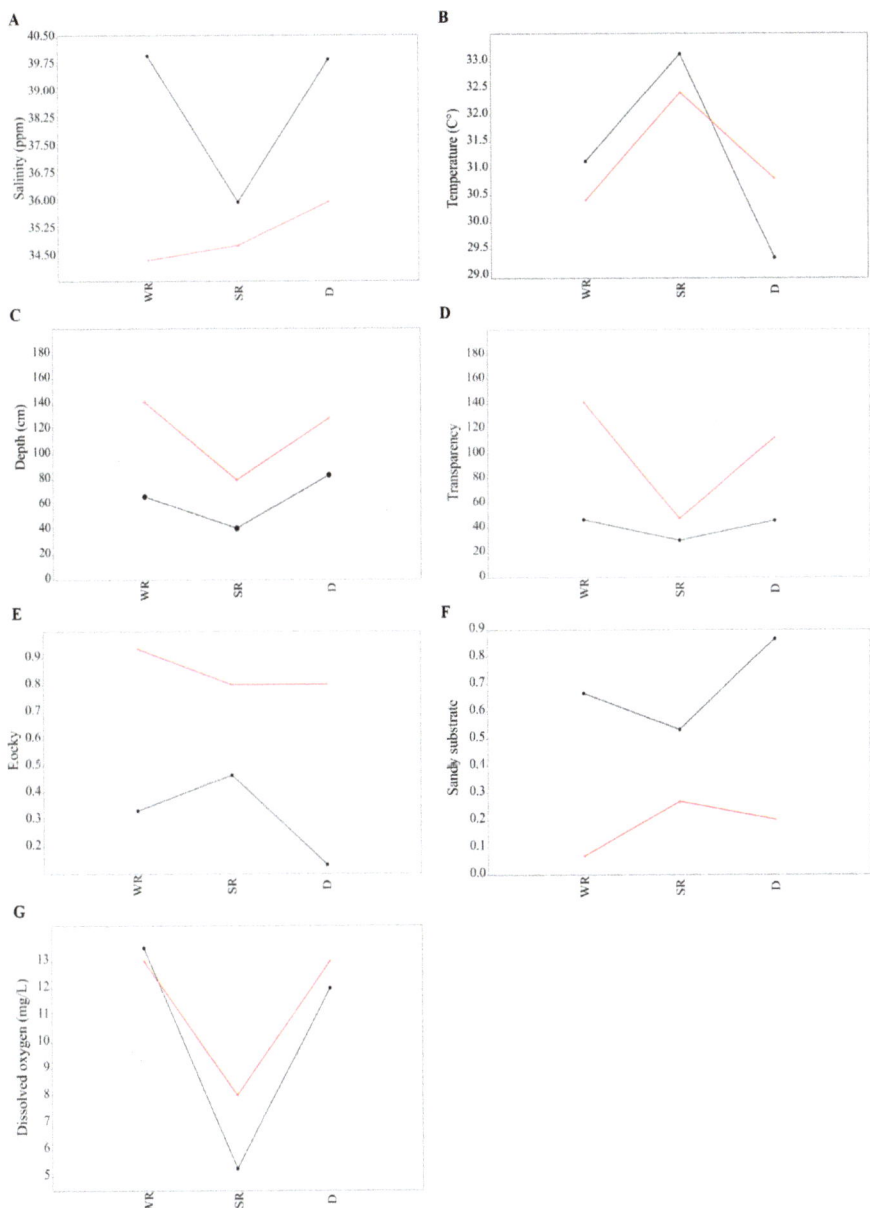

Figure 10. ANOVA of the environmental variables measured in Xpicob (black line) and Villamar (red line), Campeche, in October (2016), April (2017), and August (2017). (**A**) ANOVA of salinity, (**B**) ANOVA of temperature, (**C**) ANOVA of Depth, (**D**) ANOVA of Transparency, (**E**) ANOVA of rocky substrate, (**F**) ANOVA of sandy substrate, and (**G**) ANOVA of dissolved oxygen.

In the Canonical Correspondences analysis performed, the first three components explained 69.314% of the total variance. The first component is influenced directly by depth and indirectly by temperature. The second component is directly related to the type of sandy substrate and indirectly related to salinity (Tables 5 and 6).

Table 5. Correlation coefficient between the environmental variables recorded in Xpicob and Villamar, Campeche (2016–2017) and the biomass (in grams) of the macroalgae observed at both locations.

Environmental Variable	Axi 1	Axi 2	Axi 3	Axi 4	Axi 5	Axi 6
Salinity	6	−0.982	0.116	0.030	−0.061	0.135
Temperature	−0.222	−0.385	−0.145	−0.203	−0.507	−0.695
Depth	0.958	0.080	0.187	0.093	0.108	0.146
Transparency	0.914	−0.050	0.181	0.096	0.335	0.089
dissolved oxygen	0.375	0.419	0.269	−0.571	0.466	0.263
Rocky substrate	0.185	0.618	−0.728	0.076	−0.216	−0.029
Sandy Substrate	−0.185	−0.618	0.728	−0.076	0.216	0.029

Table 6. Percentage of variance explained in each axis of the Canonical Correspondence Analysis.

	Eigenvalue	Variance (%)	Accumulate Variance (%)
Axi 1	0.780	**35.449**	**35.449**
Axi 2	0.421	**19.146**	**54.594**
Axi 3	0.324	14.720	**69.314**
Axi 4	0.252	11.440	80.754
Axi 5	0.214	9.724	90.479
Axi 6	0.210	9.521	100

There were differences in the composition and biomass of the taxa in both locations in terms of depth. At higher depths and lower temperatures, we found specimens of *Meristotheca gelidium*, *Gracilaria flabelliformis*, *G. cervicornis*, *G. debilis*, *Halimeda opuntia*, *H. incrassata*, *H. monile*, *H. scabra*, *Udotea spinulosa*, *U. looensis*, *U. caribaea*, *U. dixonii*, *Caulerpa ashmeadii*, and *C. prolifera*, but also articulated corallines such as *Jania capillacea* and *Amphiroa fragilissima* or fleshy thalli of the genus *Laurencia*. On the other hand, in shallower areas and at higher temperatures, we found filamentous taxa such as *Ceramium corniculatum*, *Cladophora*, *Lophosiphonia*, and *Polysiphonia*, and blade taxa such as *Ulva*.

Axis two of the Canonical Correspondence Analysis revealed that the salinity and the type of substrate (sandy) are the factors that determine the biomass and the presence of the taxa at each location. In Xpicob, where the salinity is higher and the sandy substrate dominates, we mainly found filamentous specimens of the genus *Cladophora* and articulated coralline algae such as *Jania*, *Amphiroa*, and *Laurencia*. In contrast, in Villamar, where a rocky substrate and lower salinity predominate, we found species of fleshy red algae such as *Gracilaria debilis* and *Meristotheca gelidium*, calcified green algae of the genera *Halimeda*, *Udotea*, and *Penicillus*, and a few filamentous algae such as *Willeella* and *Cladophora*.

Furthermore, the separation of the summer rains season in Xpicob (XL) from its other two seasons was observed again. Similarly, summer rains season in Xpicob (XL) is ordered along the seasons in Villamar (Figure 11).

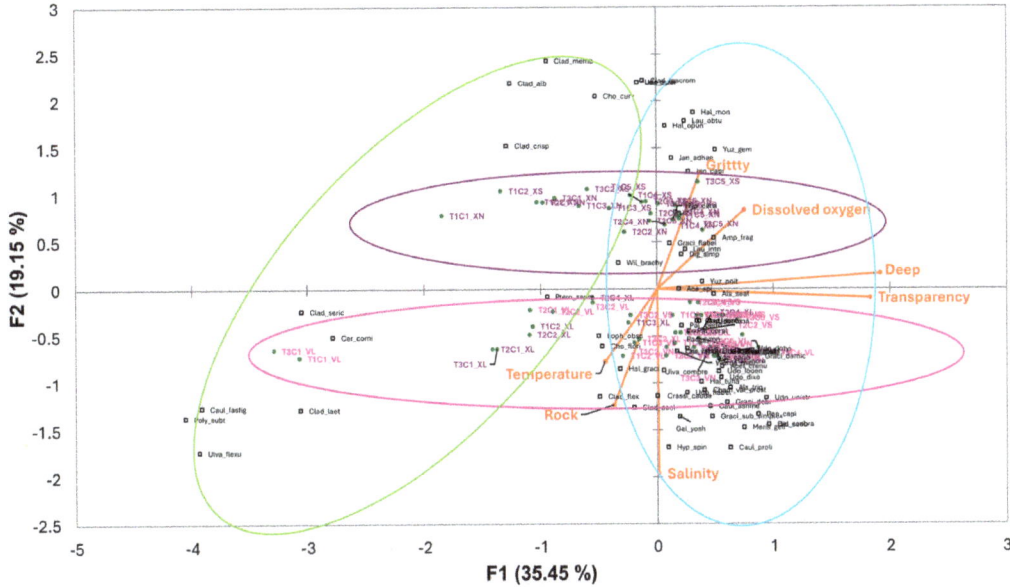

Figure 11. Canonical correspondence analysis (ACC) for environmental variables and biomass of macroalgae species by quadrant and season in Xpicob and Villamar, Campeche, collected in 2016 and 2017. Sites (quadrants) of Xpicob (purple circle). Sites (quadrants) of Villamar (pink circle). Macroalgae species located in shallow areas (green circle). Species of macroalgae located in deep areas (blue circle).

4. Discussion

4.1. Species Composition

In this study, we performed an ecological analysis of the phycoflora at two locations and identified a total of 74 taxa. These results are different from the report by Ortiz-Rosales (1988) [7], who cited 42 species for three locations in Campeche. This difference could be attributed to the sampling effort, the updated literature, and the meticulous laboratory work for the determination of the taxa.

On the coast of Campeche, there is a total of 271 known algae taxa [4–6], only 27% of which were identified in this study; this indicates that floristic studies mainly provide data on the species richness of the place, while ecological studies help understand the distribution patterns of the species in response to environmental conditions [7].

The highest species richness was observed in the class Florideophyceae (Rhodophyta), which is in keeping with previous studies in the area [3–6] reporting that the Phylum Rhodophyta and the class Florideophyceae are the best represented, in terms of the number of taxa. The success of this group of algae is attributed to their life cycle and reproduction strategies (formation and dispersal of spores) that allow them to persist at different times of the year [37,47].

4.2. Biogeographic Affinity

Lüning (1990) [48] suggests that the location of Campeche near the northern limit of the tropical Atlantic region would explain the high value of the Cheney index observed in this study (23.6), suggesting that the flora of Campeche is typically tropical. The flora of the study site included *Acanthophora spicifera, Udotea caribaea, U. cyathiformis, Halimeda incrassata, Gracilaria debilis, Hypnea cervicornis, Chondria floridana, Amphiroa fragilissima*, and *Laurencia intricata*, which are species with a tropical distribution.

The estuaries and coastal flora of the state present Cheney indices between 6.5 and 13 (tropical flora) [5,36]. In our study, we observed high index values, which are attributed to the low richness of brown algae species, with Dictyotales being the only species present. This group of brown algae is considered cosmopolitan since it can be found in both temperate and tropical zones. Furthermore, it has been observed that Dictyotales thalli remain as small branches during adverse periods [49]. In this sense, the sampling method is of great importance since direct sampling was used to evaluate subtidal populations [18]; this is a selective sampling method and, consequently, cannot ensure that all species present at the site are represented in the collection [13]. In the state of Campeche, 27 species of brown algae have been cited [5,6,50]; the Ectocarpaceae family of brown algae is composed entirely of algae of microscopic size that are mostly epiphytes [51]. This observation confirms that the type of sampling and analysis of the specimens greatly influenced the number of brown algae present in this study. In addition, environmental factors such as transparency (suspended organic matter), temperature, and depth influence the presence of brown algae. Reportedly, brown algae have their highest diversity in cold water environments such as the Pacific coastal zone of Baja California, which is influenced by the California Current (cold water) and upwelling that provide cold water that is rich in nutrients [37,52].

4.3. Spatial and Temporal Variation of Species Richness

In this work, the highest richness was observed in the dry season and the lowest in the winter rain season. This same variation in richness was observed by Mateo-Cid et al. (2013) [5] for eight locations on the coast of Campeche; Alfonso and Martinez-Daranas (2009) [53] reported a similar pattern for the northeast of Cuba. Increased richness in dry seasons and low richness in summer rains season can be attributed to the high transparency (illumination) typical of the dry months.

The classes Florideophyceae and Ulvophyceae had a different species richness at each sampling time, with Florideophyceae being the most diverse in the winter rain and dry seasons. Ulvophyceae, on the other hand, had the highest richness in the summer rain season, which can be attributed both to the intrinsic characteristics of the representatives of each Phylum at the time of sampling. Brown and green algae are annual, while red algae are both annual and perennial [54].

The taxa that were present during the three sampling periods were mainly perennial algae such as *Digenea mexicana*, *Acanthophora spicifera*, and *Gracilaria debilis*, which are characterized by a life cycle lasting more than one year. Annual algae such as *Cladophora*, *Polysiphonia*, *Lophosiphonia*, or *Ceramium* were also recorded throughout the three sampling periods, with life cycles that last less than a year with several generations [5,55].

4.4. Vertical Distribution Profile of Macroalgae

At both sampling locations at low depths (50–70 cm, the most exposed area), we found filamentous algae forming aggregates, mainly from the class Ulvophyceae, which are considered opportunistic [56]. In keeping with this observation, Lubchenco and Menge (1978) [57] indicated that the presence, and biomass, of filamentous and lamellar algae increase in areas of higher exposure to waves, limiting the establishment of other types of fleshy or calcareous algae. This result coincides with our observations in the first quadrants of the transects, where mainly Ulvophyceae and Florideophyceae filamentous algae were found; in addition to being opportunistic, these organisms (e.g., *Ceramium corniculatum*, *Polysiphonia subtilissima*, *Cladophora albida*, and *C. coelothrix*) have rapid growth rates.

Various authors [58,59] have studied the colonization and succession process of seaweeds, reporting that the colonization is initially composed of pioneer organisms with short life cycles and high reproduction rates, which modify the characteristics of the substrate. Delgado et al. (2008) [60], report that in the coastal splashing zone, there is a horizontal distribution of algae, and that the community is homogeneous and has a lower number of species, which is in keeping with what we observed in the present study: the shallow areas that are more exposed to waves have a lower diversity of algae. Furthermore, due to the sunlight available in the area, filamentous algae can grow rapidly with high photosynthetic rates per unit of biomass and rapid O_2 production rates [59].

Mathieson (1989) [61] reported a higher stability in deeper communities composed mainly of perennial algae; however, the richness was lower compared to that of shallower communities. Coleman and Mathieson (1975) [62] hypothesized that, in shallow waters with wide temperature fluctuations, the number of annual algae would be higher than that of perennial algae. This may explain the dominance of annual filamentous algae in the shallow waters that predominated in Xpicob and Villamar.

4.5. Temporal and Spatial Variation of the Biomass

In our study, the variation in total annual biomass was very marked; the annual biomass value recorded in the winter rain season in Xpicob was 10 times higher (322.24 g) than that of the summer rain season (23.43 g); this is possibly due to transparency, which is an environmental factor that varied throughout the seasons in Xpicob. At this site, the highest incidence of light was observed, which causes species of tropical affinity to reach their maximum biomass. These results coincide with what was recorded by Li-Alfaro and Zafra-Téllez (2012) [63] for the macroalgae community of Puerto Malabrigo, Peru.

In contrast, in Villamar, the biomass did not vary throughout the sampling cycle, having values between 200 g/m^2 and 250 g/m^2. These biomass values were higher than those observed in a similar study in Campeche [7] where the highest dry biomass value was 225 g/m^2.

Ortiz-Rosales (1998) [7] pointed out that *Eucheumatopsis isiformis* and *Acanthophora spicifera* were the most important species in terms of biomass (20–40%) (Champotón, Isla Arena and Sabancuy), mainly because these taxa are in deep areas (150–250 cm) with high transparency, which allows the photosynthetic rate to be higher and increases biomass.

Quirós-Rodríguez et al. (2010) [64] reported that filamentous algae present biomass values below 1, which coincides with our observation that the highest biomass values were given by leathery and fleshy species.

It should be noted that in most studies of marine macroalgae, the calculation of biomass is estimated from the cover (visually using photography). Few studies use biomass as an indicator of abundance, including those of Martinez-Daranas et al. (2016) [19] and Zúñiga-Ríos et al. (2012) [65]. The estimation of macroalgae biomass is of interest to many disciplines, and is useful in different ways, for example, it is a measure of the resource available for organisms at other trophic levels; furthermore, it provides information on the profitability of their exploitation as a natural resource [63]. Furthermore, the amount of biomass of macroalgae is related to factors such as light, temperature, nutrients, and type of substrate [19].

The clustering analysis showed the formation of three groups. Group I was formed only by Xpicob in the summer rain season; we thus infer that in this season, the number of taxa shared with the other seasons is low (four), compared to the dry and winter rain seasons with seven shared species. Furthermore, in Xpicob, the biomass values of the summer rain season were below 3 g/m^2, while in the dry and winter rain seasons, the biomass values were between 0.01 g/m^2 and 100 g/m^2.

Additionally, the environmental variables measured in Xpicob in the summer rain season were significantly different from those in the dry and winter rain seasons (which were alike). We thus infer that during the summer rain season, the environment is characterized by very particular environmental conditions. Therefore, low numbers of various species of filamentous algae, or the absence of articulated corallines, such as *Jania* or *Amphiroa*, are observed [53].

Group II of the cluster analysis includes the three climatic seasons of Villamar, even if the summer rain season had lower levels of similarity. The three climatic seasons of Villamar shared a total of 12 species. *Jania capillacea* and *Amphiroa fragilissima* were only found in the winter rain and dry seasons, but not in the summer rain season.

4.6. Macroalgal Community Attributes

4.6.1. α Diversity

Villamar was the location that presented the highest diversity. Compared to Xpicob, Villamar is deeper, which, combined with the predominance of a rocky substrate with both protected and exposed areas, allows the establishment, development, and growth of macroalgae. It has been reported that higher topographic variations result in a higher floristic diversity [66].

Rocky coasts are generally covered with vegetation formed almost exclusively by macroalgae [67]. Coasts with soft and sandy bottoms, on the other hand, have a lower abundance of macroalgae, mainly because most species are unable to attach to an unstable substrate with continuous water movement [66,67].

The highest α diversity was observed in the winter rain season and the lowest in the dry season, which contrasts with previous studies [7,37] reporting that the highest diversity was observed in the dry season when the temperature reaches the highest value, favoring the increase in species richness. In keeping with this idea, Núñez-López (1996) [68] indicated that the lowest values of richness and biomass of macroalgae are observed in the winter season when the water temperature reaches its lowest value.

Although the range of species richness between the three seasons in Xpicob is 2–8 species, there is a significant difference in the biomass of the species in the winter rain season, since at this time, the biomass was over 300 g, while in the remaining two seasons, the biomass was below 150 g.

Villamar presents an opposite pattern relative to Xpicob; the highest diversity was observed in the dry period, and the lowest in the winter rain season, coinciding with Ortiz-Rosales (1988) [7], possibly because in this season, environmental conditions such as temperature and salinity promote the increase in richness and biomass of macroalgae [69].

The Pielou evenness index presented values close to 1 in the summer rain season for Xpicob and in the dry season for Villamar, which indicates that there are no species with dominant abundance, suggesting that the macroalgae community is homogeneously distributed in these seasons. However, in the dry season of Xpicob and the winter rain season of Villamar, the Pielou evenness index was below 0.6, which, in turn, is related to alpha diversity, since in the same seasons, the diversity was lower at both locations. This observation indicates that the algae community at both sampling locations might be heterogeneously distributed, with one or several dominating species, in terms of abundance. These results coincide with the spatial behavior of macroalgae diversity in Ascensión Bay, reported by Fernández-Prieto (1988) [70].

4.6.2. β Diversity

In Xpicob, the turnover of species between seasons was higher than 0.5, which indicates that at least 50% of the species was replaced in each season. In contrast, in Villamar, the species turnover rate was below 50% in the three sampling periods.

This turnover rate is in turn related to the biological type of each species [55], since annual species that are present for a short period (short life cycle) are replaced by other species in the following season. Perennial algae, on the other hand, have life cycles that last more than one year, allowing them to be present in all seasons, thus reducing their turnover rate. In Villamar, perennial species were predominant, but in Xpicob, annual species were predominant; this could explain why the turnover rate at each location was above or below 50%.

4.7. Relationship between Environmental Factors and the Biomass of Taxa

According to our ACC results, the factors that explain the highest variation in the seasonal or temporal biomass of the taxa are depth, temperature, salinity, and type substrate (mostly sandy substrate). Environmental factors influence and determine the establishment, growth, reproduction, distribution, and abundance of algae in a specific location; similarly, the temporal fluctuations of these factors are related to the seasonal variations of algae [20].

It has been observed that temperature influences the reproduction of algae. In brown algae, for example, low temperatures cause the production of gametes, while high temperatures promote the formation of spores [11].

Salinity not only affects the distribution of algae but also their reproduction, causing the release of spores and polyspermy and influencing the viability of gametes [71].

It is important to emphasize that most environmental factors act together; for example, in tropical areas, changes in light intensity and/or temperature induce the synchronous release of gametes of holocarpic green algae (the entire thallus is transformed into gametes) [72,73].

The ACC showed that the substrate (mostly sandy) has a great influence on the biomass and composition of the macroalgae communities in both locations. Saad-Navarro and Riosmena-Rodríguez (2005) [54] in Baja California Sur, Alfonso and Martínez-Daranas (2009) [53] in Cuba, and Quirós-Rodríguez et al. (2010) [64] in Colombia concluded that climatic seasons do not determine the temporal variation of abundance and composition of algae; rather, the particular conditions and characteristics of each site influence the composition and abundance of macroalgae, mainly due to the type of substrate. McCook et al. (2001) [74] and other authors have shown that the sandy substrate negatively affects the coverage of algal associations; even in sandy bottoms with pebbles, algal richness is higher than in completely sandy or muddy bottoms [75].

Ortiz-Rosales (1988) [7] in Campeche and Delgado et al. (2008) [60] in Colombia reported that the substrate, depth, temperature, transparency, and salinity are the factors that exert the highest influence on the presence, composition, and abundance of macroalgae. It has been widely demonstrated by several reports [7,53,64,76] that the substrate greatly influences the algal composition of a given location; most macroalgae, especially Rhodophyta, adhere mainly to rocky substrates, which gives them a higher stability than sandy substrates. However, there are various macroalgae species (*Halimeda*, *Penicillus*, and *Udotea*, among others) that use rhizoids as a fixation system, which allows them to establish and develop in sandy substrates and they are thus considered substrate-forming organisms [54,76,77].

5. Conclusions

Florideophyceae stood out in terms of specific richness and dominance, with depth being the factor that favored its distribution of red algae, while for Ulvophyceae, the factor that influenced its distribution was the type of substrate. The analyzed communities present low similarity in both composition and abundance, which suggests that they are different communities, and the behavior of their attributes is different in each location. The turnover of species in both locations was high, considering that in each season, the turnover was more than 50%, which indicates that both locations have different environmental conditions in each season, causing the community to vary according.

This study demonstrates that it can be used for conservation and use studies of various species such as the agarophyte *Gracilaria debilis*.

Author Contributions: Conceptualization, methodology, C.M.H.-C., Á.C.M.-G. and L.E.M.-C.; methodology, C.M.H.-C., Á.C.M.-G., D.Y.G.-L. and L.E.M.-C.; investigation, C.M.H.-C., Á.C.M.-G. and L.E.M.-C.; formal analysis, C.M.H.-C., Á.C.M.-G. and L.E.M.-C.; writing original draft, C.M.H.-C., Á.C.M.-G., D.Y.G.-L. and L.E.M.-C.; data curation, C.M.H.-C., Á.C.M.-G., D.Y.G.-L. and L.E.M.-C.; writing—review and editing, C.M.H.-C., Á.C.M.-G., D.Y.G.-L. and L.E.M.-C.; funding acquisition, L.E.M.-C. and Á.C.M.-G. All authors have read and agreed to the published version of the manuscript.

Funding: This research was funded by Instituto Politécnico Nacional (SIP 20170696, 20180489, 20195092, 20170767, 20180491 y 20195127), for providing financial support, facilities, and necessary equipment for the development of this study.

Institutional Review Board Statement: Not applicable.

Data Availability Statement: The raw data supporting the conclusions of this article will be made available by the authors on request.

Acknowledgments: C.M.H.-C. appreciate the scholarship granted by Consejo Nacional de Humanidades Ciencias y Tecnologías (CONAHCyT) (scholarship number 704348). Á.C.M.-G., L.E.M.-C. appreciate the scholarship granted by the Commission for the Operation and Promotion of Academic Activities (COFAA) and the EDI incentives. To A. Gerardo Garduño Acosta, J. Alfredo Pérez Salgado, Itzel Gonzáles Contreras, Ángel Norberto Ocaña Valencia for their logistical support.

Conflicts of Interest: The authors declare no conflicts of interest.

References

1. Huerta-Múzquiz, L.; Garza-Barrientos, M.A. Algas marinas del litoral del estado de Campeche. *Ciencia* **1966**, *24*, 193–200.
2. Huerta-Múzquiz, L.; Mendoza-González, A.C.; Mateo-Cid, L.E. Avance de un estudio de las algas marinas de la península de Yucatán. *Phytologia* **1987**, *62*, 23–53.
3. Ortega, M.M. Observaciones del fitobentos de la Laguna de Términos, Campeche, México. *An. Inst. Biol. Ser. Bot.* **1995**, *66*, 1–36.
4. Callejas-Jiménez, M.E.; Sentíes, A.; Dreckmann, K.M. Macroalgas de Puerto Real, Faro Santa Rosalía y Playa Preciosa, Campeche, México, con algunas consideraciones florísticas y ecológicas para el estado. *Hidrobiológica* **2005**, *15*, 89–96.
5. Mateo-Cid, L.E.; Mendoza-González, A.C.; Avila-Ortiz, A.G.; Díaz Martínez, S. Algas marinas bentónicas del litoral de Campeche, México. *Acta Bot. Mex.* **2013**, *104*, 53–92. [CrossRef]
6. Mendoza-González, A.C.; Mateo-Cid, L.E.; López Garrido, P.H. Algas marinas bentónicas asociadas a pecios y otras estructuras submareales de Campeche, México. *Acta Bot. Venez.* **2013**, *36*, 119–140.
7. Ortiz-Rosales, J. Ecología de las Algas Bentónicas de tres Localidades del Litoral del Estado de Campeche, México. Master's Thesis, Universidad Autónoma de Nuevo León, Monterrey, Nuevo León, 1988.
8. Silva, P.C. Geographic patterns of diversity in benthic marine algae. *Pac. Sci.* **1992**, *46*, 429–437.
9. Lobban, C.S.; Harrison, P.J. *Seaweed Ecology and Physiology*; Cambridge University Press: Cambridge, MA, USA, 1997; 366p. [CrossRef]
10. Santelices, B.; Hommersand, M.M. *Pterocladiella*, a new genus in the Gelidiaceae (Gelidiales, Rhodophyta). *Phycology* **1997**, *36*, 114–119. [CrossRef]
11. Espinoza-Avalos, J. Fenología de macroalgas marinas. *Hidrobiológica* **2005**, *15*, 109–122.
12. Mathieson, A.C. Vertical distribution and longevity of subtidal seaweeds in Northern New England. *Bot. Mar.* **1979**, *30*, 511–520. [CrossRef]
13. Littler, M.M. Morphological form and photosynthetic performances of marine Macroalgae: Tests of a functional/form hypothesis. *Bot. Mar.* **1980**, *22*, 161–165. [CrossRef]
14. Jackelman, J.J.; Stegenga, H.; Bolton, J.J. The marine benthic flora of the Cape Hangklip area and its phytogeographical affinities. *S. Afr. J Bot.* **1991**, *57*, 295–304. [CrossRef]
15. Serviere-Zaragoza, E. Descripción y Análisis de la Ficoflora del Litoral Rocoso de Bahía de Banderas, Jalisco, Nayarit, México. Ph.D. Thesis, UNAM, Mexico City, Mexico, 1993.
16. Dawes, C. *Botánica Marina*; Editorial Limusa Ciudad de México: Mexico City, Mexico, 1986; 673p.
17. Darley, W. *Biología de las Algas. Enfoque Fisiológico*; Editorial Limusa Ciudad de México: Mexico City, Mexico, 1987; 236p.
18. Larkum, A.; Drew, W.D.; Ralph, P.J. Photosynthesis and metabolism in seagrasses at the cellular level. In *Seagrasses: Biology, Ecology and Conservation*; Larkum, A.W.D., Orth, R.J., Duarte, C.M., Eds.; Springer: Dordrecht, The Netherlands, 2006. [CrossRef]
19. Martínez-Daranas, B.; Esquivel, M.; Alcolado, P.M.; Jiménez, C. Composición específica y abundancia de macroalgas y angiospermas marinas en tres arrecifes coralinos de la plataforma Sudoccidental de Cuba (1987). *Hidrobiológica* **2016**, *26*, 323–337. [CrossRef]
20. Bula-Meyer, G. Algas marinas bénticas indicadoras de un área afectada por aguas de surgencia frente a la costa Caribe de Colombia. *An. Inst. Investig. Mar. Punta Betín* **1977**, *9*, 45–71. [CrossRef]
21. Dethier, M.N. Pattern and process in tide pool algae: Factors influencing seasonality and distribution. *Bot. Mar.* **1982**, *25*, 41–54. [CrossRef]
22. Mathieson, A.C.; Norral, T.L. Physiological studies of subtidal red algae. *J. Exp. Mar. Biol. Ecol.* **1975**, *20*, 237–247. [CrossRef]
23. Arnold, K.E.; Murray, S.N. Relationships between irradiance and photosynthesis for marine benthic green algae (Chlorophyta) of differing morphologies. *J. Exp. Mar. Biol. Ecol.* **1980**, *43*, 183–192. [CrossRef]
24. García, E. *Modificaciones al Sistema de Clasificación Climática de Köppen*; Editorial Universidad Autónoma de México, Instituto de Geografía: Mexico City, Mexico, 2004; 92p.
25. Orellana, R.; Islebe, G.A.; Espadas, C. Presente, pasado y futuro de los climas de la Península de Yucatán. In *Naturaleza y Sociedad en el Área Maya: Pasado, Presente y Futuro*; Colunga-García, M.P., Larqué-Saavedra, A., Eds.; Academia Mexicana de Ciencias Centro de Investigación Científica de Yucatán: Mexico City, Mexico, 2003; pp. 37–52.
26. Vázquez, J.; González, J. Método de evaluación de macroalgas submareales. In *Manual de Métodos Ficológicos*; Alveal, K., Ferrario, M.E., Oliveira, E.C., Sar, E., Eds.; Universidad de Concepción: Concepción, Chile, 1995; pp. 643–666.
27. Zayas, C.R.; Suárez, A.M.; Ocaña, F.A. Abundancia y diversidad de especies de fitobentos de playa Guardalavaca, Cuba. *Rev. Investig. Mar.* **2006**, *27*, 87–93.

28. Cano, M. de la C. Bases Biológicas de *Ulva fasciata* Delile, (Chlorophyta) Para su Posible Explotación, al Oeste de La Habana, Cuba. Ph.D. Thesis, Facultad de Biología, Universidad de La Habana, La Habana, Cuba, 2008.
29. Taylor, R. *Marine Algae of the Eastern Tropical and Subtropical Coasts of the America*; The University of Michigan Press: Ann Arbor, MI, USA, 1960; 870p. [CrossRef]
30. Schneider, C.W.; Searles, R.B. *Seaweeds of the Southeastern United States*; Cape Hatteras to Cape Cañaveral; Duke University Press: Durham, NC, USA, 1991; 563p. [CrossRef]
31. Littler, D.S.; Littler, M.M. *Caribbean Reef Plants. An Identification Guide to the Reef Plants of the Caribbean, Bahamas, Florida and Gulf of Mexico*; Offshore Graphics: Washington, DC, USA, 2000; 542p.
32. Dawes, C.J.; Mathieson, A.C. *The Seaweeds of Florida*; University Press of Florida: Gainesville, FL, USA, 2008; 591p.
33. Cho, T.O.; Boo, S.M.; Hommersand, M.H.; Maggs, C.A.; McIvor, L.; Fredericq, S. *Gayliella* gen. nov. in the tribe Ceramieae (Ceramiaceae, Rhodophyta) based on molecular and morphological evidence. *J. Phycol.* **2008**, *44*, 721–738. [CrossRef]
34. Wynne, M.J. A checklist of benthic marine algae of the tropical and subtropical western Atlantic: Fifth revision. *Nova Hedwig.* **2022**, *140*, 7–66.
35. Guiry, M.D.; Guiry, G.M.; Algaebase. World-Wide Electronic Publication. National University of Ireland. Galway, Ireland. Available online: http://www.algaebase.org (accessed on 28 March 2024).
36. Cheney, P. R+C/P A new and improved ratio for comparing seaweed floras. *J. Phycol.* **1977**, *13*, 12.
37. Águila-Ramírez, R.N. Variación Estacional de la Distribución de las Macroalgas en la Laguna Ojo de Liebre, B.C.S. Master's Thesis, Centro Interdisciplinario de Ciencias Marinas, Instituto Politécnico Nacional, Baja California Sur, México, 1998.
38. Curtis, J.T.; McIntosh, R.P. An upland forest continuum in the pariré-forest border region of Wisconsin. *Ecology* **1951**, *32*, 476–496. [CrossRef]
39. Zabi, S.G. Role de la biomasse dans la determination de L Importance Value pour la mise en evidence des unites de peuplements benthiques en Lagune Ebrie (Cote D'Ivoire). *Doc. Sci. Cent. Recj. Oceanogr. Abidj.* **1984**, *15*, 55–87.
40. Magurran, A. *Ecological Diversity and Its Measurement*; Princeton University Press: Princeton, NJ, USA, 1988; 179p. [CrossRef]
41. Whittaker, R.H. Evolution and measurement of species diversity. *Taxon* **1972**, *21*, 213–251. [CrossRef]
42. Hammer, O.; Harper, D.; Ryan, P. PAST: Paleontological Statistics Software Package for Education and Data Analysis. *Palaeontol. Electron.* **2001**, *4*, 1–9.
43. Brower, J.; Zar, J.; Von Ende, C. *Field and Laboratory Methods for General Ecology*, 4th ed.; Mc Graw Hill: New York, NY, USA, 1997; 273p.
44. Peet, R.K. The measurement of species individual's diversity indices applied to samples of field insects. *Ecology* **1974**, *45*, 859–861.
45. Acosta-Calderón, J.A.; Hernández-Rodríguez, C.; Mendoza-González, A.C.; Mateo-Cid, L.E. Diversity and distribution of *Udotea* genus J.V. Lamouroux (Chlorophyta, Udoteaceae) in the Yucatan peninsula littoral, Mexico. *Phytotaxa* **2018**, *345*, 179–218. [CrossRef]
46. Nuñez-Resendiz, M.L.; Dreckmann, K.M.; Wynne, M.J.; Sentíes, A. *Codiophyllum mexicanum* sp. nov. (Halymeniaceae, Rhodophyta), first record of a stalked red alga associated with sponges in the western Atlantic. *Phycologia* **2019**, *59*, 89–98. [CrossRef]
47. Mateo-Cid, L.E.; Sánchez Rodríguez, I.; Rodríguez Montesinos, E.; Casas Valdez, M. Estudio florístico de las algas bentónicas de Bahía Concepción, B.C.S., México. *Cienc. Mar.* **1993**, *19*, 41–60. [CrossRef]
48. Lüning, K. *Seaweeds: Their Environment, Biogeography and Ecophysiology*; John Wiley & Sons, Inc.: New York, NY, USA, 1990; 527p. [CrossRef]
49. Gauna, M.C.; Cáceres, E.J.; Parodi, E.R. Temporal variations of vegetative features, sex ratios and reproductive phenology in *Dictyota dichotoma* (Dictyotales, Phaeophyceae) population of Argentina. *Helgol. Mar. Res.* **2013**, *67*, 721–732. [CrossRef]
50. Maldonado-Montiel, T.J.; Chan-Keb, C.A.; Agraz-Hernández, C.M.; Peréz-Balán, R.A. *Guía para la Identificación de Algas en Sabancuy*; ECORFAM: Campeche, México, 2023; 833p. [CrossRef]
51. Mendoza-González, A.C.; Mateo-Cid, L.E.; Ortega-Murillo, M.R.; Zurita-Valencia, L.; Sánchez-Heredia, J.D.; Hernández-Casas, C.M. Nuevos registros y lista actualizada de las algas pardas (Phaeophyceae) del litoral de Michoacán, México. *Rev. Biol. Mar. Oceanogr.* **2020**, *55*, 202–216. [CrossRef]
52. Mateo-Cid, L.E.; Mendoza-González, A.C. Algas marinas bentónicas de Bahía Asunción, B.C.S., México. *Cienc. Mar.* **1994**, *20*, 41–64. [CrossRef]
53. Alfonso, Y.; Martínez-Daranas, B. Variaciones espacio-temporales en la cobertura del macrofitobentos en área costera al norte de Ciudad de La Habana, Cuba. *Rev. Investig. Mar.* **2009**, *30*, 187–201.
54. Saad-Navarro, G.; Riosmena-Rodríguez, R. Variación espacial y temporal de la riqueza florística de macroalgas en la zona rocosa de Bahía Muertos, B.C.S. México. *Cienc. Mar.* **2005**, *9*, 19–32.
55. Feldmann, J. Recherches sur la végétation marine de la Mediterranée. Côte Albères. *Rév. Algol.* **1937**, *10*, 1–339.
56. Wiencke, C.; Bischof, K. *Seaweed Biology. Novel Insights into Ecophysiology, Ecology and Utilization*; Springer: Berlin, Germany, 2012; 508p. [CrossRef]
57. Lubchenco, J.; Menge, B.A. Community development and persistence in a low rocky intertidal zone. *Ecol. Monogr.* **1978**, *48*, 67–97. [CrossRef]
58. García-Castrillo, R.G.; Lanuza Alfonso, P.; López García, P. El entorno marino de los restos arqueológicos. *Monte Buciero* **2003**, *9*, 95–108.

59. Choi, C.G.; Ohno, M.; Sohn, C.H. Algal succession on different substrata covering the artificial iron reef at Ikata in Shikoku, Japan. *Algae* **2006**, *21*, 305–310. [CrossRef]
60. Delgado, M.J.C.; Palacios, J.A.; Aguirre, N.J. Variación vertical y estacional de la comunidad de macroalgas en los costados noroccidental y nororiental del Golfo de Urabá, Caribe Colombiano. *Gestión Ambiente* **2008**, *11*, 27–42.
61. Mathieson, A.C. Phenological patterns of northern New England seaweeds. *Bot. Mar.* **1989**, *32*, 419–438. [CrossRef]
62. Coleman, D.C.; Mathieson, A.C. Investigations of New England Marine Algae. VII. Seasonal occurrence and reproduction of Marine algae near Cape Cod, Massachussetts. *Rhodora* **1975**, *77*, 76–104.
63. Li-Alfaro, G.; Zafra-Trelles, A. Composición, abundancia y diversidad de macroalgas en el litoral de puerto Malabrigo, La Libertad-Perú, 2009. *Sciendo* **2012**, *15*, 33–44.
64. Quirós-Rodríguez, J.A.; Arias-Ríos, J.E.; Ruiz Vega, R. Estructura de comunidades macroalgales asociadas al litoral rocoso del departamento de Córdoba, Colombia. *Caldasia* **2010**, *32*, 339–354.
65. Zúñiga Ríos, D.; Martínez-Daranas, B.; Alcolado, P.M. Ficoflora de los arrecifes coralinos del archipiélago Sabana-Camagüey. *Serie Oceanológica* **2012**, *11*, 57–76.
66. Seapy, K.; Littler, M.M. The distribution, abundance, community structure and primary productivity of macroorganisms from two central California rocky intertidal habitats. *Pac. Sci.* **1979**, *32*, 293–314.
67. Dawes, C.J. *Marine Botany*; John Wiley y Sons: New York, NY, USA, 1998; 496p.
68. Núñez-López, R.A. Estructura de la Comunidad de Macroalgas de la Laguna de San Ignacio, B.C.S. México (1992–1993). Master's Thesis, Centro Interdisciplinario de Ciencias Marinas, La Paz, México, 1996.
69. Álvarez-Borrego, S.; Granados-Guzmán, A. Variación espacio-temporal de temperatura en un hábitat de invierno de la ballena gris: Laguna Ojo de Liebre. *Cienc. Mar.* **1992**, *18*, 151–165. [CrossRef]
70. Fernández-Prieto, J. Sistemática y Variaciones Espacio-Temporales de las Algas Marinas Bentónicas de Bahía de la Ascensión, Quintana Roo. Bachelor's Thesis, Universidad Autónoma de Baja California, Ensenada, BC, México, 1988.
71. Santelices, B. Recent advances in fertilization ecology of macroalgae. *J. Phycol.* **2002**, *38*, 4–10. [CrossRef]
72. Clifton, K.E. Mass spawning by green algae on coral reefs. *Science* **1997**, *275*, 1116–1118. [CrossRef]
73. Clifton, K.E.; Clifton, L.M. The phenology of sexual reproduction by green algae (Bryopsidales) on Caribbean coral reefs. *J. Phycol.* **1999**, *35*, 24–34. [CrossRef]
74. McCook, L.J.; Jompa, J.; Diaz-Pulido, G. Competition between corals and algae on coral reefs: A review of evidence and mechanisms. *Coral Reefs* **2001**, *19*, 400–417. [CrossRef]
75. Hoek, C.; Colijin, F.; Cortel-Breeman, A.M.; Wanders, J.B. Algal vegetation-type along the shores of inner bays, and lagoons of Curacao, and of the Lagoon Lac (Bonaire), Netherlands Antilles. *Tweede Reeks Deel* **1972**, *61*, 1–70.
76. Garduño-Solórzano, G.; Godínez-Ortega, J.L.; Ortega, M.M. Distribución geográfica y afinidad por el sustrato de las algas verdes (Chlorophyceae) bénticas de las costas mexicanas del Golfo de México y Mar Caribe. *Bol. Soc. Bot. México* **2005**, *76*, 61–78. [CrossRef]
77. Santelices, B. *Ecología de las Algas Marinas Bentónicas*; Universidad Católica de Chile: Santiago de Chile, Chile, 1977; 384p.

Disclaimer/Publisher's Note: The statements, opinions and data contained in all publications are solely those of the individual author(s) and contributor(s) and not of MDPI and/or the editor(s). MDPI and/or the editor(s) disclaim responsibility for any injury to people or property resulting from any ideas, methods, instructions or products referred to in the content.

MDPI AG
Grosspeteranlage 5
4052 Basel
Switzerland
Tel.: +41 61 683 77 34

Diversity Editorial Office
E-mail: diversity@mdpi.com
www.mdpi.com/journal/diversity

Disclaimer/Publisher's Note: The statements, opinions and data contained in all publications are solely those of the individual author(s) and contributor(s) and not of MDPI and/or the editor(s). MDPI and/or the editor(s) disclaim responsibility for any injury to people or property resulting from any ideas, methods, instructions or products referred to in the content.